1001 WAYS TO PASS ORGANIC CHEMISTRY

A GUIDE FOR HELPING STUDENTS PREPARE FOR EXAMS

SECOND EDITION

SHELTON BANK
&
JANET F. BANK

STATE UNIVERSITY
OF NEW YORK, ALBANY

SAUNDERS GOLDEN SUNBURST SERIES

Saunders College Publishing
Harcourt Brace College Publishers

Fort Worth Philadelphia San Diego New York Orlando Austin
San Antonio Toronto Montreal London Sydney Tokyo

Printed in the United States of America

Brown: 1001 Ways to Pass Organic Chemistry, second edition: A Guide for Helping Students Prepare for Exams to accompany *Organic Chemistry*. Bank & Bank.

0-03-020692-8

567 095 7654321

A MESSAGE FOR STUDENTS

How do you study for an organic chemistry exam? That is a good question considering the vast amount of material that is usually included in organic exams. During the many years of teaching chemistry, we have found that self-tests are one of the best ways students can gauge their knowledge of the material. The 1001 questions in this book were selected to emphasize key material and to provide you with a timely assessment of your progress. The multiple-choice format is used in order to guide you in this assessment. If you cannot easily discard one or two answers, then it is not a question of understanding but rather a lack of knowledge and you should go back to the text, the lectures and the problems in the text. If you cannot distinguish between two possible answers, then you probably need to clarify a point of confusion. Having done this, you are well on the way to mastering the material. We have not included answers because we are aware of the natural inclination to look them up before ending the struggle for the correct answer. Instead we have given you **TIPS** that we hope will not only guide you to the correct answer, but also help you with other similar questions. In particular, the wrong answers in each question illustrate common misunderstandings and hopefully will uncover weaknesses that you can correct before the exam. We have tried to emphasize these wrong answers as a learning tool.

In this second edition, we have modified questions to better reflect the textbook and to remove ambiguity. We have corrected errors and revised some **TIPS** for clarity. We thank our students and colleagues for their helpful comments.

We additionally encourage you to use these questions in help sessions and study groups. A Harvard University assessment of what constitutes productive learning found that interaction among fellow students and faculty is much more effective than trying to work alone. Finally, we hope that by using these questions, you will develop ***confidence*** in your abilities in organic chemistry - an important quality of winning.

Shelton Bank
Janet F. Bank
June, 1997

ACKNOWLEDGMENTS

We are grateful for the many students at The State University at New York and at Ohio State University who have used these questions, have found errors and have made suggestions for improvements. We especially thank John Vondeling, Publisher, for his continued confidence and encouragement. Finally, we thank Sandi Kiselica for her excellent professional help and advice.

CONTENTS

CONTENTS

CONTENTS

CONTENTS

CONTENTS

CHAPTER 1 COVALENT BONDING

A. ELECTRONEGATIVITY

1. Arrange the following elements in the order of increasing electronegativity (least first).

S	Br	F	Na
I	II	III	IV

a) IV, III, I, II b) II, IV, I, III c) IV, I, II, III d) IV, II, III, I

2. Arrange the following elements in the order of decreasing electronegativity (greatest first).

O	N	C	F	Si
I	II	III	IV	V

a) IV, I, II, III, V b) IV, II, I, V, III c) I, V, IV, II, III d) IV, I, II, V, III

3. Arrange the following elements in the order of increasing electronegativity (least first).

N	S	Li	Si	C
I	II	III	IV	V

a) II, IV, III, V, I b) III, IV, II, V, I c) IV, V, III, I, II d) III, V, I, II, IV

4. Which of the following molecules has the greatest difference in electronegativity between the two different elements?

a) CO b) CS_2 c) H_2O d) CN^-

5. Arrange the following compounds in the order of increasing ionic character of the indicated bonds (least first).

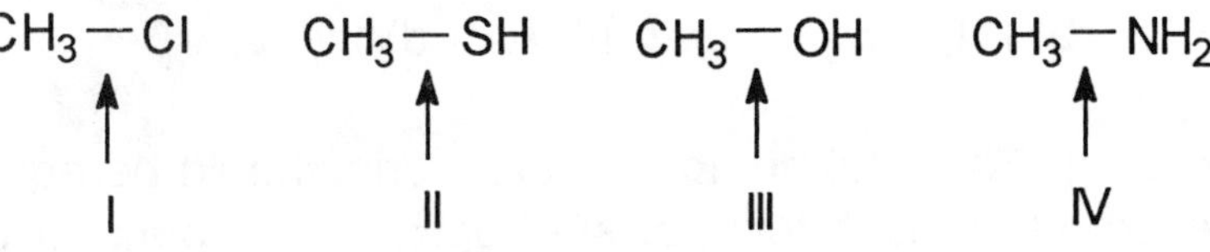

a) I, II, IV, III b) II, IV, I, III c) IV, III, I, II d) III, IV, I, II

6. Which of the following compounds are nonpolar?

H_2O	CO_2	NH_3	BF_3	$CHCl_3$
I	II	III	IV	V

a) II, III   b) II, IV   c) I, III   d) III, V

TIP A compound is nonpolar if 1) there is a balance of electronegativity for all the atoms taken as a whole or if 2) the electronegativity of the bonds is balanced. Structure I is bent, III is pyramidal, and V is tetrahedral and therefore 2) is violated. Structures II and IV violate 1) but the geometry of the molecule (linear for II, planar for IV) causes an overall balance.

B. LEWIS STRUCTURES

7. Which of the following Lewis structures are correct?

H:Ö:H:Ö:	$[H_4N]^{\oplus}$:Cl: $^{\ominus}$	$[H_3O]^{\oplus}$	CCl_4
I	II	III	IV

a) I, II, III   b) I, III, IV   c) II, III, IV   d) I, II, IV

8. Which of the following Lewis structures are incorrect?

:O::O:O:	N::N:	·Cl::Cl·	H:C::O:H (with two H on C)	H:C:N:H (with H on C and N)
I	II	III	IV	V

a) I, II, III   b) I, IV, V   c) II, III, IV   d) III, IV, V

TIP Count electrons. 1) The total number of electrons should be equal to the sum of all the valence electrons, adding or subtracting to account for charge. 2) Each atom should have a complete octet to the greatest extent possible. Structure II violates 1); structures III and IV violate 2).

9. Which of the following Lewis structures are correct?

```
H  H  H ⊖            H        H ⊖                      H
C::C:C:   H:O:::O:H   H:Si:H   H:B:H   :S::C::S:   H:N::O:H
H     H               H        H                      H
   I         II         III      IV        V           VI
```

a) II, III, IV b) I, III, IV, V c) I, II, VI d) III, IV, V

10. What is the correct Lewis structure for formaldehyde?

```
H                                       ⊕  ⊖
C::O       H:C::O:H       H:C:O:       H:C::O:
H                           H            H
 a)           b)            c)            d)
```

11. Which of the following are examples of electron configurations with completed outer shells?

sp	$1s^2$	1s 2s	$1s^22s^22p^2$	$1s^22s^23s^2$	$1s^22s^22p^6$
I	II	III	IV	V	VI

a) I, III, IV b) II, IV c) II, VI d) I, IV, VI

TIP Rule 1), completed s shells have 2 electrons.
Rule 2), completed p shells have 6 electrons.
Rule 3), the order of filling shells is 1s, 2s, 2p, 3s.

Structures I and III violate rules 1) and 2).
Structure IV violates rule 2).
Structure V violates rule 3).

12. Which of the following Lewis structures are correct?

I	II	III	IV	V
H:Ö:N̈:Ö: (with :Ö: below N)	:C̈l:H	Na ⊕ :Ö:H ⊖	H:N̈::Ö:H (H above N)	H:C̈:C:::N: (H above and below C)

a) I, II, III b) I, III, IV c) II, IV, V d) II, III, V

C. FORMAL CHARGE

13. Carbon has a formal charge in which of the following compounds and ions?

a) CO_2 b) CO_3^{-2} c) CS_2 d) CF_3^+

14. Carbon has a formal charge in which of the following ions?

a) $[(CH_3)_4N]^+$ b) $[HCO]^+$ c) $[CH_3CO_2]^-$ d) $[CH_3OH_2]^+$

15. Carbon has a formal charge in which of the following ?

a) HCO_3^- b) $CH_3CO_2^-$ c) CH_3OH d) $LiCH_3$

16. Nitrogen has a negative formal charge in which of the following compounds?

a) $NaNH_2$ b) N_2 c) NH_4Cl d) HCN

TIP The formal charge on an atom in a molecule is the difference between the number of valence electrons in the neutral atom and the number calculated by adding the number of unshared electrons plus 1/2 bonding pairs. A single bond contributes one electron, a double bond contributes two electrons and a triple bond contributes three electrons. The formal charges are:

a) $5 - [2 + 2 + 1 + 1] = -1$

b) $5 - [2 + 1 + 1 + 1] = 0$

c) $5 - [0 + 1 + 1 + 1 + 1] = +1$

d) $5 - [2 + 1 + 1 + 1] = 0$

D. BOND ANGLES AND GEOMETRY

17. Which of the following molecules have bond angles of about 109.5 degrees?

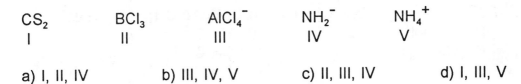

CS_2	BCl_3	$AlCl_4^-$	NH_2^-	NH_4^+
I	II	III	IV	V

a) I, II, IV   b) III, IV, V   c) II, III, IV   d) I, III, V

18. What is the bond angle for the indicated atom in the following molecules?

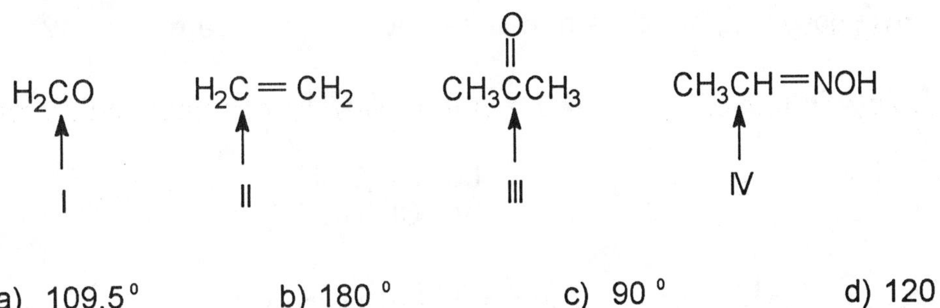

a) 109.5°   b) 180 °   c) 90 °   d) 120 °

19. Which of the following molecules have a tetrahedral shape?

BH_3	H_2O	H_3O^+	NH_4Cl	CCl_4
I	II	III	IV`	V

a) I, III, V   b) I, II, IV, V   c) III, IV, V   d) II, III, V

20. What is the molecular geometry (shape) of NO_2^+?

a) tetrahedral   b) linear   c) planar   d) trigonal

21. What is the geometry (shape) of CH_3^- ?

a) tetrahedral   b) linear   c) planar   d) trigonal

TIP Geometry is determined by the hybridization and the number of electrons around the central atom. The hybridization is sp^3 with 8 electrons and therefore the geometry is tetrahedral.

E. FUNCTIONAL GROUPS

22. Which of the following matches of names and molecules are correct?

A. ester	I. $CH_3CO_2CH_3$
B. alcohol	II. H_2CO
C. aldehyde	III. CH_3CHO
D. acid	IV. CH_3OH

a) A and III, C and II
b) B and II, D and I
c) A and I, B and IV
d) C and III, D and IV

23. Which of the following matches of names and molecules are correct?

A. alcohol	I. HCOOH
B. aldehyde	II. $(CH_3)_3COH$
C. ketone	III. CH_3OCH_3
D. acid	IV. CH_3COCH_3

a) A and II, C and IV
b) B and II, D and I
c) A and III, B and IV
d) C and III, D and IV

24. Which of the following is an acid derivative?

a) $CH_3\overset{O}{\overset{\|}{C}}CH_3$

b) $CH_3\overset{O}{\overset{\|}{C}}OCH_3$

c) CH_3OCH_3

d) $CH_3\overset{O}{\overset{\|}{C}}CH_2\overset{O}{\overset{\|}{C}}H$

25. Which functional groups are in the following molecule?

I. acid II. alcohol III. aldehyde IV. ester V. ketone

$$HO{-}CH_2{-}\underset{\displaystyle COCH_3}{\underset{|}{\overset{\displaystyle NH_2}{\overset{|}{C}}}}{-}COOH$$

a) I, II, III
b) II, III, IV
c) I, III, V
d) I, II, V

26. Which functional groups are in the following molecule? (Penicillin G)

I. acid
II. alcohol
III. aldehyde
IV. ether
V. ketone

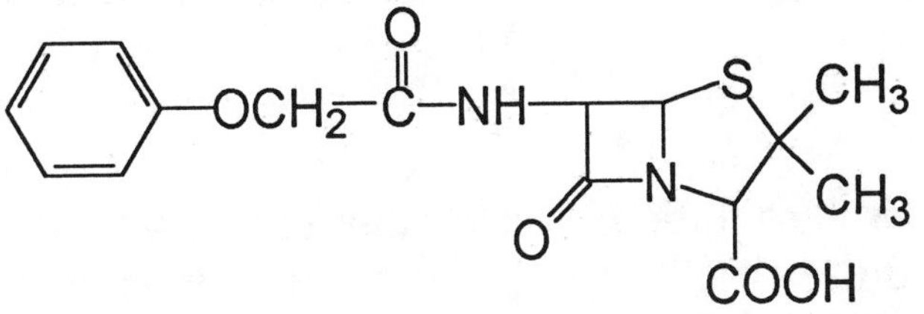

a) I, IV b) II, III, V c) III, IV, V d) I, IV, V

27. Which functional groups are in the following molecule? (Nutrasweet™, Aspartame™)

I. acid
II. alcohol
III. aldehyde
IV. ether
V. ester
VI. ketone

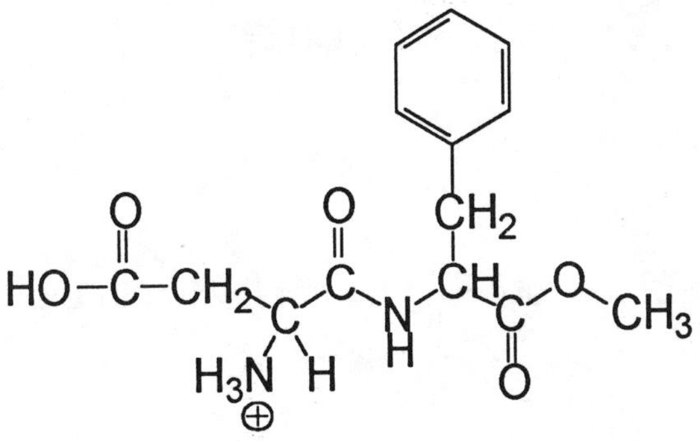

a) I, V b) I, VI c) II, IV d) III, VI

F. CONSTITUTIONAL ISOMERS

28. How many constitutional isomers are there for an alcohol having the molecular formula C_4H_9OH ?

a) 2 b) 3 c) 4 d) 5

29. How many different aldehydes or ketones can there be for the compound having the molecular formula C_3H_6O ?

a) I b) 2 c) 3 d) 4

30. How many constitutional isomers are possible for a compound having the molecular formula $C_4H_{10}O$?

a) 4 b) 5 c) 6 d) 7

TIP There are three possible ether structures: $CH_3OCH_2CH_2CH_3$, $CH_3OCH(CH_3)_2$ and $CH_3CH_2OCH_2CH_3$. There are two straight-chain alcohols: $HOCH_2CH_2CH_2CH_3$ and $CH_3CHOHCH_2CH_3$. There are two branched-chain alcohols: $(CH_3)_2CHCH_2OH$ and $(CH_3)_2COHCH_3$.

31. Which of the following do not belong in the group of constitutional isomers?

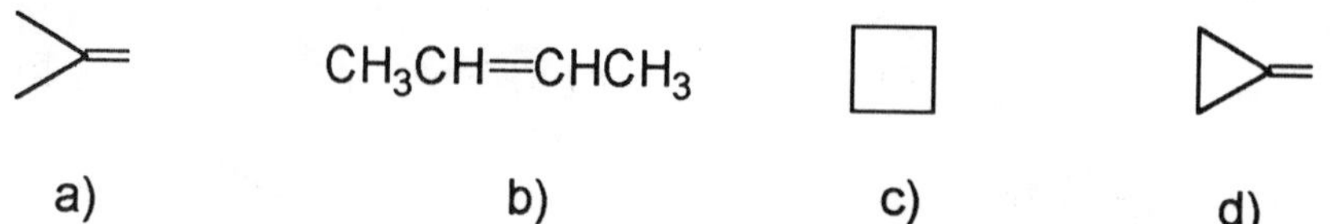

32. Which of the following do not belong in the group of constitutional isomers of C_5H_{10}?

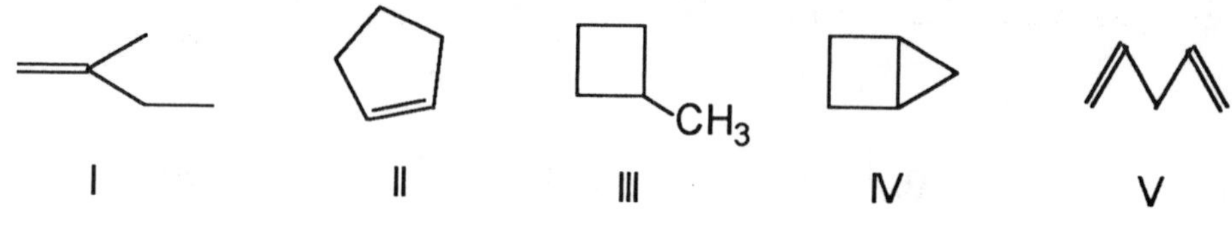

a) I, III, IV b) II, IV, V c) I, II, IV d) II, III, V

TIP Constitutional isomers must have the same molecular formula but with different arrangement of the atoms. Structures I and III are isomers that have a molecular formula of C_5H_{10} , and structures II, IV, and V are isomers that have the molecular formula of C_5H_8.

G. RESONANCE

33. Which of the following are contributing resonance structures?

I. $CH_3CH_2CH_2CH_3 \leftrightarrow (CH_3)_2CHCH_2CH_3$

II. $CH_3OCH_3 \leftrightarrow CH_3CH_2OH$

III. (methylbenzene) ↔ (methylbenzene, alternate Kekulé structure)

IV. $CH_3C(=O)CH_2^{\ominus} \leftrightarrow {}^{\ominus}O{-}C(CH_3){=}CH_2$

V. $CH_3CH(OH)CH_3 \leftrightarrow CH_3CH(\overset{\oplus}{O}H_2)CH_2^{\ominus}$

a) I, II b) III, IV c) II, V d) I, III

34. Which of the following are NOT contributing resonance structures?

I. $HN{=}C{=}O \leftrightarrow HO{-}C{\equiv}N$

II. $\overset{\ominus}{N}{=}C{=}O \leftrightarrow \overset{\ominus}{O}{-}C{\equiv}N$

III. $CH_3C(=O)CH_3 \leftrightarrow CH_3C(OH){=}CH_2$

IV. $CH_3C(=O)CH_2^{\ominus} \leftrightarrow CH_3C(O^{\ominus}){=}CH_2$

a) I, II b) II, IV c) I, III d) III, IV

35. Which of the following are NOT contributing resonance structures?

I. (methylbenzene) ↔ (methylbenzene, alternate Kekulé structure)

II. (cyclohex-1-en-1-ol) ↔ (cyclohexanone)

III. (1,4-dimethylcyclohexane) ↔ (1,3-dimethylcyclohexane)

IV. $CH_3C(=O)NH_2 \leftrightarrow CH_3C(O^{\ominus}){=}\overset{\oplus}{N}H_2$

V. $CH_3CH(OH)NH_2 \leftrightarrow CH_3CH(O^{\ominus})\overset{\oplus}{N}H_3$

a) I, IV, V b) II, III, V c) II, III, IV d) I, II, III

36. Which of the following pairs are contributing resonance structures?

A. $CH_2{=}CH\dot{C}H_2$	I. $\dot{C}H{=}CHCH_3$
B. HCl	II. $\overset{\oplus}{H}\ \overset{\ominus}{Cl}$
C. $CH_2{=}CHCH_3$	III. $CH_3CH{=}CH_2$
D. $\overset{\ominus}{C}H{=}CHCH_3$	IV. $CH{=}\overset{\ominus}{C}HCH_3$
E. $CH_3CH_2\overset{\oplus}{O}H_2$	V. $CH_3\overset{\oplus}{C}H_2\ \ OH_2$

a) A and I, B and II
b) C and III, D and IV
c) A and IV, D and I
d) B and II, E and V

37. Contributing resonance structures can have different energies, and the structure that has the lowest energy is the most important. In the following pairs of resonance structures, which structure makes a greater contribution?

$$\underset{\text{I}}{(CH_3)_2C{=}\overset{\oplus}{O}H} \longleftrightarrow \underset{\text{II}}{(CH_3)_2\overset{\oplus}{C}{-}OH}$$

$$\underset{\text{III}}{CH_3\overset{O}{\overset{\|}{C}}NH_2} \longleftrightarrow \underset{\text{IV}}{CH_3\overset{O^{\ominus}}{\overset{|}{C}}{=}\overset{\oplus}{N}H_2}$$

$$\underset{\text{V}}{CH_3\overset{O}{\overset{\|}{C}}\overset{\ominus}{C}H_2} \longleftrightarrow \underset{\text{VI}}{CH_3\overset{O^{\ominus}}{\overset{|}{C}}{=}CH_2}$$

a) I, III, V
b) II, IV, VI
c) II, III, VI
d) I, III, VI

38. In each pair of contributing structures, which is more important?

$CH_3-\overset{\ominus}{C}=\overset{\oplus}{N}=O \leftrightarrow CH_3-C\equiv\overset{\oplus}{N}-\overset{\ominus}{O}$ $\quad$ $H_2N-C\equiv N \leftrightarrow H_2\overset{\oplus}{N}=C=\overset{\ominus}{N}$

I $\quad$ II $\quad$ III $\quad$ IV

$H_2C=\overset{\oplus}{O}H \leftrightarrow H_2\overset{\oplus}{C}OH$

V $\quad$ VI

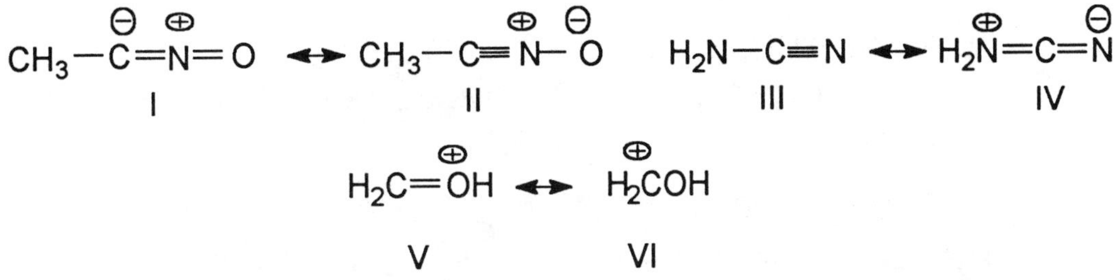

a) I, III, IV $\quad$ b) II, IV, VI $\quad$ c) I, IV, V $\quad$ d) II, III, V

TIP Rule 1), structures that have the most bonds and filled octets are the most stable. Rule 2), charge separation is unfavorable. Rule 3), with completed octets and charge separation, the most important and stable structures are those with the negative charge on the most electronegative element. Structure I violates rule 3). Structure IV violates rule 2). Structure VI violates rule 1)

39. In each pair of contributing structures, which is more important?

:C̈::Ö: ↔ :C:::O: $\quad$ Ö::Ö:Ö: ↔ :Ö:::$\overset{\oplus}{O}$:$\overset{\ominus}{Ö}$:

I $\quad$ II $\quad$ III $\quad$ IV

H:C(::O:):N̈(H):H ↔ H:C(:$\overset{\ominus}{Ö}$:)::$\overset{\oplus}{N}$(H):H

V $\quad$ VI

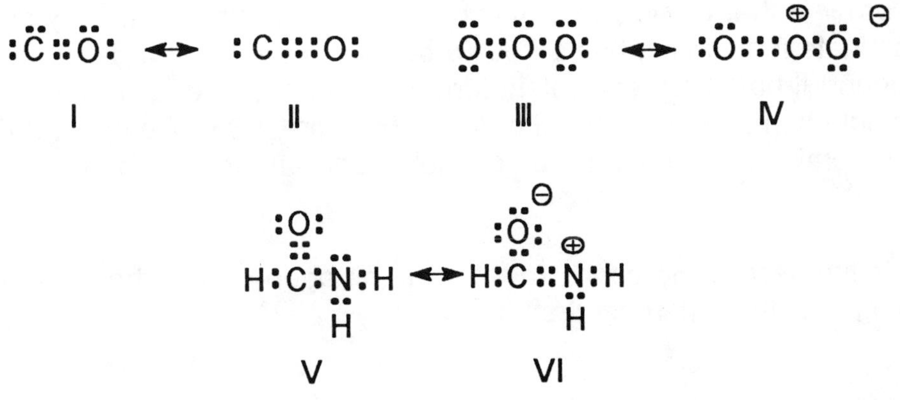

a) II, III, V $\quad$ b) I, III, V $\quad$ c) I, III, VI $\quad$ d) II, IV, V

40. Which of the following ions are stabilized by resonance?

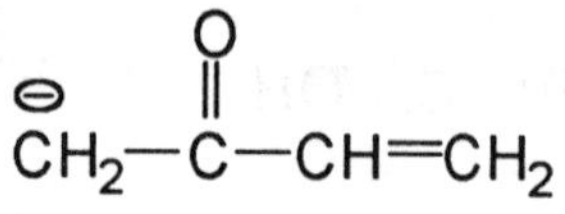

I: $\overset{\ominus}{C}H_2-C(=O)-CH=CH_2$

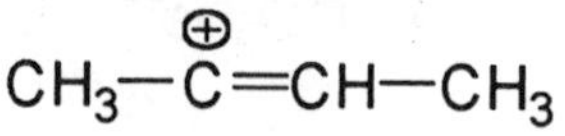

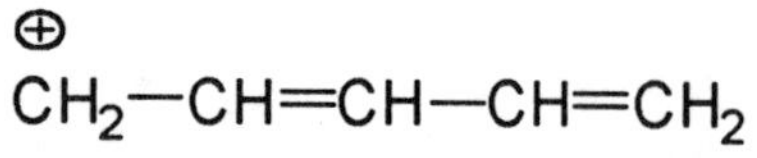

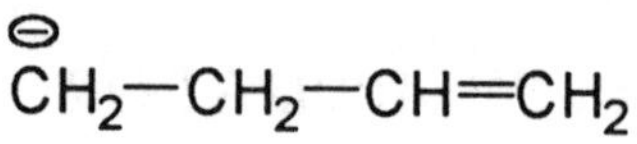

II: $CH_3-\overset{\oplus}{C}=CH-CH_3$

III: $\overset{\oplus}{C}H_2-CH=CH-CH=CH_2$

IV: $\overset{\ominus}{C}H_2-CH_2-CH=CH_2$

a) I, II b) I, III c) II, IV d) III, IV

41. What is the increasing bond order (smallest first) for the carbon-oxygen bonds in the following compounds?

I. CO_2 II. CO_3^{-2} III. CH_3OH

a) I, III, II b) II, III, I c) I, II, III d) III, II, I

TIP For single, double and triple bonds, the bond orders are 1, 2 and 3 respectively. For equivalent resonance hybrids, the bond order is the weighted average of the single, double and triple bond structures. For inequivalent structures, the bond order is that of the predominate form. Compound I has only one predominate form which has a double bond and the bond order is therefore 2. Compound II has three equivalent forms with one C-O double bond and two C-O single bonds and the bond order is therefore 1.33. Compound III has only one predominate form which has a single bond and the bond order is 1.

42. What is the decreasing bond order for the nitrogen-oxygen bond in the following structures (largest first)?

I. 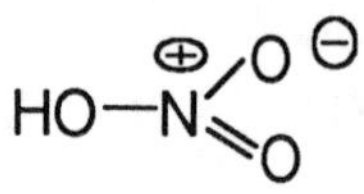 $HO-\overset{\oplus}{N}(-O^{\ominus})=O$ II. 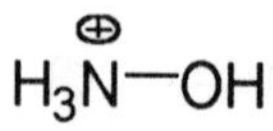 $H_3\overset{\oplus}{N}-OH$ III. 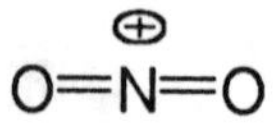 $O=\overset{\oplus}{N}=O$

a) III, I, II b) I, II, III c) II, III, I d) III, II, I

43. What is the increasing bond order for the oxygen-carbon bond in the following ions and compounds (lowest first)?

I. CO_2 II. CO_3^{-2} III. HCO_2^- IV. CH_3OH

a) IV, I, III, II b) II, III, I, IV c) I, III, II, IV d) IV, II, III, I

H. ORBITALS and HYBRIDIZATION

44. What is the order of increasing energy for the following orbitals?

I. sp^3 II. sp^2 III. sp IV. s V. p

a) V, III, II, I, IV b) IV, V, III, II, I c) I, II, III, IV, V d) IV, III, II, I, V

45. Which are the correct orbital hybridizations for the carbon atoms in the following structures?

$H_2C=O$	$H_2C=CH_2$	$CH_3^{\oplus}$	$HC\equiv N$	$O=C=O$
sp	sp^2	sp^3	sp	sp
I	II	III	IV	V

a) I, II, III b) II, III, IV c) II, IV, V d) I, IV, V

46. What are the correct orbital descriptions for the indicated atoms?

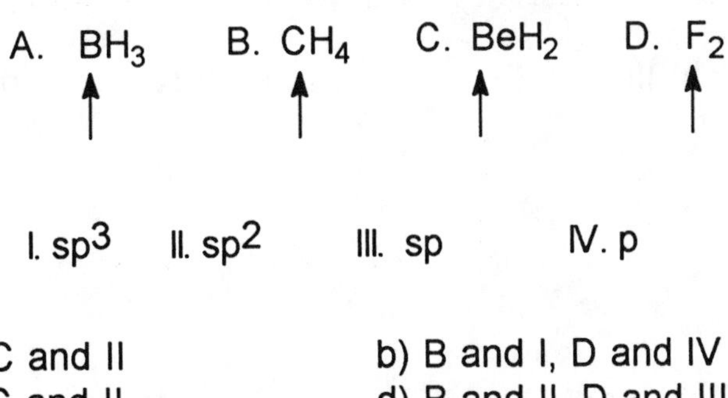

a) A and I, C and II
b) B and I, D and IV
c) B and I, C and II
d) B and II, D and III

47. What are the correct orbital hybridizations for carbon in the following?

A. CH_4 B. $\overset{\oplus}{CH_3}$ C. $\overset{\ominus}{CH_3}$

I. sp II. sp^2 III. sp^3

a) A and III, B and II
b) B and III, C and II
c) B and II, C and I
d) A and I, B and III

48. When forming molecular orbitals from the following atomic orbitals, what is the order of decreasing strength for the resulting bond (strongest first)?

I. sp^3 II. sp^2 III. sp IV. p

a) I, II, III, IV b) IV, III, II, I c) III, II, I, IV d) IV, III, I, II

TIP The s-orbital has a lower energy than the p-orbital, and therefore the greater the s contribution of a hybrid orbital, the lower the energy. The s contribution in I is 25%, in II is 33%, in III is 50%, and in IV is 0%.

49. What is the order of increasing bond length (shortest and strongest first) derived from the following molecule orbitals?

sp^3— sp^3	sp^2 — sp^2	sp — sp	p—p
I	II	III	IV

a) IV, III, II, I b) III, II, I, IV c) II, III, I, IV d) I, II, III, IV

50. Which carbon-hydrogen bond is the strongest?

51. Arrange the following in decreasing order of s-character for the indicated bonds.

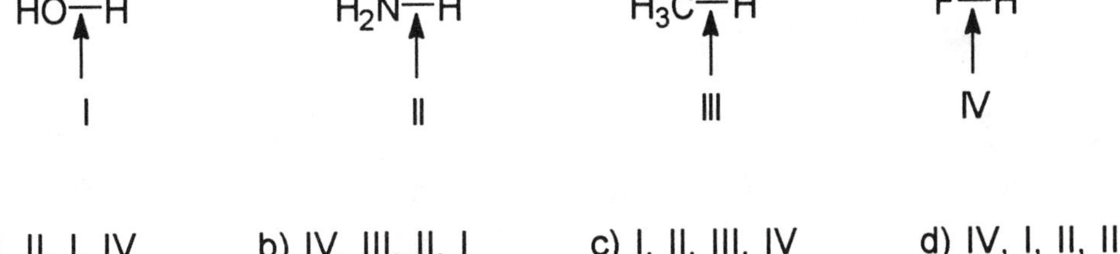

a) III, II, I, IV b) IV, III, II, I c) I, II, III, IV d) IV, I, II, III

52. What is the hybridization of the carbon-hydrogen bond in the methyl cation (CH_3+) ?

a) sp^3 b) sp^2 c) sp d) p

53. What is the hybridization of the boron-hydrogen bond in BH_4^- ?

a) sp^3 b) sp^2 c) sp d) p

TIP First, count the electrons around the central atom in the Lewis structure. Second, count the number of attached atoms. Third, count the number of lone pairs of electrons. In BH_4^- there are 8 electrons around boron, each hydrogen contributes 1, boron contributes 3, and 1 from the negative charge. There are 4 atoms attached to boron and there are no lone pairs of electrons. Therefore, for boron, the 1s orbital and the 3p orbitals are filled and the hybridization is sp^3.

54. What is the hybridization of the nitrogen-boron bond in $H_3N - BF_3$?

a) p - sp^3 b) sp^3 - sp^2 c) sp^3 - sp^3 d) p - p

55. Which compound has the most s-character in the carbon-hydrogen bond?

a)

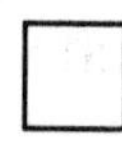

b)

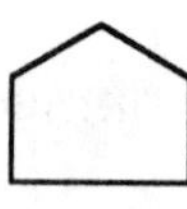

c)

CHAPTER 2 ALKANES and CYCLOALKANES

A. NOMENCLATURE AND STRUCTURE

56. Which of the following is NOT an IUPAC name for an isomer of a compound having the formula C_4H_9Br ?

a) 1-bromobutane
b) 1-bromo-2-methylpropane
c) 2-bromobutane
d) 1-bromo-3-methylpropane

57. What is the IUPAC name for the compound having the following structure?

Cl

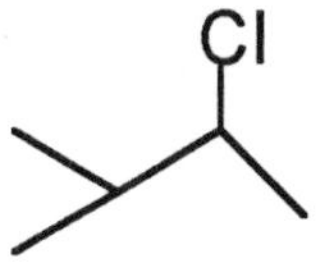

a) 2-methyl-3-chloropentane
b) 2-methyl-3-chlorobutane
c) 1,1-dimethyl-2-chloropropane
d) 2-chloro-3-methylbutane

58. What is the IUPAC name for the compound having the following structure?

$$CH_3-C(CH_2CH_3)_2-CH_2-CH_3$$

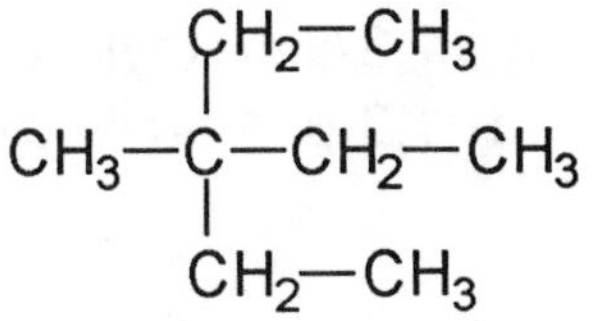

a) 2,2-diethylbutane
b) 2,2,2-triethylethane
c) 3-methyl-3-ethylpentane
d) 3-ethyl-3-methylpentane

59. Which of the following structures have the correct names?

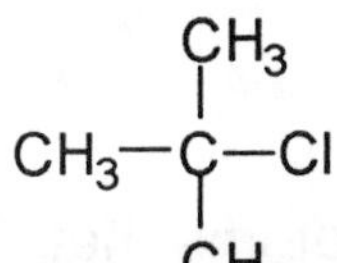

I. $CH_3-C(CH_3)_2-Cl$ — I. tert-butyl chloride

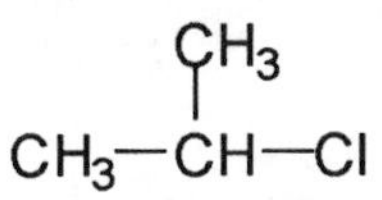

II. $CH_3-CH(CH_3)-Cl$ — II. isobutyl chloride

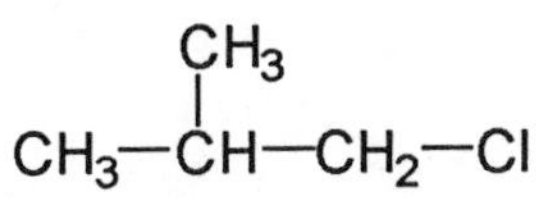

III. $CH_3-CH(CH_3)-CH_2-Cl$ — III. sec-butyl chloride

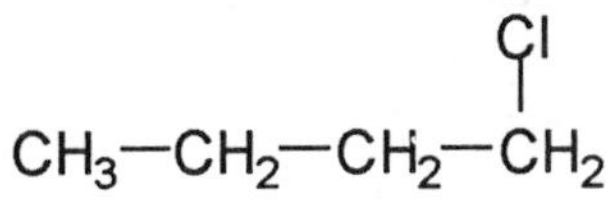

IV. $CH_3-CH_2-CH_2-CH_2Cl$ — IV. n-butyl chloride

a) I and II
b) II and III
c) III and IV
d) I and IV

60. Which of the following structures have the correct names?

$CH_3-C(CH_3)_2-OH$	$CH_3-CH(CH_3)-OH$	$CH_3-CH(CH_3)-CH_2-OH$	$CH_3-CH_2-CH(OH)-CH_3$
I. tert–butyl alcohol	II. sec–butyl alcohol	III. isobutyl alcohol	IV. butane–3–ol

a) I and II b) I and III c) II and IV d) III and IV

61. What is the correct IUPAC name for the following structure?

H_3C–◇–CH_3

a) 1,2-dimethylcyclobutane
b) 1,3-dimethylcyclobutane
c) 1,4-dimethylcyclobutane
d) 1,3-dimethylbicyclo[1.1.0]butane

62. What is the correct IUPAC name for the following structure?

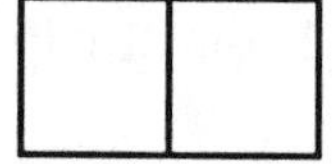

a) bicyclo[2. 1. 0]hexane
b) bicyclo[2. 2. 1]hexane
c) bicyclo[2. 2. 0]hexane
d) bicyclo[3. 1. 0]hexane

TIP Step one is to identify the bridgehead atoms; they are the carbon atoms that are part of two or more rings. Step two is to count the number of atoms that connect the bridgehead atoms in each ring. These numbers are put in brackets in decreasing order. The total of these numbers, plus 2 for the bridgehead atoms, determines the name of the alkane. The correct name is c).

63. What is the correct IUPAC name for the following structure?

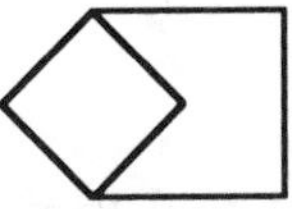

a) bicyclo[2. 1. 1]hexane b) bicyclo[2. 2. 1]hexane
c) bicyclo[2. 2. 0]hexane d) bicyclo[3. 1. 0]hexane

64. Which of the following is a secondary alcohol having the molecular formula $C_4H_{10}O$?

a) $(CH_3)_2CHCH_2OH$ b) $CH_3CH_2CHOHCH_3$ c) HO CH_3 d) $(CH_3)_2CHOH$

65. Which of the following is a primary alcohol having the molecular formula C_4H_8O ?

a) OH b) $-CH_2OH$ c) -OH d) OH

66. What is the formula for isobutylbromide?

a) $CH_3CH_2CH_2CH(Br)CH_3$ b) $(CH_3)_2CHCH_2Br$
c) $Br(CH_3)CHCH_2CH_3$ d) $(CH_3)_2CHCH_2CH_2$ Br

67. How many branched-chain isomers are there of C_6H_{14}?

a) 2 b) 3 c) 4 d) 6

TIP Because of the branch, hexane is eliminated. Pentane has two places for methyl substitution (at 2 and 3). Butane must be substituted with 2 methyl groups and the only possibilities are at 2,2 and 2,3. Any substituted propane is more correctly named as a pentane or butane. So 2 plus 2 is 4, answer c).

68. How many total isomers are there for dimethylcyclopentane (include cis,trans) ?

a) 3 b) 5 c) 7 d) 9

69. How many isomeric alkanes are there that have the molecular formula C_6H_{14}?

a) 3 b) 5 c) 6 d) 7

70. Which of the following is not a conformer of C_5H_{12} ?

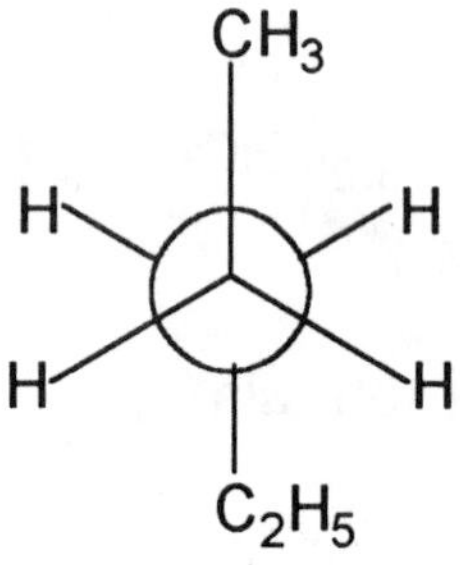

a) 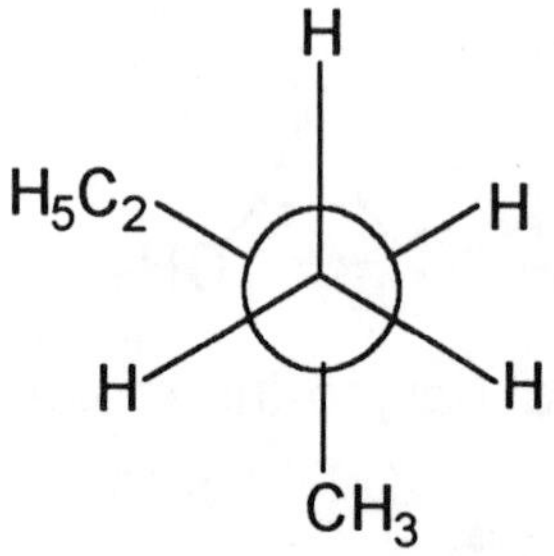 b) 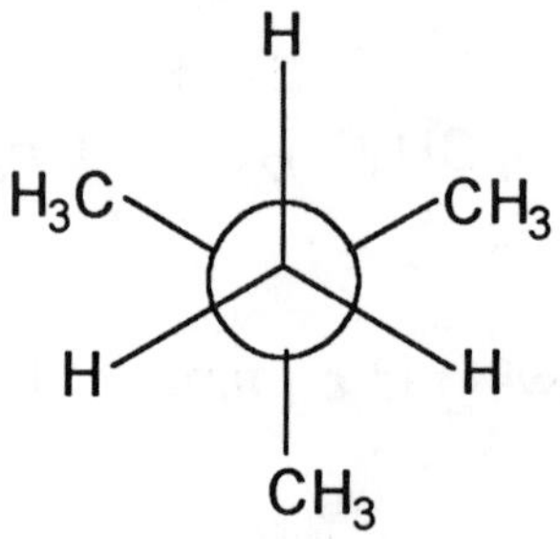 c) 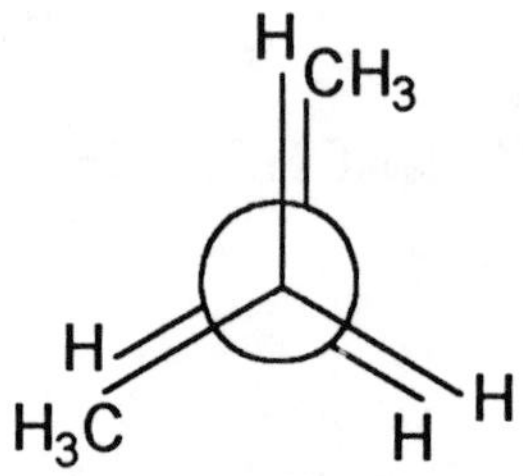 d)

71. Which of the following is a conformer of isobutane?

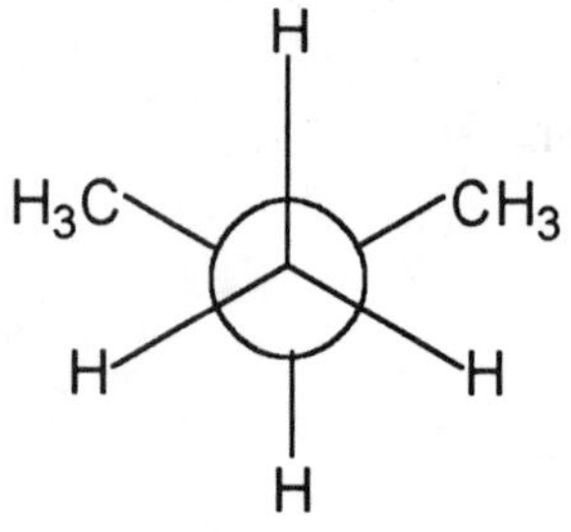

a) 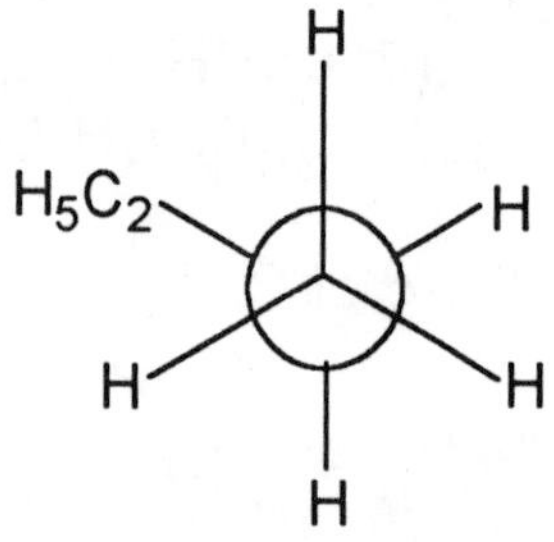 b) 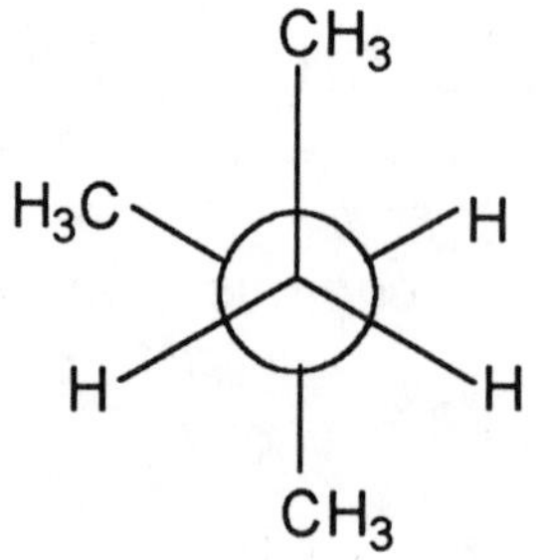 c) 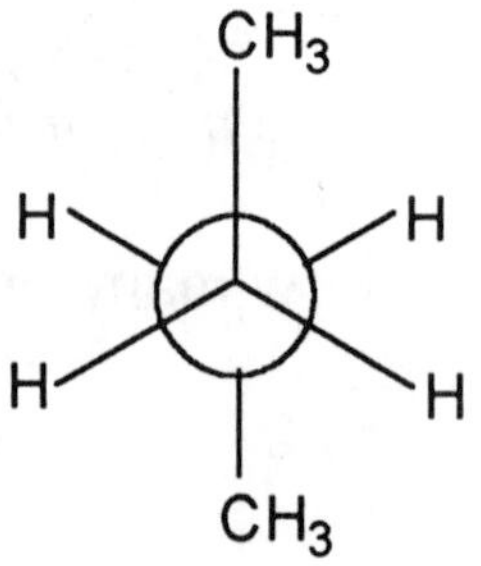 d)

72. Which of the following are gauche forms?

CH_3 H_3C H H H H I
CH_3 H H H H CH_3 II
H H CH_3 H_3C H H III
H H CH_3 H CH_3 H IV
H CH_3 H_3C H H H V

a) I and IV b) II and III c) II and IV d) III and V

73. Which of the following are anti forms?

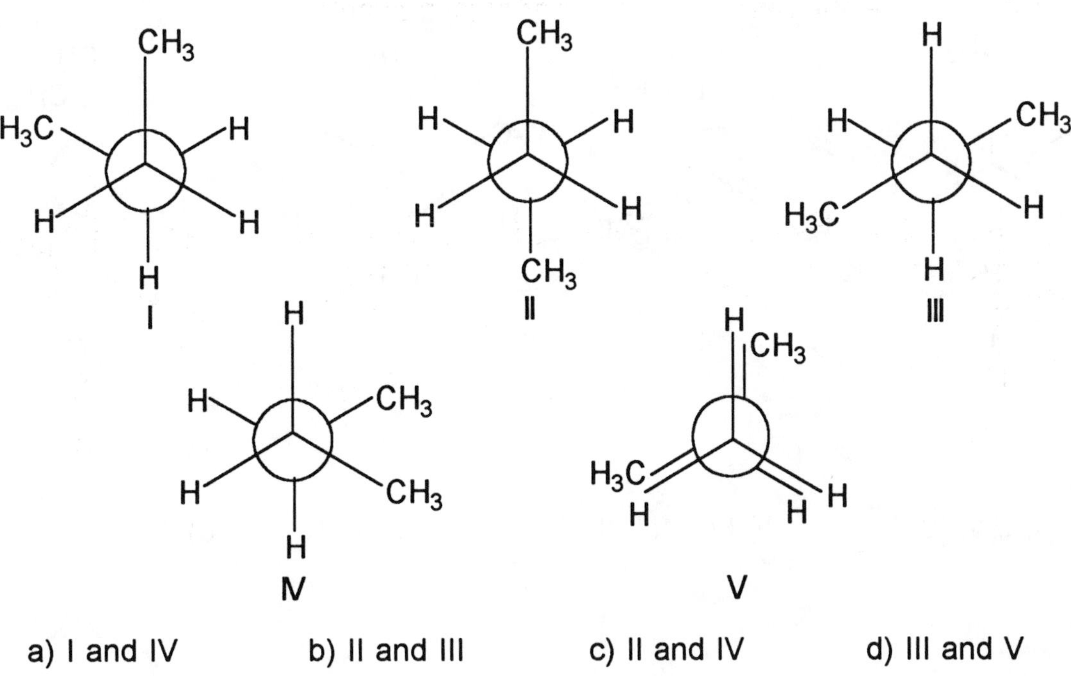

a) I and IV b) II and III c) II and IV d) III and V

74. Which of the following conformers has the highest energy?

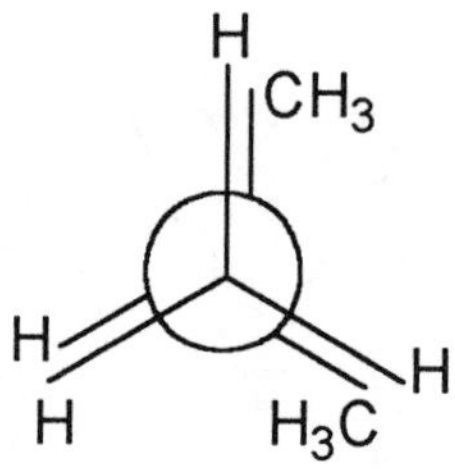

a) 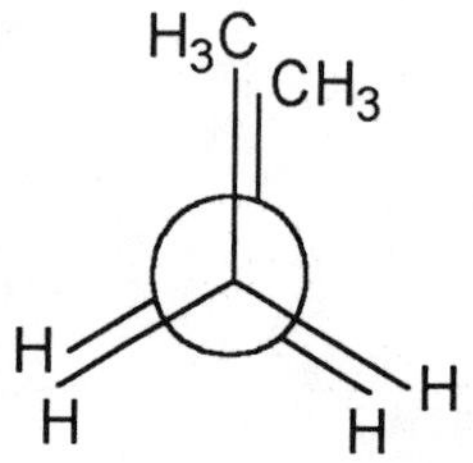 b) 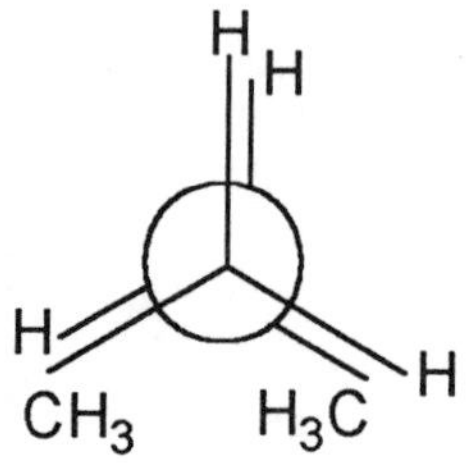 c) 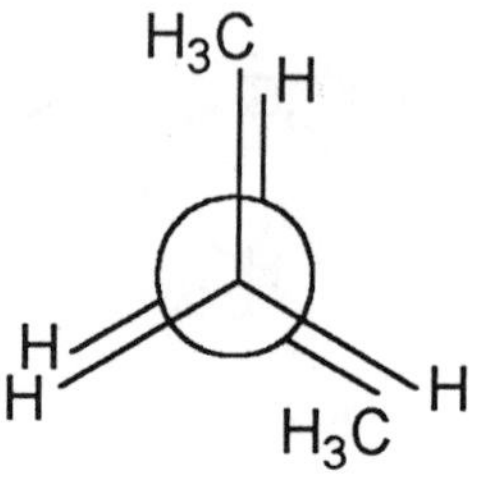 d)

TIP Each of these structures is an eclipsed form and the highest energy structure will have the largest groups eclipsed. Since methyl is larger than H, the methyl-methyl interaction in answer b has the highest energy,

75. Arrange the conformers of 2,3-dimethylbutane in the order of increasing stability (least first).

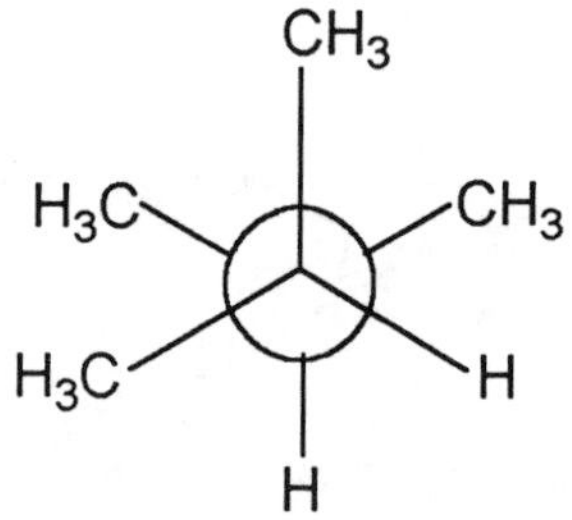

I 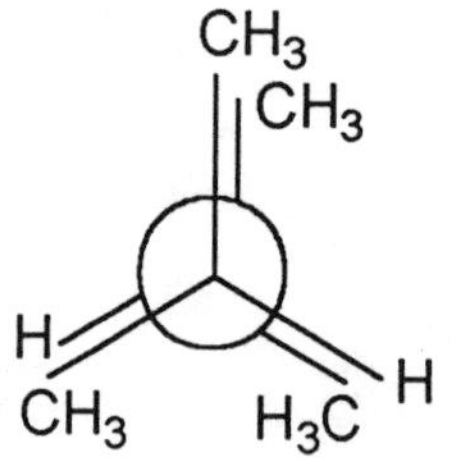 II 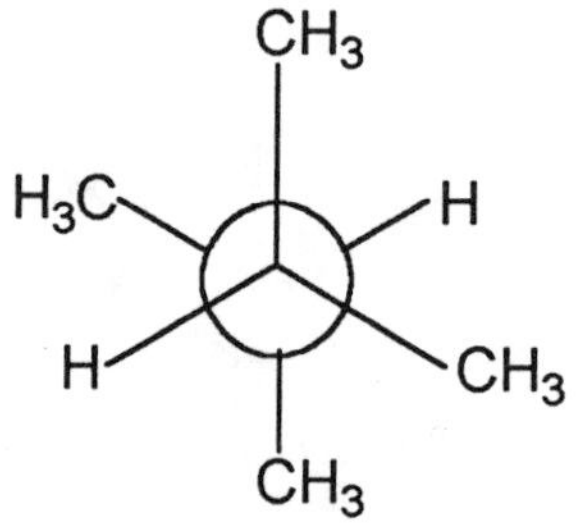 III 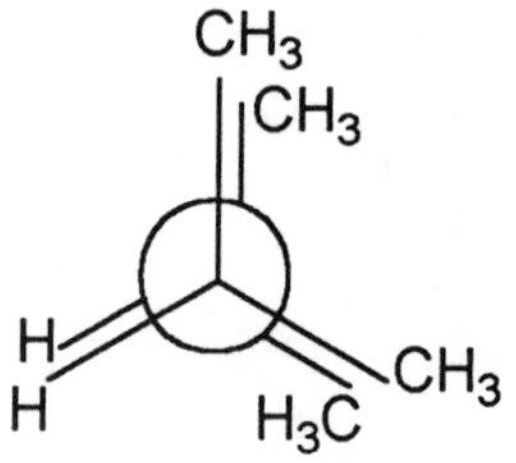 IV

a) I, II, IV, III b) IV, I, III, II c) II, IV, III, I d) IV, II, I, III

76. For methylcyclohexane, the axial conformer is less stable than the equatorial because of which of the following gauche interactions?

a) one methyl-methylene b) two methyl-methylene
c) one methyl-methyl d) two methyl-methyl

77. How many gauche interactions are there for trans-1,3-dimethylcyclohexane?

a) 1 b) 2 c) 3 d) 4

78. How many gauche interactions are there for 1,1-dimethylcyclohexane?

a) 1 b) 2 c) 3 d) 4

79. Which of the following is the most stable chair form of trans-1-isopropyl-3-methylcyclohexane?

a) axial isopropyl, equatorial methyl b) axial isopropyl, axial methyl
c) equatorial isopropyl, axial methyl d) equatorial isopropyl, equatorial methyl

TIP For a trans-1,3 compound, one substituent must be equatorial and the other substituent must be axial. The most stable form will be the one with the largest group (in this case isopropyl) in the equatorial position.

80. How many cis isomers are there for dimethylcyclobutane?

a) 1 b) 2 c) 3 d) 4

81. How many trans isomers are there for dimethylcyclopentane?

a) 1 b) 2 c) 3 d) 4

82. Which of the conformations of cyclohexane is most stable?

a) boat b) chair c) twist boat d) planar

83. What is the order of increasing energy for the different conformers of cyclohexane (least first)?

I.boat II. chair III. twist boat IV. planar

a) I, III, IV, II b) IV, III, I, II c) II, III, I, IV d) II, I, IV, III

TIP The planar form has all 12 hydrogen atoms eclipsed and is therefore the least stable. The boat form has eight hydrogen atoms eclipsed and also has the so-called flagpole interaction. Some of these high energy interactions are reduced in the twist boat. Nevertheless, the chair form with no eclipsed interactions is the most stable.

84. Which of the following are more stable?

cis- or trans-1,4-dimethylcyclohexane cis- or trans-1,3-dimethylcyclohexane

a) cis, cis b) cis, trans c) trans, cis d) trans, trans

85. Which of the following are more stable?

cis- or trans- 1,2-dimethylcyclohexane cis- or trans-1,3-dimethylcyclohexane

a) cis, cis b) cis, trans c) trans, cis d) trans, trans

86. In the most stable conformation of trans 1,4-dimethylcyclohexane, what positions do the methyl groups occupy?

a) axial,axial b) equatorial, axial
c) equatorial, equatorial d) axial, equatorial

87. Which of the following positions is the bridgehead carbon atom?

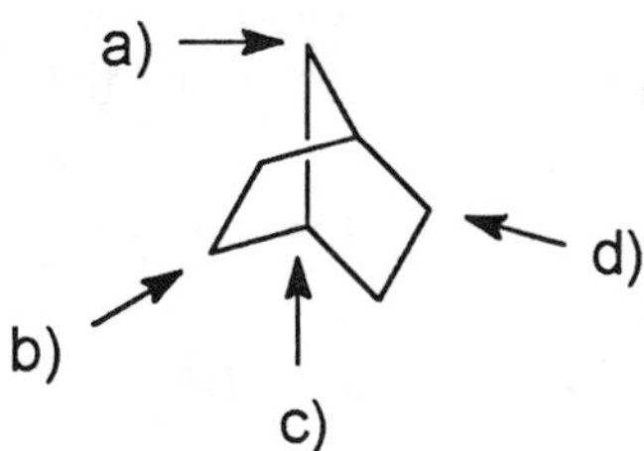

88. Which isomer of dimethylbicyclo [2. 2. 2] octane can have cis and trans forms?

a) 1,2 b) 2,3 c) 1,3 d) 1,4

TIP There is only one substituent at the bridgehead carbon which is numbered 1. This position bisects all other substitution positions and therefore all are equivalent. For the 2 or 3 position, substitution is possible at two different positions and therefore cis and trans forms are possible.

B. PROPERTIES and REACTIONS

89. What is the order of increasing boiling point (lowest first) for the following forms of pentane?

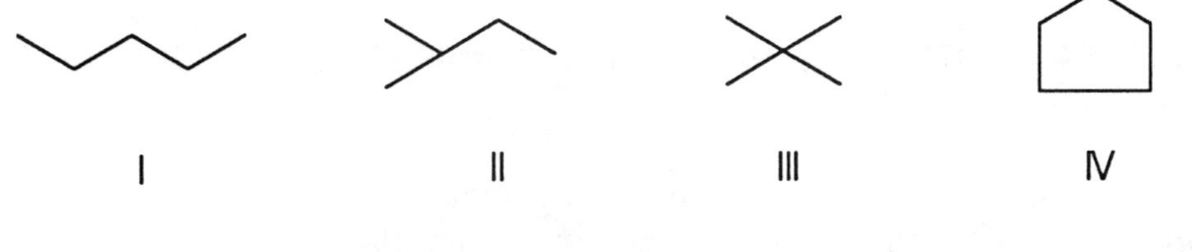

a) III, II, I, IV b) IV, II, III, I c) I, II, III, IV d) III, IV, II, I

90. Which fuel gives the highest heat upon combustion (per mole)?

a) natural gas b) gasoline c) LPG d) all are the same

91. Which of the following contribute most to the strain in cyclopentane?

a) eclipsed hydrogen atoms
b) bent bonds or angle strain
c) transannular non-bonded interaction
d) all of these

92. Arrange the following attractive forces in the order of increasing energy.

I. covalent bonds II. hydrogen bonds III. ionic bonds IV. dispersion forces

a) IV, II, I, III b) II, IV, I, III c) III, II, IV, I d) II, I, III, IV

93. A compound with the molecular formula C_6H_{14} reacted with chlorine in the presence of light to give 5 monochlorinated products. The compound is which of the following?

a) n-hexane
b) 2,3-dimethylbutane
c) 2-methylpentane
d) 3-methylpentane

94. A hydrocarbon gave only one monochlorinated product when reacted with chlorine and heat. What is the hydrocarbon?

a) neopentane
b) isobutane
c) 2,3-dimethylbutane
d) 2,2,3,3-tetramethylbutane

95. In which of the following molecules is abstraction of a primary, a secondary and a tertiary hydrogen possible?

a) neopentane b) isobutane c) isopentane d) cyclopentane

96. Light-catalyzed chlorination of a hydrocarbon is a good synthetic procedure for which of the following compounds?

I. 2-chlorobutane
II. neopentyl chloride
III. 2-chloro-2-methylpropane
IV. cyclohexyl chloride

a) II,IV b) I,II c) III,IV d) II,III

TIP Chlorination is not a selective reaction, so if there are 2 or more kinds of hydrogens, there will be a mixture of products which is very difficult to separate. The starting materials for I (butane) and III (2-methylpropane) have two kinds of hydrogens, but the starting materials for II (neopentane) and IV (cyclohexane) have only one.

97. When 2-methylbutane and bromine react with light, what is the order of increasing amounts of products having bromine in the indicated positions (least first)?

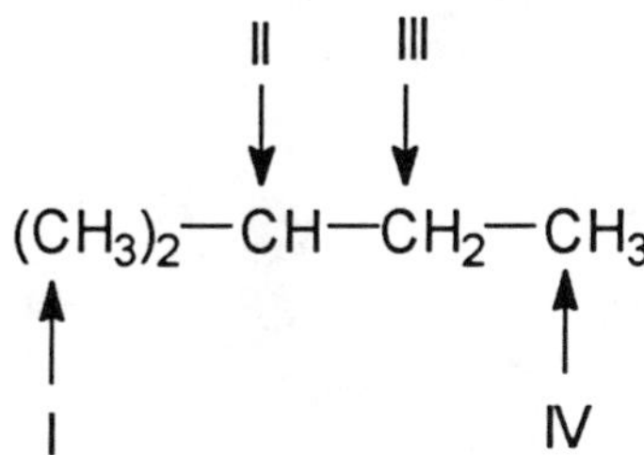

a) I,II,III,IV b) IV, II, III, I c) I, IV, III, II d) IV, I, III, II

98. When 2-methylpentane and fluorine react, what is the ratio of the products, 1-fluoro-2-methylpentane to 2-fluoro-2-methylpentane?

a) 9:1 b) 2:1 c) 3:1 d) 1:400

99. What is the most likely ratio of ethyl chloride to methyl chloride when one mole each of methane and ethane are reacted with chlorine and heat?

a) 1:1 b) 6:4 c) 400:1 d) 4:6

CHAPTER 3 ACIDS and BASES

100. Which of the following are acid-base reactions according to the Bronsted-Lowry theory?

I. $CH_2{=}CH_2 + \overset{\oplus}{H} \rightleftharpoons CH_3\overset{\oplus}{C}H_2$ II. $CH_3\overset{\ominus}{}\overset{\oplus}{Li} + H_2O \rightleftharpoons CH_4 + \overset{\oplus}{Li}\overset{\ominus}{OH}$

III. $AlCl_3 + \overset{\ominus}{Cl} \rightleftharpoons \overset{\ominus}{AlCl_4}$ IV. $HCCl_3 + \overset{\ominus}{OH} \rightleftharpoons \overset{\ominus}{C}Cl_3 + H_2O$

a) I, II, III b) II, III, IV c) I, III, IV d) I, II, IV

101. Which of the following are acid-base reactions according to the Bronsted-Lowry theory?

I. $Br_2 + FeBr_3 \rightleftharpoons \overset{\ominus}{FeBr_4} + \overset{\oplus}{Br}$

II. $AlCl_3 + \overset{\ominus}{Cl} \rightleftharpoons \overset{\ominus}{AlCl_4}$

III. $CH_3NH_2 + HCl \rightleftharpoons CH_3\overset{\oplus}{N}H_3\overset{\ominus}{Cl}$

IV. $CH_3NH_2 + BF_3 \rightleftharpoons CH_3\overset{\oplus}{N}H_2\overset{\ominus}{B}F_3$

V. $(CH_3)_3COH + H_2SO_4 \rightleftharpoons (CH_3)_3\overset{\oplus}{C} + H_2O + \overset{\ominus}{HSO_4}$

a) I, II b) III, V c) IV, V d) II, III

102. Which are correctly identified as Bronsted-Lowry acids in the following reactions?

$$\underset{I}{CH_3NH_2} + \underset{II}{H_2O} \rightleftharpoons \underset{III}{CH_3\overset{\oplus}{N}H_3} + \underset{IV}{\overset{\ominus}{OH}}$$

$$\underset{V}{CH_3OH} + \underset{VI}{HCl} \rightleftharpoons \underset{VII}{CH_3\overset{\oplus}{O}H_2} + \underset{VIII}{\overset{\ominus}{Cl}}$$

$$\underset{IX}{C_2H_5OH} + \underset{X}{\overset{\oplus}{Na}\overset{\ominus}{H}} \rightleftharpoons \underset{XI}{C_2H_5\overset{\ominus}{O}\overset{\oplus}{Na}} + \underset{XII}{H_2}$$

a) II, VI, VII,X b) II, III, VII, IX c) III, VIII, IX, XII d) V, VII, IX, XI

TIP To determine if a species is a Bronsted-Lowry acid or conjugate acid, check the reaction to see if the species has donated a proton. Thus II is correctly identified as an acid as it donates a proton to the Bronsted-Lowry base I to give the conjugate base IV. Also, VII is a conjugate acid as it donates a proton to the conjugate base VIII to give the base V in the reverse reaction.

103. Which are correctly identified as Bronsted-Lowry bases in the following reactions?

$$\underset{I}{CH_3NH_2} + \underset{II}{H_2O} \rightleftharpoons \underset{III}{CH_3\overset{\oplus}{N}H_3} + \underset{IV}{\overset{\ominus}{O}H}$$

$$\underset{V}{CH_3OH} + \underset{VI}{HCl} \rightleftharpoons \underset{VII}{CH_3\overset{\oplus}{O}H_2} + \underset{VIII}{\overset{\ominus}{Cl}}$$

$$\underset{IX}{C_2H_5OH} + \underset{X}{\overset{\oplus}{Na}\overset{\ominus}{H}} \rightleftharpoons \underset{XI}{C_2H_5\overset{\ominus}{O}\overset{\oplus}{Na}} + \underset{XII}{H_2}$$

a) I, IV, V, XII   b) III, VIII, IX, XII   c) II, V, VIII, XI   d) IV, V, VIII, XI

104. Which of the following are acid-base reactions according to the Lewis theory?

$$\text{I. } CH_3\overset{\overset{O}{\|}}{C}CH_3 + \overset{\ominus}{C}H_3 \rightleftharpoons CH_3\overset{\overset{O^{\ominus}}{|}}{C}(CH_3)_2$$

$$\text{II. } CH_3CO_2H + H_2O \rightleftharpoons CH_3C\overset{\ominus}{O_2} + H_3\overset{\oplus}{O}$$

$$\text{III. } NH_3 + BF_3 \rightleftharpoons \overset{\oplus}{N}H_3\overset{\ominus}{B}F_3$$

$$\text{IV. } CH_3OCH_3 + BF_3 \rightleftharpoons (CH_3)_2\overset{\oplus}{O}\overset{\ominus}{B}F_3$$

a) I, II   b) III, IV   c) IV   d) I, II, III, IV

105. Indicate which are acids in the following reactions.

$$\underset{I}{NH_3} + \underset{II}{HCl} \rightleftharpoons \overset{\oplus}{NH_4}\overset{\ominus}{Cl}$$

$$\underset{III}{NH_3} + \underset{IV}{BF_3} \rightleftharpoons \overset{\oplus}{N}H_3\overset{\ominus}{B}F_3$$

$$\underset{V}{AlCl_3} + \underset{VI}{CH_3CH_2OH} \rightleftharpoons CH_3CH_2H\overset{\oplus}{O}\overset{\ominus}{Al}Cl_3$$

a) I, IV, V   b) II, IV, V   c) II, IV, VI   d) I, IV, VI

106. Identify the conjugate acids in the following reactions.

$$CH_3OH + HCl \rightleftharpoons \underset{I}{CH_3\overset{\oplus}{O}H_2} + \underset{II}{\overset{\ominus}{Cl}}$$

$$C_2H_5OH + \overset{\oplus}{Na}\overset{\ominus}{H} \rightleftharpoons \underset{III}{C_2H_5\overset{\ominus}{O}\overset{\oplus}{Na}} + \underset{IV}{H_2}$$

$$C_2H_5\overset{\ominus}{O}\overset{\oplus}{Na} + H_2O \rightleftharpoons \underset{V}{C_2H_5OH} + \underset{VI}{NaOH}$$

a) II, IV, V b) II, III, VI c) I, IV, V d) I, III, V

107. Identify the conjugate bases in the following reactions.

$$C_6H_{11}\text{-}NH_2 + HBr \rightleftharpoons \underset{I}{C_6H_{11}\text{-}\overset{\oplus}{N}H_3} + \underset{II}{\overset{\ominus}{Br}}$$

$$C_2H_5NH_2 + H_2CO_3 \rightleftharpoons \underset{III}{C_2H_5\overset{\oplus}{N}H_3} + \underset{IV}{HC\overset{\ominus}{O}_3}$$

$$(CH_3)_2NH + CH_3CO_2H \rightleftharpoons \underset{V}{(CH_3)_2\overset{\oplus}{N}H_2} + \underset{VI}{CH_3C\overset{\ominus}{O}}$$

$$CH_3OH + H_3PO_4 \rightleftharpoons \underset{VII}{CH_3\overset{\oplus}{O}H_2} + \underset{VIII}{H_2P\overset{\ominus}{O}_4}$$

a) I, III, V, VII b) II, IV, VI, VIII c) I, III, VI, VII d) II, IV, V, VIII

TIP A conjugate base is the product species derived from loss of a proton in the reaction of an acid. Thus II is the conjugate base of the acid HBr, IV is the conjugate base of carbonic acid, VI is the conjugate base of acetic acid, and VIII is the conjugate base of phosphoric acid.

108. Put the following compounds in the order of increasing acidity (least first).

CH_4	NH_3	H_2O	HF	HCl
I	II	III	IV	V

a) II, I, III, IV, V b) II, III, I, V, IV c) I, III, II, V, IV d) I, II, III, IV, V

109. What is the order of increasing basicity for the following compounds and ions (weakest first) ?

H_2O	OH^-	Cl^-	NH_3
I	II	III	IV

a) I, III, IV, II b) III, I, IV, II c) IV, I, II, III d) III, IV, I, II

110. Arrange the following compounds and ions in the order of increasing acidity (weakest first).

CH_3OH	OH^-	H_3O^+	CH_3COOH
I	II	III	IV

a) III, II, I, IV b) I, IV, III, II c) II, I, IV, III d) IV, I, II, III

111. Which of the following ions is the strongest base?

a) $CH_2{=}\overset{\ominus}{C}H$ b) $CH{\equiv}\overset{\ominus}{C}$ c) $CH_3CH_2\overset{\ominus}{O}$ d) $CH_3\overset{\ominus}{C}H_2$

112. Arrange the following substances in the order of increasing acidity according to the Bronsted-Lowry theory (weakest first).

NH_4^+	H_2O	H_3O^+	CH_5^+
I	II	III	IV

a) II, IV, I, III b) IV, III, II, I c) II, I, III, IV d) IV, II, I, III

TIP Loss of a proton from I, III, and IV lead to neutral molecules, whereas loss of a proton from II leads to a negatively charged species which is of higher energy than the neutral molecules. Thus II is the weakest acid. Protonation of methane demands expansion of the octet and is therefore very energetically unfavorable. Therefore, IV is a facile proton donor and is the strongest acid. Considering I and III, a positive charge on the more electronegative atom, O, is more unstable than on N, so III is a stronger acid than I.

113. Arrange the following substances in the order of increasing acidity according to the Lewis theory (weakest first).

BH_3	NH_3	CH_3^+
I	II	III

a) I, II, III b) II, I, III c) III, II, I d) II, III, I

114. Arrange the following in the order of decreasing basicity (strongest first).

CH_3OH	CH_3CH_3	CH_3NH_2	$CH_2{=}CH_2$
I	II	III	IV

a) III, I, IV, II b) I, III, IV, II c) II, III, IV, I d) III, I, II, IV

115. Arrange the following anions in the order of increasing basicity (weakest first).

$CH_3COO^{\ominus}$	$CH_3O^{\ominus}$	$NH_2^{\ominus}$	$HSO_4^{\ominus}$
I	II	III	IV

a) III, II, I, IV b) I, IV, II, III c) IV, I, II, III d) II, III, I, IV

116. What is the approximate pH of 2 grams of NaOH in 100 ml water?

a) 2.5 b) 1.7 c) 12.7 d) 10

117. What is the approximate pH of 36.5 grams of HCl in 1 liter of water?

a) 1 b) 0.1 c) 2 d) 2.1

118. Which of the following solutions of 1 mole each in water would have the highest pH?

a) HI b) HCl c) HF d) CH_3COOH

TIP The stronger the acid, the larger the Ka, and since pKa = - log Ka, the stronger the acid, the smaller the pKa. Moreover, the pH and the pKa are linearly related, so the lowest value of pH will correspond to the strongest acid. The equimolar solution with the highest pH therefore is of the weakest acid, d).

119. Which of the following compounds are Lewis bases?

$$\underset{\text{I}}{CH_3\ddot{N}H_2} + \underset{\text{II}}{CH_3CH_2\ddot{Cl}:} \longrightarrow CH_3\overset{\oplus}{N}H_2CH_2CH_3 + :\ddot{Cl}:^{\ominus}$$

$$\underset{\text{III}}{(CH_3)_3CCl} + \underset{\text{IV}}{AlCl_3} \longrightarrow (CH_3)_3\overset{\oplus}{C} + \overset{\ominus}{AlCl_4}$$

$$\underset{\text{V}}{H-\overset{\overset{\displaystyle O}{\|}}{C}-H} + \underset{\text{VI}}{:NH_3} \longrightarrow H-\underset{\underset{\displaystyle \oplus NH_3}{|}}{\overset{\overset{\displaystyle :\ddot{O}:^{\ominus}}{|}}{C}}-H$$

a) I, III, VI b) I, IV, V c) II, IV, V d) II, IV, V

120. Which of the following species can be both Lewis acids and Lewis bases?

H_2O	$HC\equiv CH$	CCl_4	$CH_3-\overset{\overset{\displaystyle O}{\|}}{C}-CH_3$
I	II	III	IV

a) I, III, IV b) I, II, IV c) II, III, IV d) I, IV

TIP A Lewis acid is an electron acceptor and a Lewis base is an electron donor. For a molecule to function as both, there have to be reasonable reactions of both types. The easiest to see as both is H_2O, which by loss of a proton is an electron acceptor and therefore a Lewis acid. But reaction at the lone pair of electrons on oxygen involves electron donation and H_2O now acts as a Lewis base. For II, the pi bonds can act as electron donors, and since the C-H bond is acidic, both Lewis acid and Lewis base reactions are reasonable. A similar case is made for IV, where reaction at the oxygen electron pair is a Lewis base reaction and reaction at the C-H bonds is a Lewis acid reaction. While a basic reaction from an electron pair on chlorine is reasonable, there is no electron pair accepting reaction that leads to stable products.

121. Which of the following species can be both Bronsted-Lowry acids and Bronsted-Lowry bases?

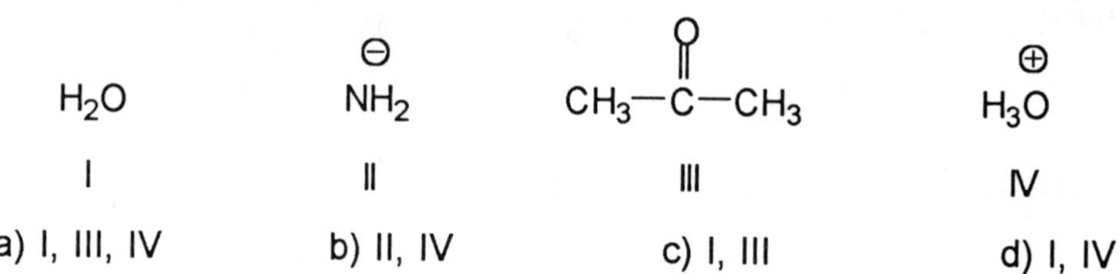

a) I, III, IV   b) II, IV   c) I, III   d) I, IV

122. Which of the following species is not a Lewis acid?

$FeCl_3$	HCl	CCl_4	$HClO_4$
I	II	III	IV

a) I, III, IV   b) III   c) II, III, IV   d) I, IV

123. What is the role of ethyl chloride in the following reaction?

$$CH_3CH_2Cl + AlCl_3 \longrightarrow CH_3\overset{\oplus}{C}H_2 + \overset{\ominus}{AlCl_4}$$

a) Lewis acid   b) Bronsted-Lowry acid
c) Lewis base   d) Bronsted-Lowry base

124. What is the product of the reaction of propanide ion with deuterium oxide?

$$CH_3CH_2\overset{\ominus}{C}H_2 + D_2O \longrightarrow ?$$

$CH_3CH_2CH_2OD$	$CH_3CH_2CH_3$	$CH_3CH_2CH_2D$	$CH_3CH_2CD_3$
a)	b)	c)	d)

125. At which site is the preferential attack by a Bronsted-Lowry acid on mescaline?

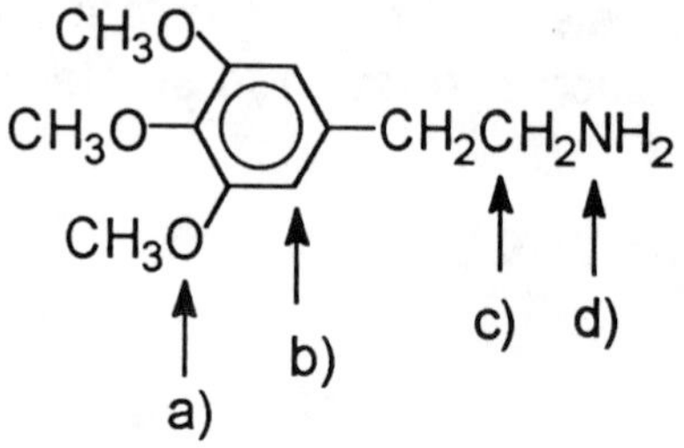

CH_3O, CH_3O, CH_3O (ring) —$CH_2CH_2NH_2$

a) (methoxy O)   b) (ring)   c) (CH_2)   d) (NH_2)

CHAPTER 4 ALKENES I

A. NOMENCLATURE

126. What is the correct name for the following structure?

$$CH_2{=}CH{-}CH_2Cl$$

a) vinyl chloride b) 1-chloro-2-propene c) 3-propylchloride d) allyl chloride

127. What is the correct name for the following structure?

$-CH_3$

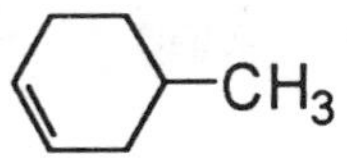

a) 1-methyl-4-cyclohexene
b) 1-methyl-3-cyclohexene
c) 4-methylcyclohexene
d) 3-methylcyclohexene

128. What is the correct name for the following structure?

H_3C, H on one carbon; H, CH_2CH_3 on the other: $(H_3C)(H)C{=}C(H)(CH_2CH_3)$

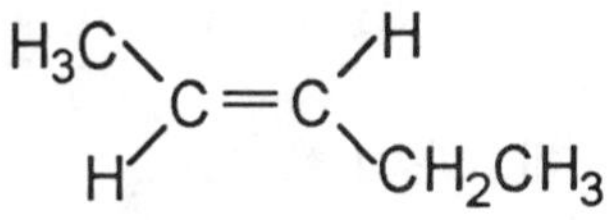

a) 2-pentene
b) trans-1-methyl-2-ethylethene
c) cis-2-pentene
d) trans-2-pentene

TIP The steps for naming are as follows. First, look for the longest chain. The name will come from that with the ending, ene. Second, number the double bond with the lowest possible number. Third, number and name other substituents. Fourth, use cis and trans where appropriate to designate groups on the same (cis) or opposite (trans) sides of the double bond. Thus for the first step, pentene; for the second step, 2; for the third step, no other substituents; and for the fourth step, trans.

129. What is the correct name for the following structure?

Cl
CH_3C H
C=C
H H CH_3

a) trans-2-chloro-3-pentene
b) cis-2-chloro-3-pentene
c) trans-4-chloro-2-pentene
d) cis-4-chloro-2-pentene

130. What is the correct name for the following structure?

H H
C=C
H_3C CH_2Cl

a) cis-1-methyl-3-chloroethylene
b) cis-1-chloro-2-butene
c) 4-chloro-2-butene
d) cis-3-methylethylene chloride

131. What is the correct name for the following structure?

$-CH_2-CH=CH_2$

a) 3-cyclopentyl-2-butene
b) allyl cyclopentane
c) vinylcyclopentane
d) 1-cyclopentyl-2-methylethylene

132. What is the correct structure for methylenecyclopentane?

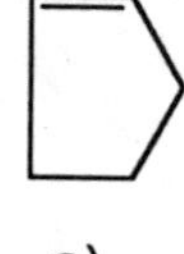
a)

$-CH_3$
b)

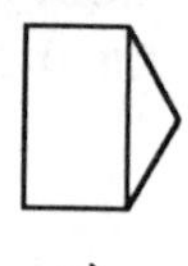
c)

$=CH_2$
d)

133. What is the correct structure for vinylcyclobutane?

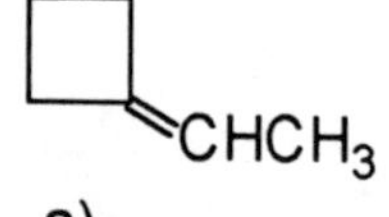

a)

CH_3
b)

CH_2
c)

$CH=CH_2$
d)

B. STRUCTURE

134. Which compound contains the shortest carbon-carbon bond distance?

a) ethyne b) ethylene c) cyclohexane d) ethane

135. Which compound contains the shortest carbon-carbon single bond?

a) pentane b) 1,4-pentadiene c) pentene d) 1,3-pentadiene

136. Which molecule contains the strongest carbon-carbon single bond?

a) ethane b) propene c) 1,3-butadiene d) cyclohexene

137. In the following alkene, what is the order of decreasing bond strength (strongest first) for the indicated carbon-hydrogen bond?

$CH_2{=}CH{-}CH_2{-}CH{=}CH{-}CH_2{-}CH_3$

(arrows: I at CH_2 (C3), II at CH (C4), III at CH_2 (C6), IV at CH_3)

a) II, I, III, IV b) II, IV, III, I c) II, III, I, IV d) I, II, III, IV

TIP Carbon-hydrogen bond strengths follow from the hybridization of the carbon and decreases in the order of $sp>sp^2>sp^3$. A second factor is whether the radical so formed is resonance stabilized and therefore the C-H bond is weaker. The carbon in II is sp^2 hybridized as opposed to sp^3 for the others, and is therefore the strongest. The radical derived from breaking the carbon-hydrogen bond in I is resonance stabilized by two adjacent pi bonds. The radical from III is stabilized by one pi bond. The radical from IV is not stabilized. So, the order is II, IV, III, I.

138. What is the hybridization for the 3,4 carbon-carbon bond in cyclohexene?

a) sp, sp b) sp^2, sp^3 c) sp^3, sp^3 d) s, sp^3

139. What is the correct hybridization for the indicated carbon atoms?

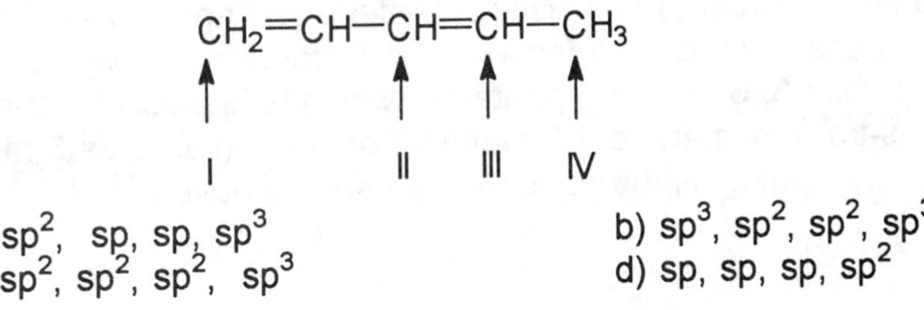

a) sp^2, sp, sp, sp^3
b) sp^3, sp^2, sp^2, sp^3
c) sp^2, sp^2, sp^2, sp^3
d) sp, sp, sp, sp^2

140. What is the correct description for the hybridized bonds (C-C, C-H) in ethylene?

a) sp-sp, sp-s b) sp^2-sp^2, sp^2-s c) sp^2-sp^2, sp^3-s d) Π, s

141. What is the correct representation for the Π bonding molecular orbital of ethylene?

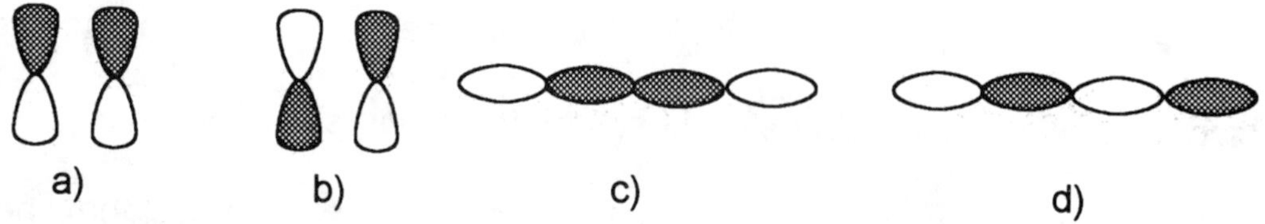

C. CIS - TRANS ISOMERIZATION

142. Which of the following alkenes show cis-trans isomerization?

a) 2-methyl-2-hexene
b) 3-methyl-3-hexene
c) 3-ethyl-3-methyl-1-pentene
d) 2-chloropropene

143. Which of the following alkenes does not show cis-trans isomerization?

a) 2,6-dimethyl-2,6-octadiene
b) 2-pentene
c) 3-hexene
d) 3-methylcyclohexene

144. How many total isomers can there be for a methyl-substituted pentene?

a) 5 b) 8 c) 9 d) 12

TIP First, determine how many methyl-substitutions are possible and where. Then for each of these isomers, determine how many show cis-trans isomerization. Then do the numbers. For a pentene, substitution can be only at carbon 2 and 3, since at 1 it would be a hexene and at 4 it would be the same as at 2. Substitution at carbon 2 would allow the following alkenes: 2-methyl-1-pentene, 2-methyl-2-pentene (no cis-trans isomers for either because both have identical groups on one of the carbons), 4-methyl-2-pentene (cis and trans), 4-methyl-1-pentene (no cis-trans isomers). So far we have 5 isomers. Moving on to substitution at carbon 3, 3-methyl-2-pentene (which is identical to 3-methyl-3-pentene) has cis-trans isomers, but 3-methyl-1-pentene (same as 3-methyl-5-pentene) does not. Adding all the possible isomers we have 8.

145. How many total stereoisomers are possible for 2,4-heptadiene?

a) 4 b) 8 c) 2 d) 6

146. What is the order of increasing priority in the EZ system for the following groups (lowest first)?

NH_2	OH	Cl	CH_3
I	II	III	IV

a) IV, II, I, III b) III, I, IV, II c) IV, I, II, III d) I, IV, III. II

147. Place the following groups in the order of increasing priority in the EZ system (lowest first).

$CH_3O-C(=O)-$	$CH_3-C(=O)-$	$HC(=O)-$	$-C(=O)OH$
I	II	III	IV

a) III, II, IV, I b) II, I, IV, III c) IV, I, III, II d) III, I, II, IV

148. What is the correct order of increasing priority in the EZ system for the following groups (lowest first)?

I. vinyl II. ethyl III. isopropyl IV. tert. butyl

a) II, I, III, IV b) III, IV, I, II c) II, III, I, IV d) IV, III, I, II

TIP In assigning priorities when the atoms connected to the double bond are the same, we use the next atoms along the chain. For tert. butyl, isopropyl, and ethyl, the connecting atom is carbon for all, but the next atoms are 3 carbons, 2 carbons and one, respectively. Double bonds are treated as if each bond was separately bonded to another carbon atom. To determine the priority, break the Π bond and mentally place a carbon atom at each end. Therefore vinyl is of a higher priority than isopropyl, but less than tert. butyl.

149. Which of the following is a designated Z structure?

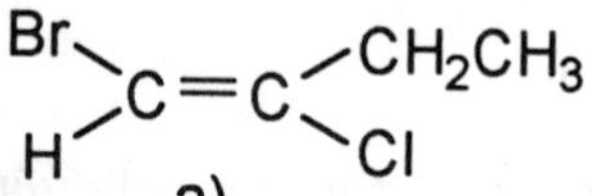

a) $\mathrm{(Br)(H)C{=}C(CH_2CH_3)(Cl)}$

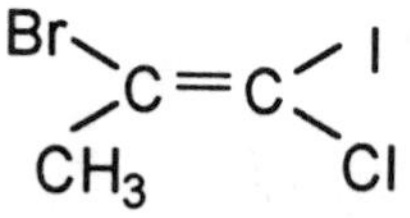

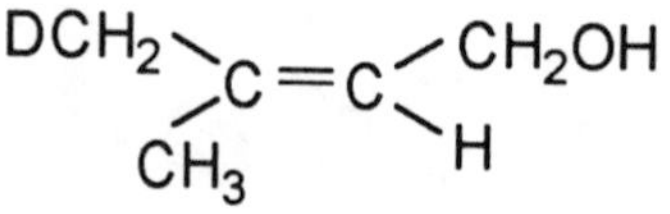

b) $\mathrm{(CH_3)(H)C{=}C(CH_2CH_3)(CH(CH_3)_2)}$

c) $\mathrm{(Br)(CH_3)C{=}C(I)(Cl)}$

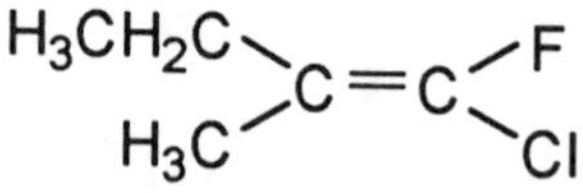

d) $\mathrm{(H_3CH_2C)(H_3C)C{=}C(F)(Cl)}$

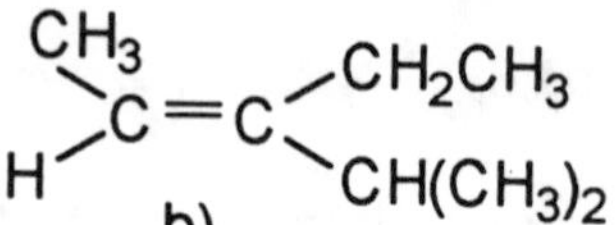

150. Which of the following is a designated Z structure?

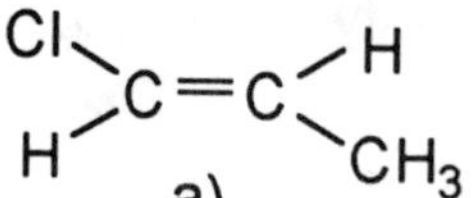

a) $\mathrm{(Cl)(H)C{=}C(H)(CH_3)}$

b) $\mathrm{(ClCH_2)(CH_3)C{=}C(CH_3)(CH_2CH_3)}$

c) $\mathrm{(DCH_2)(CH_3)C{=}C(CH_2OH)(H)}$

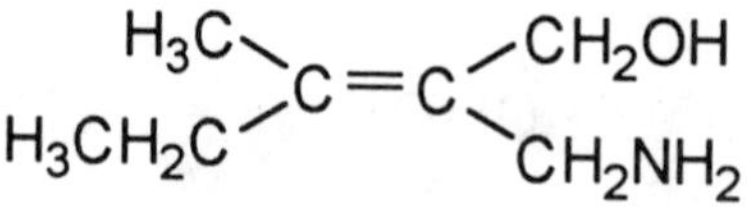

d) $\mathrm{(H_3C)(H_3CH_2C)C{=}C(CH_2OH)(CH_2NH_2)}$

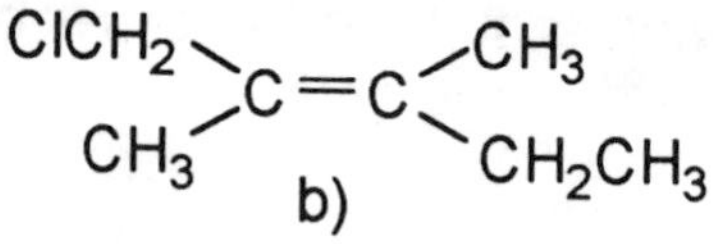

151. Which of the following is a designated Z form?

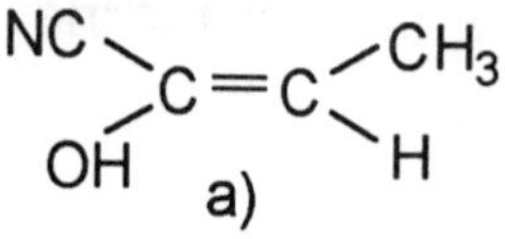

a) $\mathrm{(NC)(OH)C{=}C(CH_3)(H)}$

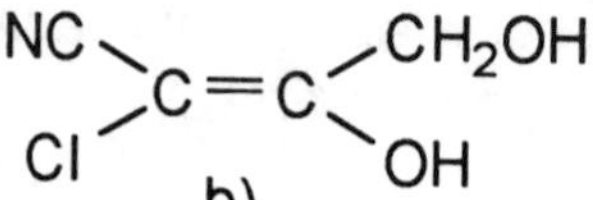

b) $\mathrm{(NC)(Cl)C{=}C(CH_2OH)(OH)}$

c) $\mathrm{(CH_3)(SH)C{=}C(CH_2OH)(H)}$

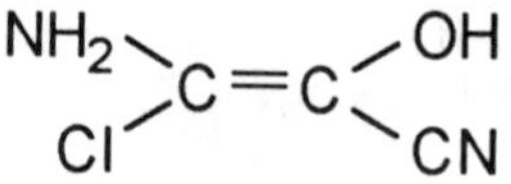

d) $\mathrm{(NH_2)(Cl)C{=}C(OH)(CN)}$

152. Which of the following is a designated E form?

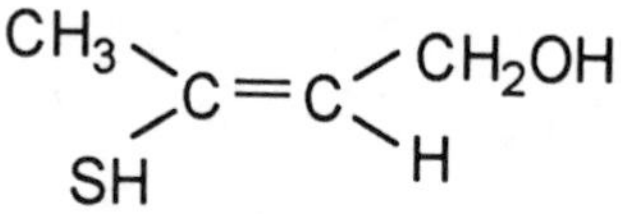

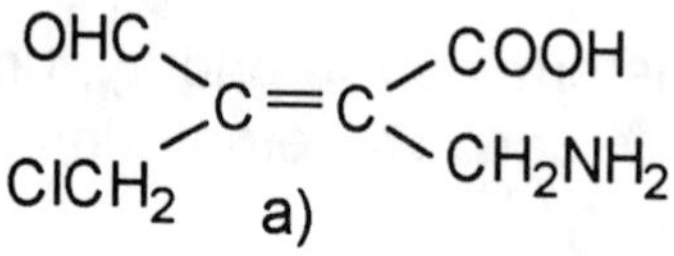

a) $\mathrm{(OHC)(ClCH_2)C{=}C(COOH)(CH_2NH_2)}$

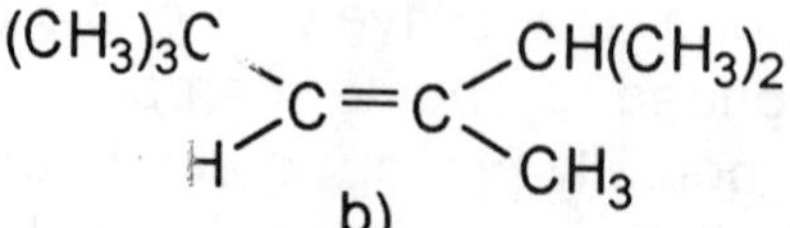

b) $\mathrm{((CH_3)_3C)(H)C{=}C(CH(CH_3)_2)(CH_3)}$

c) $\mathrm{(CH_3)(H)C{=}C(H)(H)}$

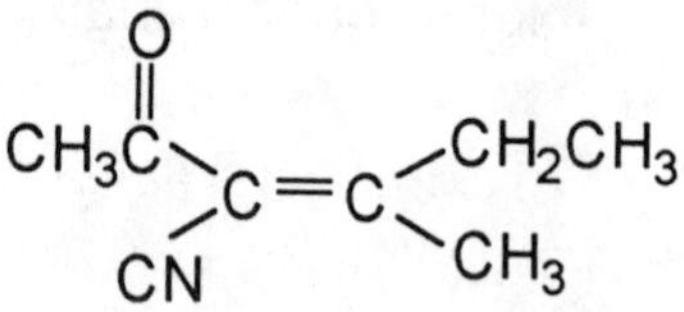

d) $\mathrm{(CH_3C(=O))(CN)C{=}C(CH_2CH_3)(CH_3)}$

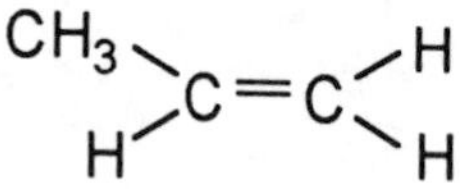

153. What is the correct structure for Z-2-bromo-3-chloro-2-pentene?

a) Br(CH$_3$)C=C(Cl)CH$_2$CH$_3$ (Br and Cl on the same side; CH$_3$ and CH$_2$CH$_3$ below)

b) Br(Cl)C=C(CH$_3$)CH$_2$CH$_3$ (Br and CH$_3$ on the same side; Cl and CH$_2$CH$_3$ below)

c) CH$_3$(Br)C=C(Cl)CH$_2$CH$_3$ (CH$_3$ and Cl on the same side; Br and CH$_2$CH$_3$ below)

d) H(CH$_3$)C=C(Br)CH$_2$Cl (H and Br on the same side; CH$_3$ and CH$_2$Cl below)

154. Which carbon atoms in butene lie in a straight line?

C=C—C—C

I II III IV

a) I, II b) I, III c) I, II, III d) I, II, III, IV

155. Which carbon atoms in cyclohexene lie in the same plane?

V IV

CH$_3$

I II III

a) II, IV b) I, II, IV c) I, II, III, IV d) I, II, III, IV, V

TIP Both carbon atoms of the double bond are sp^2 hybridized and the atoms bound to them must lie in a plane for maximum pi bonding. So six atoms are in the same plane. The sixth atom not shown is the hydrogen on the carbon atom labeled IV.

156. Trans-cyclooctene is less stable than cis-cyclooctene principally for which of the following reasons?

a) steric interactions of the alkyl chain
b) hydrogen-hydrogen interactions
c) bending of the carbon-carbon sigma bond
d) twisting of the carbon-carbon pi bond

D. DEHYDROHALOGENATION REACTION

157. Which halides on dehydrohalogenation give as a major product 3,4-dimethyl-3-heptene?

I. 1-chloro-3,4-dimethylheptane
II. 2-chloro-3,4-dimethylheptane
III. 3-chloro-3,4-dimethylheptane
IV. 4-chloro-3,4-dimethylheptane
V. 5-chloro-3,4-dimethylheptane

a) I, II, IV   b) II, III   c) III, IV   d) III, IV, V

158. Which halide can give only one product upon dehydrohalogenation?

a) 3-bromopentane
b) 2-chloro-4,4-dimethylpentane
c) 3-chloro-2,2-dimethylhexane
d) 2-chloro-3,3-dimethylpentane

159. Which halide on dehydrohalogination gives an alkene which will show cis-trans isomerization?

a) butylbromide
b) cyclohexylbromide
c) 2-bromo-4,4-dimethylpentane
d) 2-bromo-3,3-dimethylbutane

160. Which halide gives the highest yield of isobutene upon dehydrohalogenation?

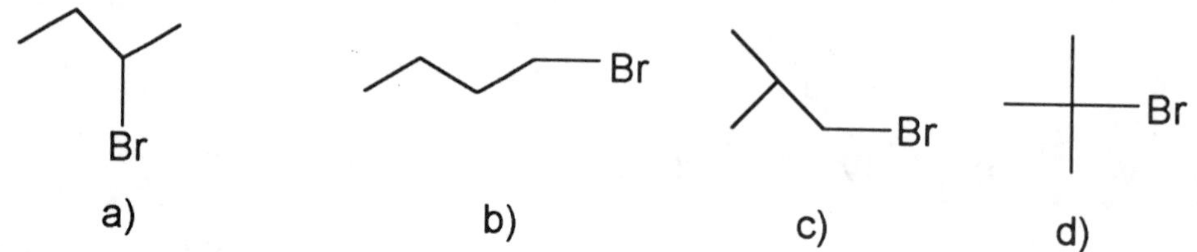

TIP For the elimination reaction, the order of reactivity is tertiary>secondary>primary. Moreover, for the competing substitution reaction, the order is exactly opposite. For compounds a) and b) the dehydrohalogenation products are not isobutene. Answer d) is tertiary and answer c) is primary, therefore, d) will give the highest yield of isobutene.

161. Which halide gives the highest yield of 2-methyl-2-butene upon dehydrohalogenation with base?

a) 1-chloro-2-methylbutane
b) 1-chloro-3-methylbutane
c) 2-chloro-3-methylbutane
d) 2-chloro-2-methylbutane

162. How many trans isomers are there for an alkene with the formula C_4H_7Br?

a) 2 b) 3 c) 4 d) 6

163. Which of the following halides will give the highest yield of cyclohexene by elimination?

a) bromocyclohexane b) chlorocyclohexane
c) fluorocyclohexane d) iodocyclohexane

TIP In the elimination reaction, the units of H-X are lost and so the leaving group ability of X is critical. Leaving group ability is solely determined by the basicity of X, so the stronger the acid, HX, the weaker the base, X, and the better the leaving group ability. Since HI is the strongest acid, iodide is the best leaving group and d) will give the highest yield.

164. How many total alkenes are possible for the product of the following reaction?

$$CH_3CH_2\underset{\underset{Br}{|}}{\overset{\overset{CH_3}{|}}{C}}CH_2CH_2CH_3 \xrightarrow[\text{heat}]{\text{KOH}}$$

a) 2 b) 4 c) 5 d) 7

165. How many total alkenes are possible for the product of the following reaction?

$$CH_3\overset{\overset{CH_3}{|}}{C}H\underset{\underset{Br}{|}}{C}HCH_2CH_2CH_3 \xrightarrow[\text{heat}]{\text{KOH}}$$

a) 2 b) 3 c) 4 d) 5

CHAPTER 5 ALKENES II

A. REACTION MECHANISMS

166. Which of the following alkenes has the highest heat of hydrogenation?

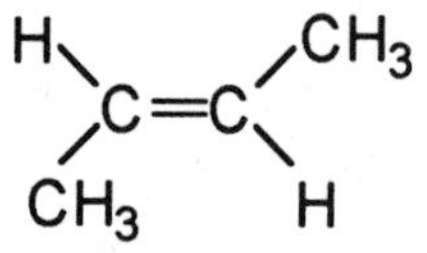

H CH$_3$
C=C
CH$_3$ H
a)

H H
C=C
CH$_3$ CH$_3$
b)

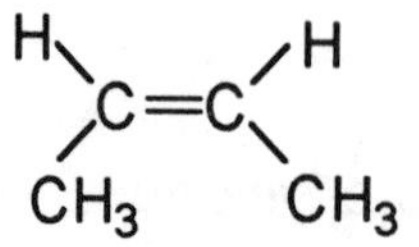

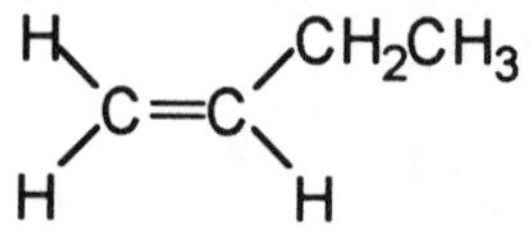

H CH$_2$CH$_3$
C=C
H H
c)

CH$_3$ CH$_3$
C=C
CH$_3$ CH$_3$
d)

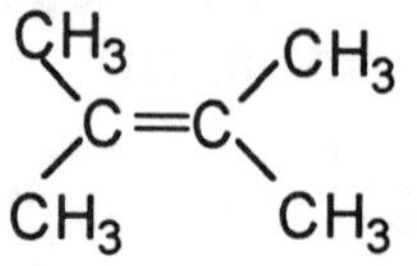

167. Arrange the following isomers of hexadiene in the order of increasing stability (least stable first).

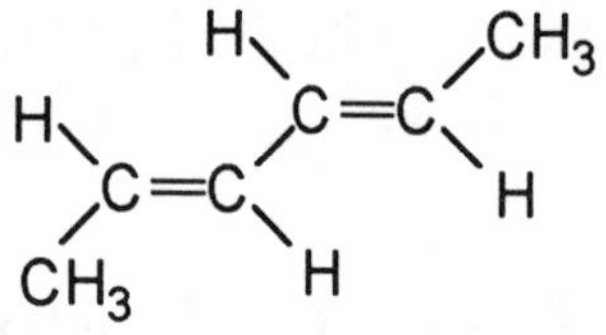

H CH$_3$
C=C
H H
C=C
CH$_3$ H
I

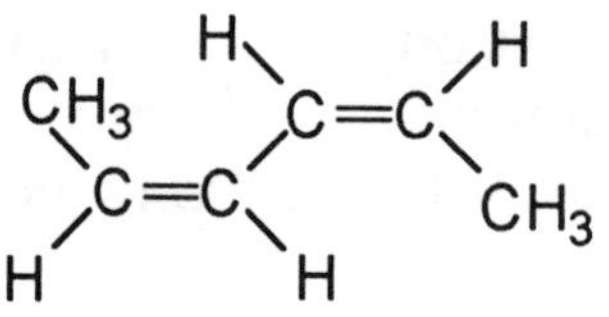

H H
CH$_3$ C=C
C=C CH$_3$
H H
II

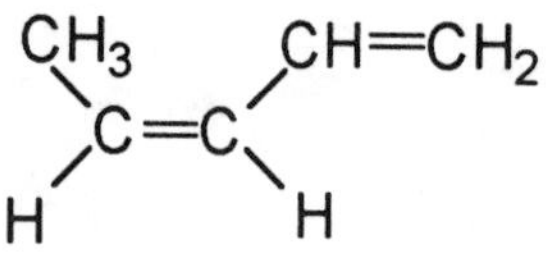

CH$_3$ CH=CH$_2$
C=C
H H
III

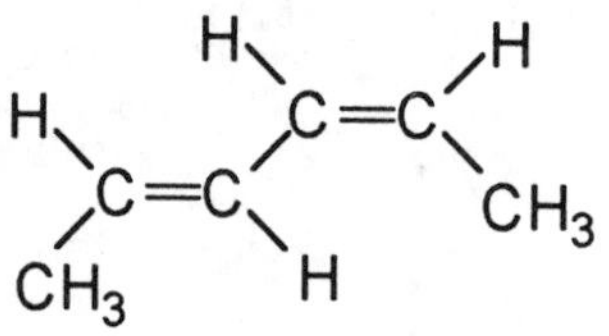

H H
H C=C
C=C CH$_3$
CH$_3$ H
IV

a) II, IV, III, I b) III, II, IV, I c) I, IV, II, III d) IV, II, III, I

TIP Two factors contribute to stability: the degree of alkene substitution (the more highly substituted is more stable); and the trans or cis geometry (the trans isomer is more stable than the cis). The III isomer is only monosubstituted at one double bond (all the others are disubstituted at both double bonds) and is therefore the least stable. The II isomer has a cis-cis geometry, the IV isomer has cis-trans and the I isomer has trans-trans, so the order is II, IV, I.

168. Which are the allylic hydrogen atom positions in the following molecule?

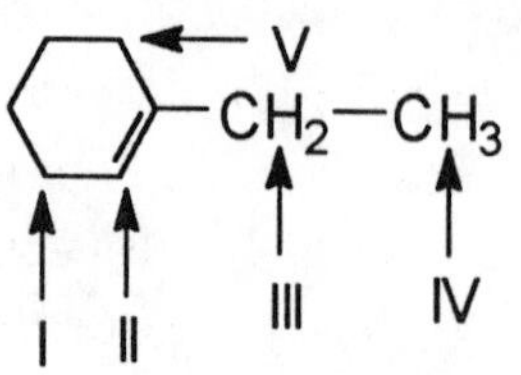

V
CH$_2$–CH$_3$
I II III IV

a) I, V b) II, III c) IV d) I, III, V

169. Which of the following reactions with alkenes involve a carbocation intermediate?

I. Br_2 / H_2O II. HBr III. NBS IV. $KMnO_4$

a) I, II b) III, IV c) II, III d) I, IV

170. Which of the following carbocations are most likely to rearrange?

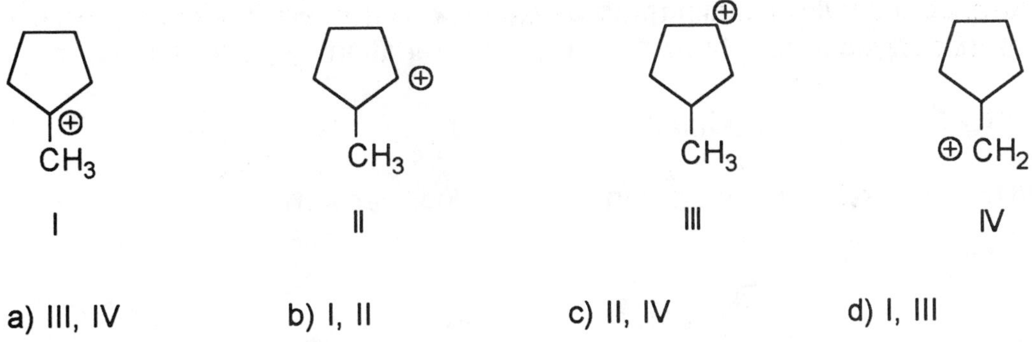

a) III, IV b) I, II c) II, IV d) I, III

171. Arrange the following carbocations in the order of increasing stability (least stable first).

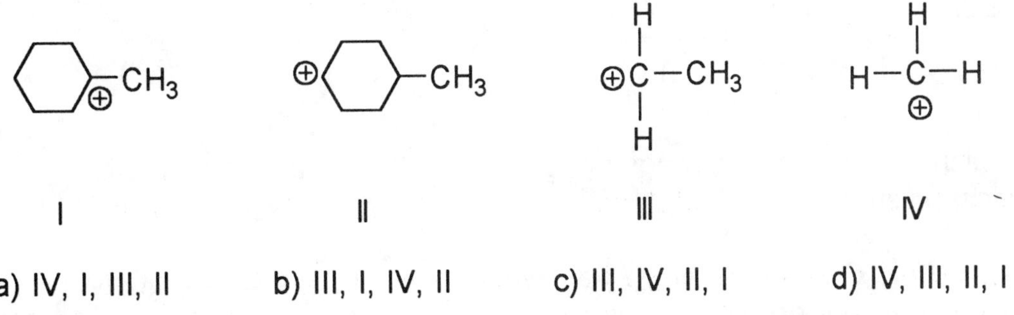

a) IV, I, III, II b) III, I, IV, II c) III, IV, II, I d) IV, III, II, I

172. Which of the following substituents can stabilize a carbocation?

I. CH_3 II. C_2H_5 III. H IV. Cl V. $(CH_3)_3C$

a) I, II, V b) III, IV c) I, II, III d) I, II

TIP According to Coulomb's Law, factors that concentrate charge are destabilizing, but factors that disperse charge are stabilizing. A carbocation is electron deficient and is therefore stabilized by electron donating groups. Alkyl groups are electron donating and thus methyl, ethyl and tert-butyl substituents can stabilize a carbocation. In contrast, chlorine is electron withdrawing and so is destabilizing. Hydrogen is neutral.

173. Rearrangements involving carbocations are common for which of the following reactions?

I. hydration of alkenes
II. catalytic hydrogenation of alkenes
III. dehydrohalogenation in base
IV. electrophilic additions to alkenes

a) I, II, III b) I, III, IV c) III, IV d) I, IV

174. Which of the following reactions with alkenes involve a free radical mechanism?

I. hydrobromination with peroxides
II hydroboration
III. high temperature bromination
IV. oxidation by potassium permanganate

a) I, III b) II, III c) I, IV d) III, IV

175. What is a likely intermediate in the following reaction?

1-methylcyclohexene ($-CH_3$) + HBr + H_2O_2 →

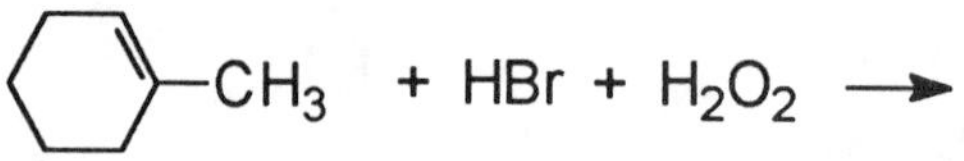

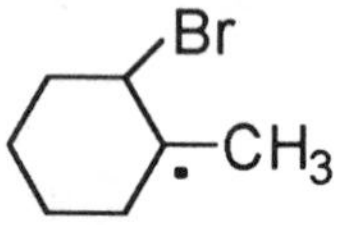

a) 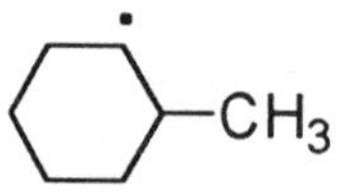 b) 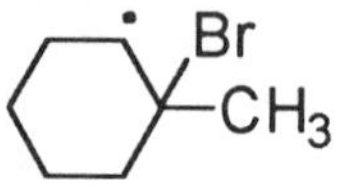 c) 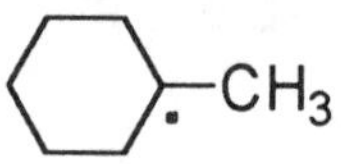 d)

176. Which of the following reagents react with alkenes by a free radical mechanism?

I. HCl / H_2O
II. HBr / peroxides
III. $Hg(OAc)_2$ / CH_3OH
IV. B_2H_6 followed by H_2O_2 and base
V. CCl_4 with heat and peroxides

a) I, IV b) II, V c) I, III d) II, IV

177. What is a likely side product when trans-2-butene reacts with HCl in ethanol?

a) 2-chloro-3-ethoxybutane
b) sec-butyl ethyl ether
c) dibutyl ether
d) diisobutyl ether

TIP The intermediate in the reaction is the 2-butyl cation which can collapse with all the nucleophiles present in solution. In addition to chloride ion which reacts with the cation to give the major product, the solvent, ethanol, provides another nucleophile. Reaction of ethanol with the carbocation followed by loss of a proton from oxygen leads to sec-butyl ethyl ether.

B. KEY REACTIONS

178. Which of the following reactions involve addition to an alkene in a Markovnikov orientation?

I. catalytic hydrogenation
II. hydrohalogenation with peroxides
III. acid-catalyzed hydration
IV. electrophilic addition of HX

a) I, II b) III, IV c) I, III, IV d) II, III

179. Which of the following reagents add to an alkene in a Markovnikov orientation?

I. HCl in H_2O
II. $Hg(OAc)_2$ in methanol
III. B_2H_6 followed by H_2O_2 and base
IV. H_2O, H_2SO_4

a) I, II, IV b) II, IV c) I, II d) I, III, IV

180. Which of the following reagents add to an alkene in an anti-Markovnikov orientation?

a) water / sulfuric acid
b) HCl in acetic acid
c) $Hg(OAc)_2$
d) HBr / peroxides

181. Which of the following reactions with alkenes follow Markovnikov's rule?

I. electrophilic addition of unsymmetrical reagents
II acid-catalyzed hydration
III. hydroboration and oxidation
IV. free radical addition of HBr

a) I, II b) II, IV c) III, IV d) II, III

182. Which of the following are examples of syn addition to alkenes?

I. catalytic hydrogenation
II. addition of HBr
III. hydration
IV. hydroboration

a) I, II b) II, III c) III, IV d) I, IV

183. Which of the following reagents react with alkenes with syn addition?

I. BH_3 / H_2O_2 II. $Hg(OAc)_2$, $NaBH_4$

III. $KMnO_4$, H_2O IV. O_3, $(CH_3)_2S$

a) I, II b) I, III c) III, IV d) II, IV

184. Which of the following reactions involve an addition in an anti orientation to cyclopentene?

I. electrophilic addition of bromine

II. reaction with potassium permanganate and water

III. hydroboration IV. addition of HOCl

a) I, II b) III, IV c) II, IV d) I,IV

TIP Electrophilic addition of bromine (or chlorine) to an alkene leads to the bridged bromonium ion. Attack by a nucleophile in the second step occurs on the opposite face, leading to overall anti addition. Thus the addition of bromine (I) and HOCl (IV) will follow this route. The other two reactions do not involve electrophilic additions and in fact both have syn addition.

185. Which of the following reagents give an electrophilic addition to alkenes?

I. H_2 and Pt II. HCl III. 10% H_2SO_4

IV. Br_2 in CCl_4 V. O_3

a) II, III, IV b) I, II c) II, III d) I, IV, V

186. Which of the following reagents give an electrophilic addition to alkenes?

I. Br_2 in H_2O II. $Hg(OAc)_2$

III. HCl IV. H_2 and Pt

a) II, III b) I, II, III c) I, IV d) II, IV

187. According to the mechanism proposed for electrophilic addition of bromine to alkenes, which of the following could not be products from the bromination of 1-pentene in methanol?

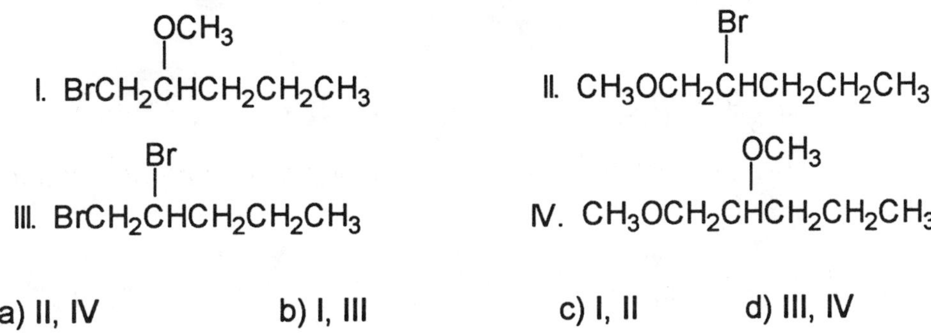

a) II, IV b) I, III c) I, II d) III, IV

TIP The first step involves electrophilic addition of bromine to the double bond at carbon atom 1 to give the more stable carbocation at carbon atom 2. The second step involves attack at carbon atom 2 by an available nucleophile. In the solvent, methanol, there are two possible nucleophiles, bromide and the lone pair of electrons on oxygen in methanol. Each possibility for step two leads to III and I. Product II involves addition of bromine in step one to the wrong carbon atom and product IV erroneously involves electrophilic attack by methanol in step one.

C. PREPARATIONS

188. Which reagents could be used to prepare a diol from an alkene?

I. H^+ and H_2O II. cold, dilute $KMnO_4$ III. OsO_4 and Na_2SO_3 in H_2O

IV. hot, concentrated $KMnO_4$ V. O_3 followed by $(CH_3)_2S$

a) I, III b) II, III c) IV, V d) II, V

189. Which reagents could be used to prepare cis-1,2-dihydroxycyclohexane from cyclohexene?

a) H^+ and H_2O b) CH_3COOH and H_2O
c) cold, dilute $KMnO_4$ d) O_3, then $(CH_3)_2S$

190. Which of the following alkenes will yield a ketone when reacted with ozone followed by reduction with dimethyl sulfide?

$CH_3CH{=}CH_2$ I

$CH_3CH{=}CHCH_3$ II

$(CH_3)_2C{=}CH_2$ III

$(CH_3)_2C{=}C(CH_3)_2$ IV

(1-methylcyclohexene)–CH_3 V

a) II, III, V b) I, III, V c) III, IV, V d) I, II, III

191. Which of the following alkenes will yield an aldehyde when reacted with ozone followed by reduction and hydrolysis with dimethyl sulfide?

(1,2-dimethylcyclohexene, CH_3, CH_3) I

$CH_2{=}CHCH_3$ II

$(CH_3)_2C{=}C(CH_3)_2$ III

$(CH_3)_2{=}CH_2$ IV

(cyclohexenyl)–$C{=}CH_2$ V

a) I, II, IV b) II, IV, V c) III, IV, V d) I, II, V

TIP In the reaction of ozone with alkenes followed by reduction/hydrolysis, the carbon-carbon double bond is cleaved and a carbon-oxygen double bond is made at each end. Thus to obtain an aldehyde (R-CHO), the original alkene must have at least one hydrogen atom on the alkene double bond. If not, a ketone will be the product, as in structures I and III. The others have the necessary hydrogen atoms to make aldehydes.

192. Which alcohol is prepared from the reaction of 1-methylcyclohexene with borane followed by hydroperoxide in base?

a) 1-methylcyclohexanol
b) trans-2-methylcyclohexanol
c) cis-2-methylcyclohexanol
d) cis and trans-2-methylcyclohexanol

193. What is a product when 3-methyl-1-pentene is reacted with ozone and then with dimethylsulfide?

a) 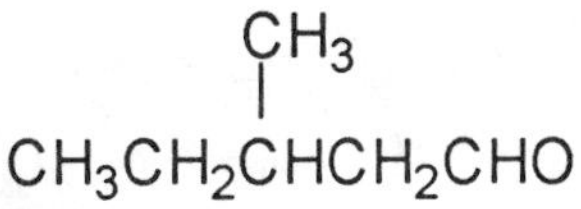 $CH_3CH_2CH(CH_3)CH_2CHO$

b) 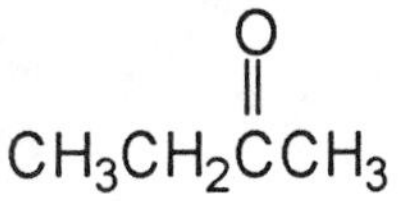 $CH_3CH_2C(=O)CH_3$

c) 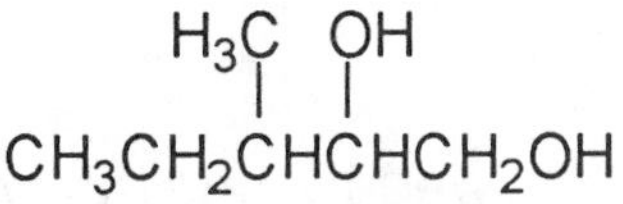 $CH_3CH_2CH(CH_3)CH(OH)CH_2OH$

d) 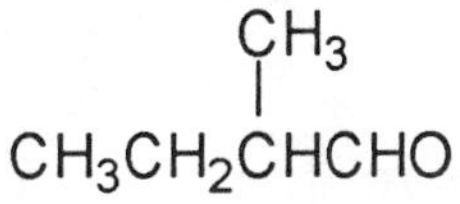 $CH_3CH_2CH(CH_3)CHO$

D. STEREOCHEMICAL CONSIDERATIONS

194. In the product from the following reaction, how many deuterium atoms in the equatorial position are there for the most stable conformer?

1,2-dideuteriocyclohexene $\xrightarrow[Pt]{H_2}$

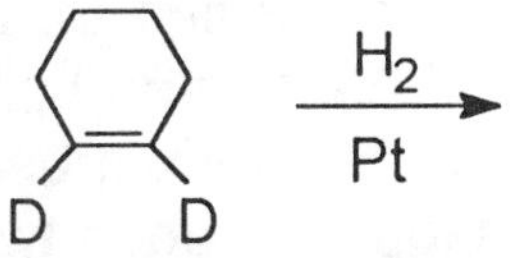

a) 0 b) 1 c) 2 d) 3

TIP Hydrogenation on the platinum metal surface (Pt) leads to cis addition of the hydrogen atoms to the alkene. In 1,2 substituted cyclohexenes, if the substituents are cis, one is in an axial position and one is in an equatorial position. Since addition of hydrogen is on the same side (cis), one of the deuterium atoms remains in the equatorial position.

195. In the product from the following reaction, how many deuterium atoms in the equatorial position are there for the most stable conformer?

1,2-dimethylcyclohexene $\xrightarrow[Pt]{D_2}$

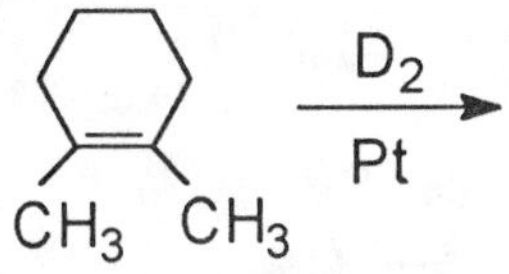

a) 0 b) 1 c) 2 d) 3

196. In the product from the following reaction, how many deuterium atoms in the equatorial position are there for the most stable conformer?

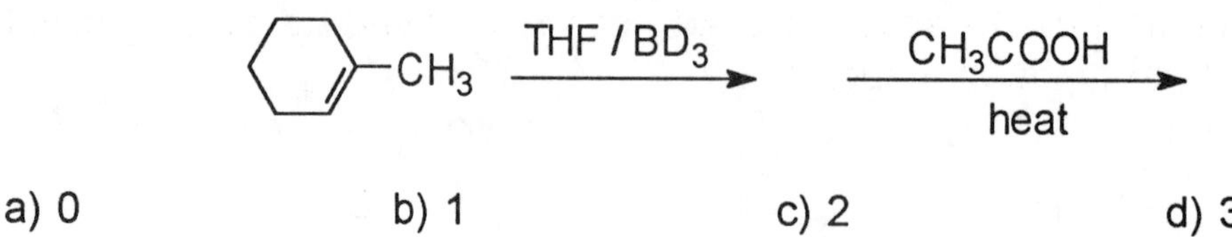

a) 0 b) 1 c) 2 d) 3

197. 1-Methylcyclohexene is treated with THF / BD_3 followed by heating with CH_3COOD. How many equatorial deuteriums are there in the most stable conformer of the product?

a) 0 b) 1 c) 2 d) 3

TIP Borane reacts by a cis and anti-Markovnikov addition. The first step involves addition of deuterium to the carbon atom 2 and BD_2 to the carbon atom 1 in a cis configuration. Then conversion of the carbon-boron bond to the final product, in this case forming a carbon-deuterium bond, occurs with retention of configuration. Thus this two-step sequence adds two atoms of deuterium with defined cis stereochemistry. As in question 194, one deuterium is axial and one is equatorial.

E. "ROADMAP" REACTIONS

198. What is the product from the following series of reactions starting with cyclopentane?

I. free radical chlorination

II. reaction with potassium hydroxide in refluxing ethanol

III. reaction with bromine in carbon tetrachloride

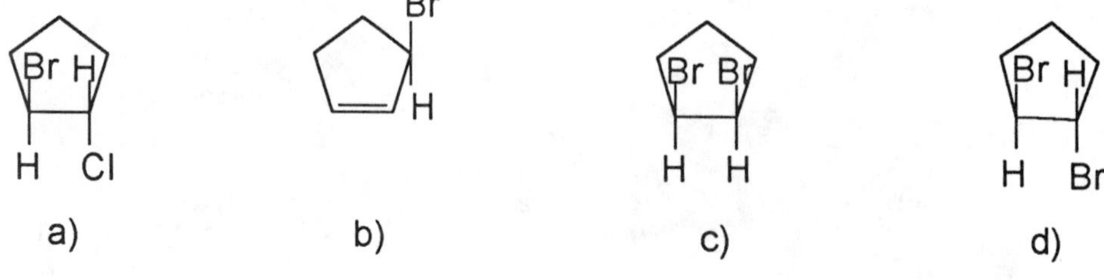

199. When a 1:1 mixture of 2,3-dimethylbutane and bromine is heated, the product, A, is easily separated by fractional distillation. When A is treated with potassium hydroxide in refluxing ethanol, the product, B, is isolated and dissolved in alcohol. Reaction of B with ozone followed by dimethylsulfide yields C. What is the structure of C?

a) $(CH_3)_2C(OH)-C(OH)(CH_3)_2$

b) $(CH_3)_2CHCH(CH_3)CH_2OH$

c) $(CH_3)_2C(OH)CH(CH_3)_2$

d) $(CH_3)_2C{=}O$

TIP Most organic preparations require more than one step. "Roadmap" questions are devised to help you think in terms of a series of reactions, starting at one point and going by a roadmap to the desired product. In this roadmap prep, the first step is a free radical bromination to give 2-bromo-2,3-dimethyl butane. The second step signals an E2 elimination to give B which is 2,3-dimethyl-2-butene (the Saytzeff product). As in question 191, the carbon-carbon double bond is cleaved to give a carbon-oxygen double bond at both ends. In this case the final product is two moles of acetone (d).

200. Compound A has the molecular formula, $C_{12}H_{18}$, and undergoes catalytic hydrogenation to give $C_{12}H_{24}$. What is the correct combination of rings and double bonds for compound A?

	rings	double bonds
a)	0	4
b)	3	1
c)	1	3
d)	4	0

201. Compound A has the molecular formula, C_8H_{14}, and reacts with hydrogen and platinum to give compound B, C_8H_{16}. Compound A has a dipole moment, but compound B does not. Which of the following structures is a possibility for A?

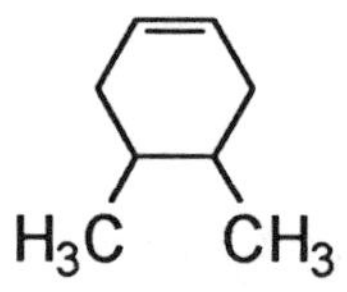

a) H₃C CH₃ (cyclohexene ring bearing two methyl groups)

b) $CH_3CH_2CH{=}CHCH_2CH{=}CHCH_3$

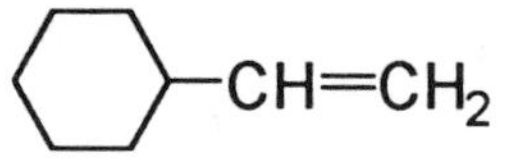

c) (cyclohexyl)$-CH{=}CH_2$

d) $CH_2{=}CHCH_2CH_2CH_2CH_2CH{=}CH_2$

202. Cis-1,4-dimethylcyclohexane was heated with one equivalent of bromine to give a monobrominated product, A. Compound A was heated with potassium ethoxide in ethanol followed by reaction with cold, dilute potassium permanganate to give two products, B and C, Which of the following are the most reasonable structures for compounds B and C?

I. 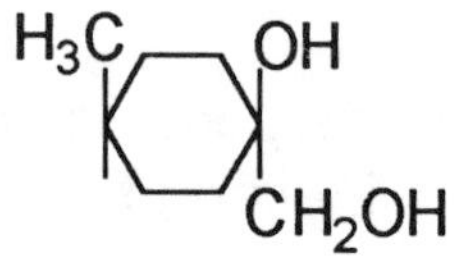 H_3C, OH, CH_2OH

II. 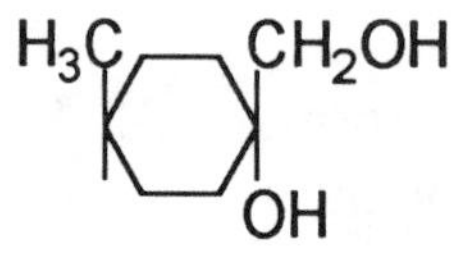 H_3C, CH_2OH, OH

III. 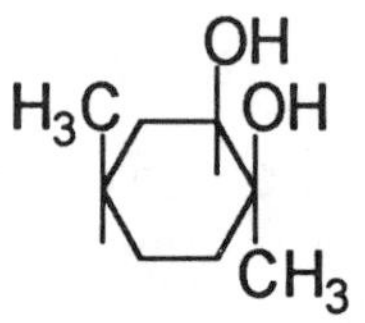 H_3C, OH, OH, CH_3

IV. 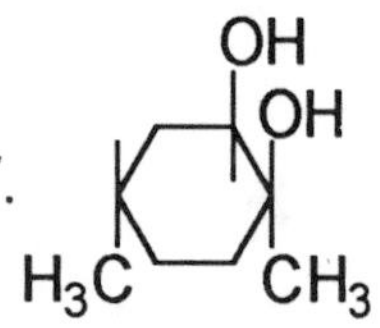 OH, OH, H_3C, CH_3

V. 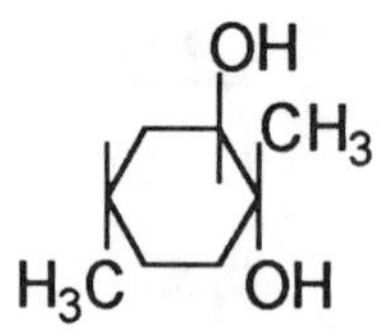 OH, CH_3, H_3C, OH

VI. 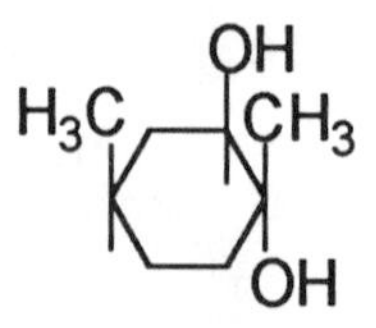 H_3C, OH, CH_3, OH

a) I, II b) III, IV c) V, VI d) III, V

TIP The free radical bromination leads to the tertiary 1-bromo-1,4-dimethylcyclohexane as a mixture of cis and trans since configuration is lost at the free radical intermediate step. Nevertheless, the elimination reaction leads to 1,4-dimethyl-1-hexene as the only product. The hydroxylation with potassium permanganate involves cis addition of the two hydroxyl groups but the 4-methyl group can be either cis or trans to these. Therefore the correct answer is b), compounds III and IV.

CHAPTER 6 ALKYNES

A. NOMENCLATURE and STRUCTURE

203. Which of the following structures have the correct IUPAC name?

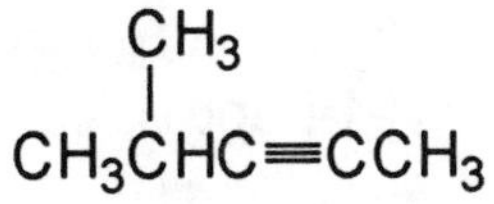

I: $CH_3CH(CH_3)C{\equiv}CCH_3$ — 4-methyl-2-pentyne

II: $CH_3C{\equiv}CCH_2OH$ — 4-ol-2-butyne

III: $BrC{\equiv}CBr$ — dibromoacetylene

IV: $CH_3CH(C{\equiv}CCH_3)CH_2CH_3$ — 4-methyl-2-hexyne

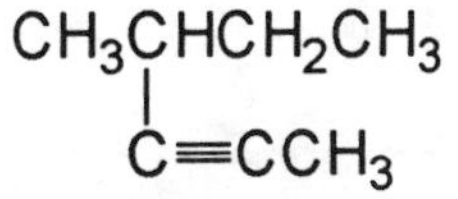

a) I, II b) II, IV c) II, III d) I, IV

204. Which of the following structures have the correct IUPAC name?

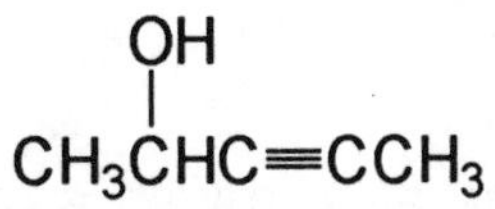

I: $CH_3CH(OH)C{\equiv}CCH_3$ — 3-pentyn-2-ol

II: $CH_2{=}CHCH_2C{\equiv}CH$ — 1-penten-4-yne

III: $HC{\equiv}CCH_2CH_2CH_2OH$ — 1-yne-5-pentanol

IV: $BrCH_2CH(CH_3)C{\equiv}CH$ — 1-bromo-2-methyl-butyne

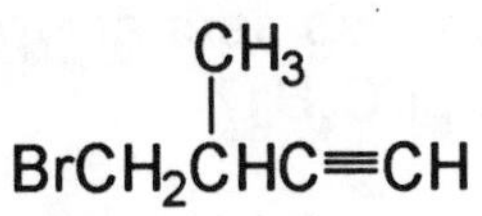

a) I, II b) II, IV c) III, IV d) II, III

205. Which of the following structures has the correct IUPAC name?

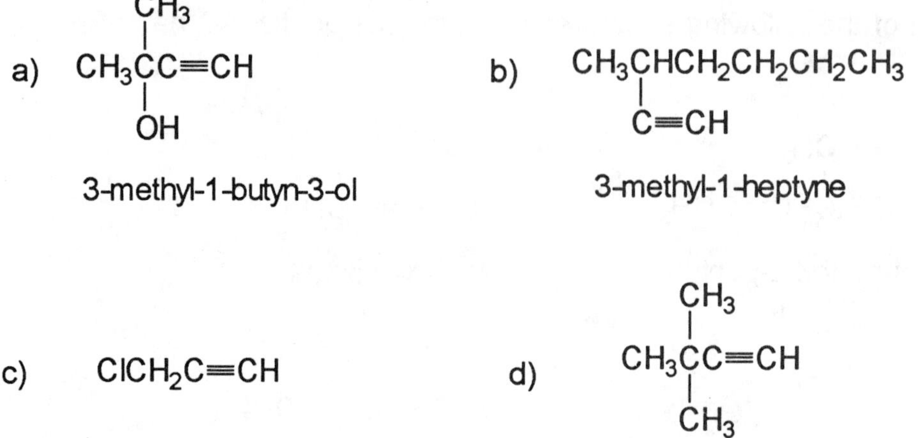

206. Which of the following structures have the correct common name?

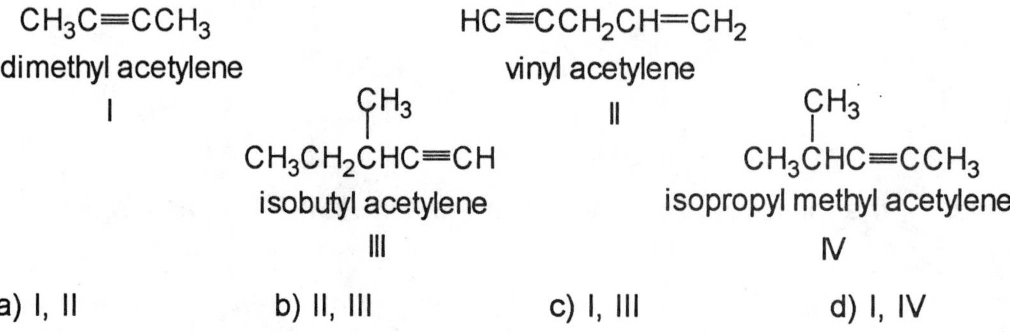

a) I, II b) II, III c) I, III d) I, IV

207. How many structural isomers are possible for an alkyne with a molecular formula of C_5H_8?

a) 2 b) 3 c) 4 d) 5

208. How many structural isomers are possible for a straight-chain compound with a molecular formula of C_4H_6?

a) 2 b) 3 c) 4 d) 5

209. How many structural isomers are possible for a compound with a molecular formula of C_3H_4?

a) 0 b) 2 c) 3 d) 4

210. What is the order of decreasing basicity (strongest first) for the following ions?

$CH_3C{\equiv}C^{\ominus}$	$CH_3^{\ominus}$	$CH_2{=}CH^{\ominus}$	$CH_3CH_2^{\ominus}$	$H_2N^{\ominus}$
I	II	III	IV	V

a) I, III, IV, II, V   b) IV, II, III, V, I   c) II, IV, III, I, V   d) V, I, III, IV, II

211. What is the order of increasing acidity (weakest first) for the following compounds?

$HC{\equiv}CH$	NH_3	$(CH_3)_2CHOH$	H_2O
I	II	III	IV

a) II, I, III, IV   b) I, II, III, IV   c) IV, I, III, II   d) III, II, I, IV

212. What is the order of decreasing stability (most first) for the following ions?

$CH_3C{\equiv}C^{\ominus}$	$CH_3^{\ominus}$	$CH_2{=}CH^{\ominus}$	$CH_3CH_2^{\ominus}$
I	II	III	IV

a) I, III, IV, II   b) II, IV, III, I   c) IV, II, III, I   d) I, III, II, IV

TIP Two factors contribute to anion stability: 1) hybridization and 2) less important, inductive effects. For 1) the greater the s character of the orbital, the greater the electronegativity of the carbon atom ($sp>sp^2>sp^3$ or I >III>II and IV). for 2) a methyl group is electron donating relative to hydrogen (II>IV).

213. Which of the following equations are favored to the right?

I. $CH_3CH_2C{\equiv}C^{\ominus} + CH_3CH_3 \rightleftharpoons CH_3CH_2C{\equiv}CH + CH_3CH_2^{\ominus}$

II. $CH_3C{\equiv}CH + CH_3CH_2CH_2CH_2^{\ominus} \rightleftharpoons CH_3C{\equiv}C^{\ominus} + CH_3CH_2CH_2CH_3$

III. $CH_3C{\equiv}CH + CH_2{=}CH^{\ominus} \rightleftharpoons CH_3C{\equiv}C^{\ominus} + CH_2{=}CH_2$

IV. $CH_3C{\equiv}CH + {}^{\ominus}OH \rightleftharpoons CH_3C{\equiv}C^{\ominus} + H_2O$

a) I, II   b) II, III   c) III, IV   d) I, IV

214. Which of the following reactions are favored to the right?

I. $CH_3CH_2C{\equiv}C^{\ominus} + CH_3CH_3 \rightleftharpoons CH_3CH_2C{\equiv}CH + CH_3CH_2^{\ominus}$

II. $CH_3C{\equiv}CH + CH_3CH_2CH_2CH_2^{\ominus} \rightleftharpoons CH_3C{\equiv}C^{\ominus} + CH_3CH_2CH_2CH_3$

III. $CH_3C{\equiv}CH + CH_2{=}CH^{\ominus} \rightleftharpoons CH_3C{\equiv}C^{\ominus} + CH_2{=}CH_2$

IV. $CH_3C{\equiv}C^{\ominus} + H_2O \rightleftharpoons CH_3C{\equiv}CH + OH^{\ominus}$

a) I, II, III b) I, II,IV c) I, III, IV d) II, III, IV

TIP The stability of the anion will determine the position of equilibria. In addition to hybridization and inductive effects (see question 212), the relative electronegativity of the atoms is a factor and is usually the dominate term. Thus except for IV, the equilibria can be predicted by the effect of hybridization as in Q. 212. For IV, the greater electronegativity of oxygen overrides the hybridization of carbon.

215. Which of the following pairs of structures represent tautomers?

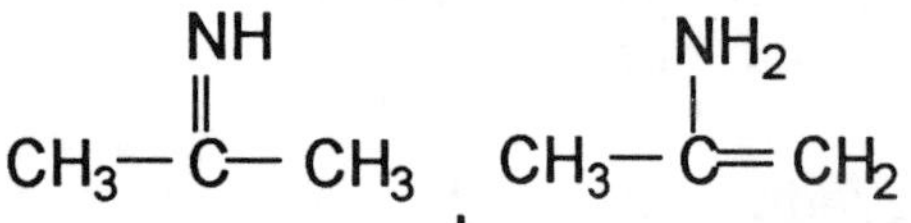

I. $CH_3-C(=NH)-CH_3$ and $CH_3-C(NH_2)=CH_2$

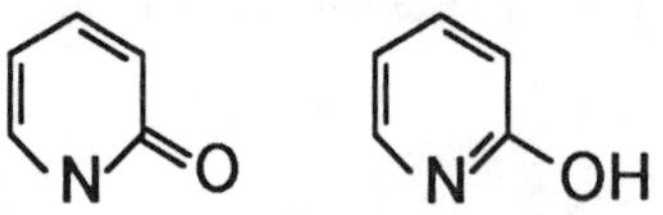

II. 2-pyridone (N, C=O) and 2-hydroxypyridine (N, OH)

III. 1,4-benzoquinone (O, O) and hydroquinone (OH, OH)

IV. $CH_3-C(O^{\ominus})=NH$ and $CH_3-C(=O)-NH^{\ominus}$

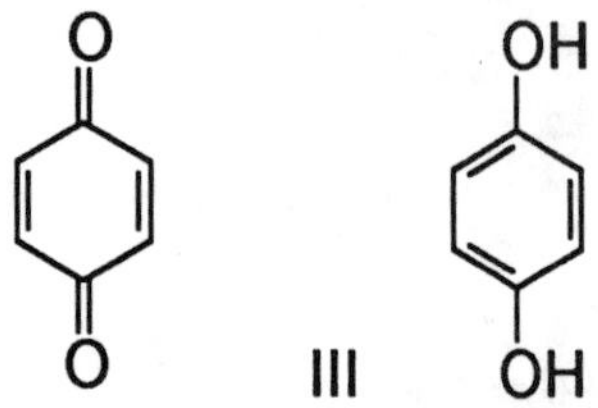

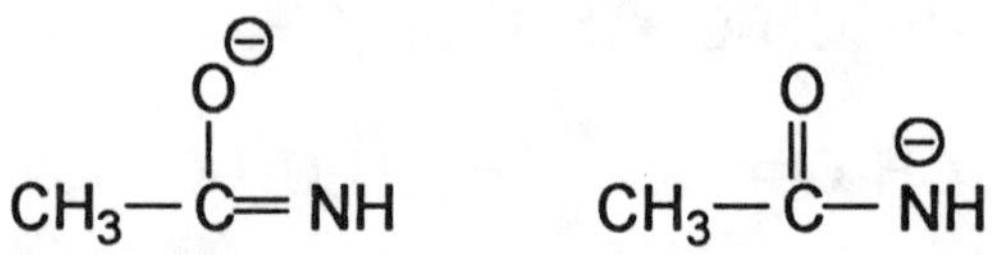

a) I, II b) III, IV c) I, III d) II, IV

216. Which of the following pairs of structures represent tautomers?

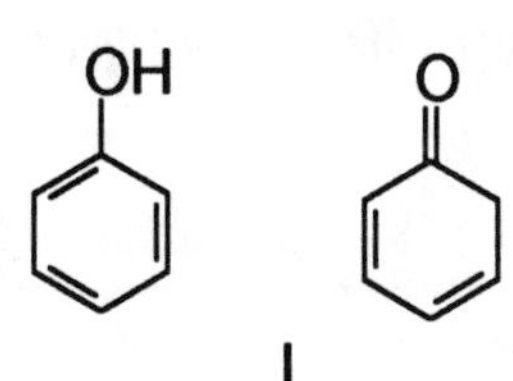

I

$O^{\ominus}$ O $\ominus$

II

$CH_3-CH=CH(OH)$ $CH_3-CH(OH)=CH_2$

III

$CH_3-C(=O)-CH_3$ $CH_3-CH(OH)=CH_2$

IV

a) I, III b) II, IV c) III, IV d) I, IV

217. Propene and propyne can be distinguished by which reagent?

a) Br_2 in CCl_4 b) dilute $KMnO_4$

c) concentrated H_2SO_4 d) $Ag(NH_3)OH$

218. Which of the following compounds forms a precipitate with $Ag(NH_3)_2OH$?

a) $CH_3CH_2C{\equiv}CCH_3$ b) $CH_3CH_2C{\equiv}CH$

c) $CH_2{=}CHCH_2CH_3$ d) $CH_3CH{=}CHCH_3$

B. PREPARATION

219. What are the best conditions for preparing 2-pentyne from 2-chloropentane?

a) $\xrightarrow{Cl_2}\ \xrightarrow{2\,NaH}$ b) $\xrightarrow{NaOH}\ \xrightarrow{Cl_2}\ \xrightarrow{2\,Na\,/\,NH_3(l)}$

c) $\xrightarrow{H_2SO_4}\ \xrightarrow{HCl}\ \xrightarrow{NaH}$ d) $\xrightarrow{NaOCH_3}\ \xrightarrow{Br_2\,/\,CCl_4}\ \xrightarrow{NaNH_2\,/\,NH_3(l)}$

TIP The best conditions are first an elimination to give the alkene, followed by addition of halogen, then two dehydrohalogenations with strong base to give the alkyne. a) is wrong because the first step gives a hopeless mixture and the second step does not eliminate halogens. The same is true for c). b) and d) are ok until the last step where b) has Na / NH_3 which is a reducing agent not a strong base.

220. What are the best conditions for preparing dideutero acetylene from ethylene?

a) $\xrightarrow{Cl_2}\ \xrightarrow[D_2O]{H_2SO_4}\ \xrightarrow[PdCl_2\ CuCl_2]{CH_3COOD}$ b) $\xrightarrow{HBr}\ \xrightarrow{D_2SO_4}\ \xrightarrow{D_2O}$

c) $\xrightarrow{Br_2\,/\,CCl_4}\ \xrightarrow[NH_3(l)]{NaNH_2}\ \xrightarrow[D_2O]{NaH}$ d) $\xrightarrow[Pd\ CuCl_2]{CH_3COOD}$

221. Which alkyne is prepared when 2-bromo-4-methylpentane is treated with potassium hydroxide, then with bromine in carbon tetrachloride, and then with sodamide in liquid ammonia?

a) $CH{\equiv}CCH_2CH(CH_3)_2$ b) $CH_3CH_2CH(CH_3)_2$

c) $(CH_3)_2CHCH_2C{\equiv}CCH_2CH(CH_3)_2$ d) $CH_3C{\equiv}CCH(CH_3)_2$

222. What is the best starting material for the preparation of cyclohexyl acetylene?

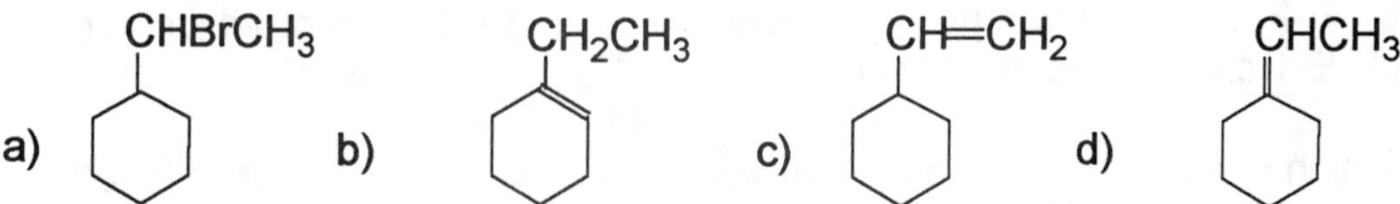

C. REACTIONS

223. What is the product when 3-hexyne is reacted with lithium in liquid ammonia?

a) cis-3-hexene

b) $CH_3(CH_2)_3C^{\ominus}{\equiv}C\ Li^{\oplus}$

c) trans-3-hexene

d) diethylacetylene

224. What is the product from the following reaction?

$$HC{\equiv}\overset{\ominus}{C}\ \overset{\oplus}{Na} + CH_3CH_2OD \longrightarrow$$

a) $HC{\equiv}CH$

b) $CH_3CH_2OC{\equiv}CH$

c) $HC{\equiv}CD$

d) $CH_3CH_2OCH{=}CHD$

225. What is the best method for converting acetylene to trans-3-hexene?

a) $\xrightarrow{NaNH_2}\ \xrightarrow{2\ C_2H_5Br}\ \xrightarrow{2\ Na\ /\ NH_3(l)}$

b) $\xrightarrow{2\ Na\ /\ NH_3(l)}\ \xrightarrow{NaNH_2}\ \xrightarrow{2\ C_2H_5Br}$

c) $\xrightarrow{NaNH_2}\ \xrightarrow{2\ C_2H_5Br}\ \xrightarrow[Pt]{H_2}$

d) $\xrightarrow{NaNH_2}\ \xrightarrow{C_2H_5I}\ \xrightarrow{CH_3(CH_2)_3^{\ominus}\ Li^{\oplus}}\ \xrightarrow{CH_3Br}\ \xrightarrow{Na\ /\ NH_3(l)}$

TIP The first step is alkylation of both ends of acetylene to give a structure with 6 carbon atoms. b) is wrong because the steps are out of order. d) is wrong because the alkylated product has only 5 carbons. Both a) and c) give 3-hexyne, but the reducing conditions in c) result in hexane.

226. What is the product from the reaction of 1-pentyne with one equivalent of DCl?

a) E-2-chloro-1-deuteriopent-1-ene
b) Z-2-chloro-1-deuteriopentene
c) E-1-chloro-2-deuteriopentene
d) Z-1-chloro-2-deuteriopentene

227. Which oxidizing conditions would convert acetylene to vinylacetylene?

a) $CuCl_2$ + O_2 in pyridine
b) $CuCl_2$ + NH_4Cl + HCl
c) ozone + Zn + CH_3COOH
d) $KMnO_4$

228. What are the best conditions for converting 4-methyl-2-pentyne to trans-4-methyl-2-pentene?

a) H_2 + Pt b) Lindlar's catalyst c) Na in liquid ammonia d) H_2 + Ni

229. Which compound is not in the product mixture from the following reaction?

$$CH_3C{\equiv}\overset{\ominus}{C}\ \overset{\oplus}{Li} + (CH_3)_2CHBr \longrightarrow$$

a) $CH_3C{\equiv}CCH(CH_3)_2$
b) $CH_3C{\equiv}CH$
c) $(CH_3)_2CHCH_2C{\equiv}CH$
d) $CH_2{=}CHCH_3$

TIP What is at stake here is the competition between substitution and elimination. In this case both are possible since the anion is a strong base and the substrate allows both. a) is the substitution product, b) and d) are the two elimination products.

230. What is the product from the following reaction?

$$CH_3CH_2C{\equiv}CH \xrightarrow{KMnO_4}$$

a) $CH_3CH_2\underset{\displaystyle\overset{|}{HO}}{CH}{-}\underset{\displaystyle\overset{|}{OH}}{CH_2}$ (HO on the CH, OH on the CH_2)
b) 2 CH_3COOH
c) CH_3CH_2COOH + HCOOH
d) $CH_3CH_2\overset{O}{\overset{\|}{C}}H$ + $H\overset{O}{\overset{\|}{C}}H$

231. What is the product from the following reaction?

$$\text{C}_6\text{H}_{11}\text{–C}\equiv\text{CH} \xrightarrow[\text{H}_2\text{SO}_4\ +\ \text{CH}_3\text{CO}_2\text{H}]{\text{HgSO}_4}$$

a) $\text{C}_6\text{H}_{11}\text{–C(OCOCH}_3\text{)=CH}_2$ (cyclohexyl–C(=CH₂) bearing $\text{O}\overset{\text{O}}{\overset{\|}{\text{C}}}\text{CH}_3$)

b) $\text{C}_6\text{H}_{11}\text{–CH=CH O}\overset{\text{O}}{\overset{\|}{\text{C}}}\text{CH}_3$

c) $\text{C}_6\text{H}_{11}\text{–C(COOCH}_3\text{)=CH}_2$ (cyclohexyl–C(=CH₂) bearing $\overset{\text{O}}{\overset{\|}{\text{C}}}\text{OCH}_3$)

d) $\text{C}_6\text{H}_{11}\text{–CH=CH}\overset{\text{O}}{\overset{\|}{\text{C}}}\text{OCH}_3$

232. What is the product of the reaction of propyne with mercuric sulfate and sulfuric acid?

a) CH_3CH_2CHO b) $CH_3\overset{O}{\overset{\|}{C}}CH_3$ c) $CH_3\overset{OH}{\overset{|}{C}}HCH_2OH$ d) $H_2C{=}CHCH_2OH$

233. Which of the following is not a product from the oxidation reaction?

$$CH_3C\equiv CCH_2C\equiv CCH(CH_3)_2 \xrightarrow{\text{hot KMnO}_4}$$

a) CH_3CO_2H b) $(CH_3)_2CHCOOH$ c) $\underset{\text{COOH}}{\overset{\text{COOH}}{|}}$ d) $CH_2(COOH)_2$

TIP Oxidation with hot $KMnO_4$ leads to bond cleavage at the triple bond to give a carboxylic acid at each end. Since there are two triple bonds, there are three acids and one is a diacid formed from the middle section. The diacid must have 3 carbon atoms, not 2.

234. What are the best conditions for the following conversion?

$$C_6H_5\text{-}CH_2CH_3 \rightarrow C_6H_5\text{-}C{\equiv}CH$$

a) $\xrightarrow[\text{light}]{2\ Br_2}$ $\xrightarrow{NaNH_2}$ b) $\xrightarrow[\text{heat}]{Cl_2}$ $\xrightarrow{NaOCH_2CH_3}$

c) $\xrightarrow{NBS}$ $\xrightarrow{KOt\text{-}C_4H_9}$ d) $\xrightarrow[\text{heat}]{2\ Br_2}$ $\xrightarrow[CH_3COOH]{Zn}$

235. What are the best conditions for converting propyne to the following product?

$$(H_3C)(D)C{=}C(H)(D)$$

a) $\xrightarrow{NaH}$ $\xrightarrow{D_2O}$ $\xrightarrow[H_2]{Pd/CaCO_3}$ b) $\xrightarrow[D_2]{Pd/CaCO_3}$

c) $\xrightarrow{(sia)_2BH}$ $\xrightarrow{CH_3COOD}$ d) $\xrightarrow{(sia)_2BD}$ $\xrightarrow{CH_3COOH}$

236. What are the best conditions for converting propyne to the following product?

$$(H_3C)(D)C{=}C(H)(H)$$

a) $\xrightarrow{NaH}$ $\xrightarrow{D_2O}$ $\xrightarrow[H_2]{Pd/CaCO_3}$ b) $\xrightarrow[D_2]{Pd/CaCO_3}$

c) $\xrightarrow{(sia)_2BH}$ $\xrightarrow{CH_3COOD}$ d) $\xrightarrow{(sia)_2BD}$ $\xrightarrow{CH_3COOH}$

237. What are the best conditions for converting propyne to the following product?

$$\mathrm{H_3C}(\mathrm{H})\mathrm{C{=}C}(\mathrm{D})(\mathrm{H})$$

(H₃C and D on the same side; H and H on the same side)

a) $\xrightarrow{NaH}$ $\xrightarrow{D_2O}$ $\xrightarrow[H_2]{Pd/CaCO_3}$ b) $\xrightarrow[D_2]{Pd/CaCO_3}$

c) $\xrightarrow{(sia)_2BH}$ $\xrightarrow{CH_3COOD}$ d) $\xrightarrow{(sia)_2BD}$ $\xrightarrow{CH_3COOH}$

238. What are the best conditions for converting propyne to the following product?

$$\mathrm{H_3C}(\mathrm{H})\mathrm{C{=}C}(\mathrm{H})(\mathrm{D})$$

(H₃C and H on the same side; H and D on the same side)

a) $\xrightarrow{NaH}$ $\xrightarrow{D_2O}$ $\xrightarrow[H_2]{Pd/CaCO_3}$ b) $\xrightarrow[D_2]{Pd/CaCO_3}$

c) $\xrightarrow{(sia)_2BH}$ $\xrightarrow{CH_3COOD}$ d) $\xrightarrow{(sia)_2BD}$ $\xrightarrow{CH_3COOH}$

TIP Questions 235-238 involve the introduction of deuterium. Both the regioselectivity and the stereoselectivity can be controlled by using the right conditions. For question 235, addition of D_2 in a cis configuration is done by using Lindlar's catalyst, answer b). The hydrogen attached to the triple bond can be replaced by deuterium by first forming the anion with NaH and then reacting this with D_2O. Then addition of hydrogen in a cis configuration leads to the answer in question 237, answer a). The remaining two questions (236 and 238) start with hydroboration because here what is needed is to add B-H in an anti Markovnikov orientation so that the replacement of C-B by C-H (answer c in 236) or by C-D (answer d in 238) is stereospecific.

239. Triclene is a commercial solvent which is 1,1,2-trichloroethane. It can be prepared from acetylene by which of the following sequences?

a) $\xrightarrow{Cl_2}$ $\xrightarrow[Pt]{H_2}$

b) $\xrightarrow{HCl}$ $\xrightarrow{Cl_2}$

c) $\xrightarrow{Cl_2}$ $\xrightarrow{\text{base}}$

d) $\xrightarrow{Cl_2}$ $\xrightarrow{\text{base}}$ $\xrightarrow[Pt]{H_2}$

240. Propene is reacted with bromine to give product A. Product A is further reacted with sodamide to give product B. Product B is treated with sodium hydride then with methyl iodide to give product C. What is the most reasonable structure for product C?

a) $CH_3CH_2CH_2C{\equiv}CH$

b) $CH_3CH_2C{\equiv}CH$

c) $CH_3C{\equiv}CCH_3$

d) $CH_3CH_2C{\equiv}CCH_3$

241. Cyclopentyl acetylene is reacted with sodium hydride and then with ethyl bromide to give compound A. Compound A is reacted with lithium in liquid ammonia to give compound B. What is the most reasonable structure for compound B?

a) cyclopentyl–$C(CH_2CH_3){=}CH_2$

b) cyclopentyl–$CH{=}CHCH_2CH_3$ (both H on the same side)

c) cyclopentyl–$CH{=}CHCH_2CH_3$ (H on opposite sides)

d) cyclopentyl–$CH_2CH_2CH_2CH_3$

TIP The first step involves the alkylation with ethyl bromide, but the second step is a stereospecific reduction. Answer a) is wrong because the alkylation takes place on the wrong carbon atom. Answer b) is wrong because the reduction gives a trans configuration and not cis. Answer d) is wrong because the reduction does not go all the way to an alkane.

CHAPTER 7 CONJUGATED DIENES

A. STRUCTURE

242. Which of the following compounds are conjugated?

I. 1,2-pentadiene
II. 1,3-pentadiene
III. 1,4-pentadiene
IV. 2,3-pentadiene
V. 2,4-pentadiene

a) I, IV b) II, V c) III, V d) II, III

243. Which of the following compounds are conjugated?

a) I, II b) III, IV c) I, II, III d) II, III

244. Which of the following compounds are conjugated?

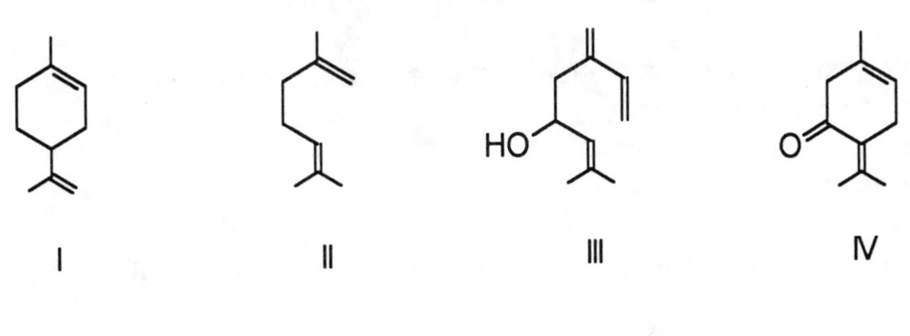

a) I, II b) III, IV c) II, III d) I, IV

245. How many conjugated bonds are there in the following structure of hematin?

a) 12 b) 9 c) 11 d) 13

246. How many nodes are there for the highest occupied molecular orbital for 1,3-butadiene?

a) 1 b) 2 c) 3 d) 4

247. Which statement is true for the pentadienyl cation?

a) The highest occupied molecule has one node.
b) There are 6 pi electrons
c) There are 3 molecular orbitals.
d) There are two resonance forms.

248. What species is best described by the following orbital energy diagram?

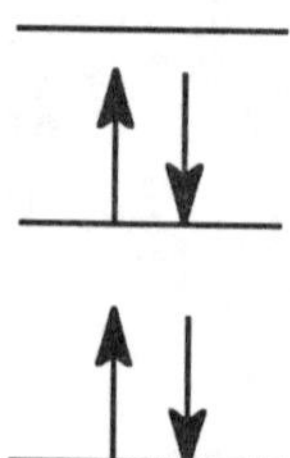

a) allyl cation b) allyl anion c) butadiene d) butadiene cation

249. What species has this description for the highest occupied molecular orbital?

a) allyl cation b) butadiene dianion c) butadiene d) butadiene dication

250. What species is best described by the following orbital energy diagram?

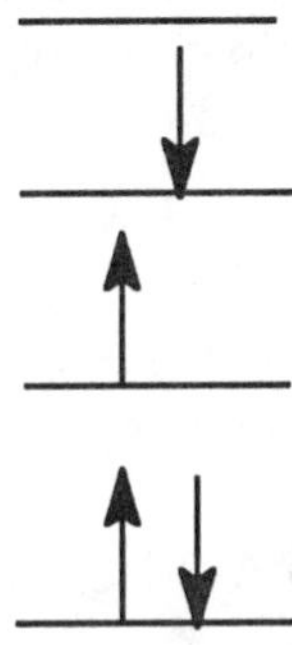

a) allyl cation b) allyl anion c) butadiene d) excited butadiene

TIP (Questions 246-250) The number of Π molecular orbitals is determined by the number of p atomic orbitals in the system. Thus allyl has three molecular orbitals and butadiene four. The most bonding orbital has no nodes and the number of nodes increases as the energy of the orbital increases. The highest occupied molecular orbital for butadiene (the second) thus has one node (question 246). The number of electrons placed in the molecular orbitals is determined by the charge on the species. Each nucleus contributes one positive charge. The allyl cation has two electrons, the allyl anion has four and the neutral butadiene has four. (The pentadienyl cation has four electrons in two orbitals and the highest occupied orbital has one node-question 247.) In the ground state for molecules, the electrons are paired in orbitals with opposite spins (question 248-three orbitals, two occupied with paired electrons). For question 250, there are four orbitals (a butadiene system) and the electrons are unpaired, one electron in the second orbital and one in the third, so this is an excited state.

251. Which of the following species has 2 nodes in the highest occupied molecular orbital?

a) 1,3-butadiene
b) 1,3-butadienyl dianion
c) 1,3-butadienyl dication
d) 1,3-pentadienyl cation

TIP Two nodes indicates that the third molecular orbital is occupied. For butadiene with four electrons, levels one and two are occupied; for the dication with two electrons, level one is occupied. The pentadienyl cation has five nuclei and four electrons with a charge of +1 so levels one and two are occupied (see question 247). The butadiene dianion has four nuclei and six electrons with a charge of -2, thus for six electrons in levels one, two and three, the highest occupied molecular orbital has two nodes.

252. What is the order of increasing heat of hydrogenation for the following compounds (lowest first)?

I. 1-hexene II. cis-2-hexene III. trans-2-hexene IV. 1,3-hexadiene

a) III, II, I, IV b) III, II, IV, I c) IV, III, II, I d) II, I, IV, III

253. What is the order of increasing heat of hydrogenation for the following compounds (lowest first)?

a) I, III, II, IV b) III, I, II, IV c) IV, I, II, III d) I, II, III, IV

254. What is the order of increasing heat of hydrogenation for the following compounds (lowest first)?

I: cyclohexyl–$CH{=}C{=}CH_2$

II: cyclohexylidene$=CH-CH{=}CH_2$

III: cyclohexenyl–$CH_2-CH_2-CH_3$

IV: cyclohexenyl–$CH{=}CH-CH_3$

a) I, II, IV, III b) II, IV, III, I c) III, IV, I, II d) III, II, IV, I

TIP The greater the stability, the lower the heat of hydrogenation. Two factors contribute to the stability. First and more important, conjugated systems are more stable than nonconjugated systems. Second, more highly substituted double bonds are more stable than less highly substituted. Thus III has the lowest heat of hydrogenation as it is conjugated and the most substituted. I has the highest heat of hydrogenation as it is nonconjugated and it is the least substituted. Since II is conjugated and IV is not, the correct order is d).

255. Which of the following are examples of allylic carbocations?

I: $CH_3-\overset{\oplus}{C}H-CH{=}CH_2$

II: cyclohexylidene$=\overset{\oplus}{C}H$

III: $CH_2{=}CH-CH_2-\overset{\oplus}{C}H_2$

IV: cyclohexenyl–$\overset{\oplus}{C}H_2$

V: $\oplus C(CH_3)_3$

a) II, V b) II, IV c) I, III, IV d) I, IV

256. Which of the following are examples of allylic carbocations?

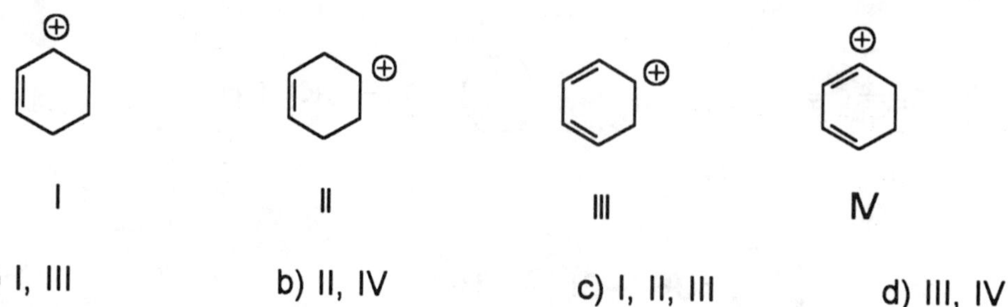

a) I, III   b) II, IV   c) I, II, III   d) III, IV

B. REACTIONS

257. What is the major factor favoring 1,2-additions to conjugated dienes?

a) alkene stability   b) Markovnikov addition
c) allylic cation stability   d) steric effects

258. What is the major factor favoring 1,4-additions to conjugated dienes?

a) alkene stability   b) Markovnikov addition
c) allylic cation stability   d) steric effects

259. What is the principal product from the reaction of 1,3-pentadiene with bromine at low temperature?

a) $CH_3-\underset{Br}{CH}-CH_2-\underset{Br}{CH}-CH_3$ b) $CH_3-\underset{Br}{CH}-\underset{Br}{CH}-CH=CH_2$

c) $CH_3-CH=CH-\underset{Br}{CH}-\underset{Br}{CH_2}$ d) $CH_3-\underset{Br}{CH}-CH=CH-\underset{Br}{CH_2}$

260. What is the principal product from the reaction of 1,3-pentadiene with bromine at high temperature?

a) $CH_3-CH(Br)-CH_2-CH(Br)-CH_3$

b) $CH_3-CH(Br)-CH(Br)-CH=CH_2$

c) $CH_3-CH=CH-CH(Br)-CH_2(Br)$

d) $CH_3-CH(Br)-CH=CH-CH_2(Br)$

261. What is a major product from the following reaction at low temperature?

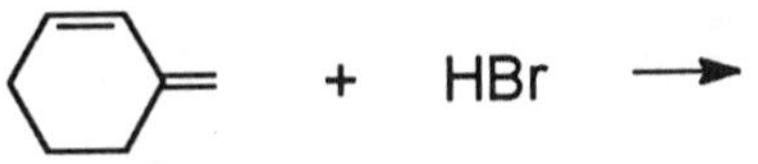

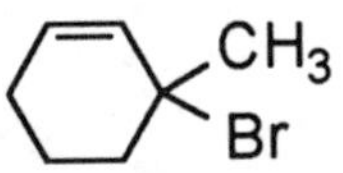

(1-methylenecyclohex-2-ene) + HBr →

a) 1-bromo-1-methylcyclohex-2-ene (CH_3, Br)

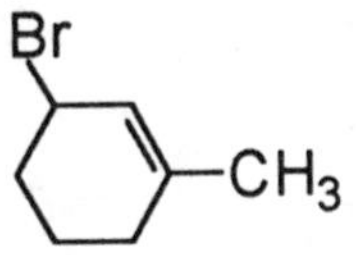

b) 3-bromo-1-methylcyclohexene (Br, CH_3)

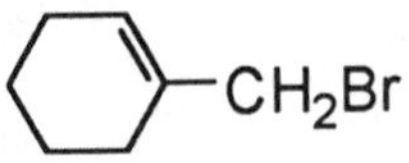

c) 1-(bromomethyl)cyclohexene (CH_2Br)

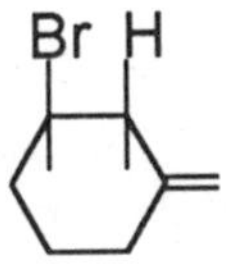

d) (Br H adduct of 3-methylenecyclohexene)

TIP Electrophilic reaction of a conjugated system at low temperatures is dominated by allylic cation stability and leads to 1,2-addition. Addition of a proton at carbon atom 1 of the external double bond gives an allylic cation with charge delocalized at a secondary and a tertiary position. Under kinetic control (low temperature) the more stable tertiary position is attacked by the nucleophile (Br -) to give a).

262. What is a major product from the following reaction at high temperature?

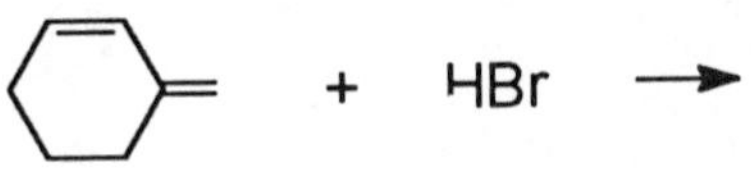

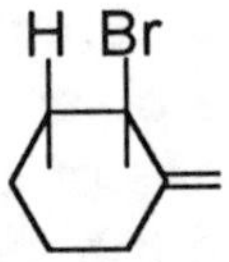

(1-methylenecyclohex-2-ene) + HBr →

a) (H Br adduct of 3-methylenecyclohexene)

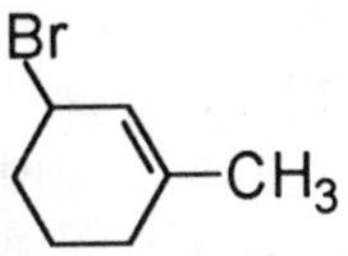

b) 3-bromo-1-methylcyclohexene (Br, CH_3)

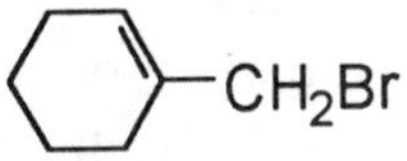

c) 1-(bromomethyl)cyclohexene (CH_2Br)

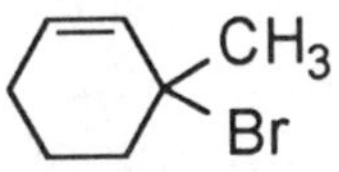

d) 1-bromo-1-methylcyclohex-2-ene (CH_3, Br)

263. What is the principal product from the reaction of 1,3-cyclopentadiene with bromine?

a) 3,4-dibromocyclopentene
b) 3,5-dibromocyclopentene
c) 2,3-dibromocyclopentene
d) a mixture of a) and b)

264. What is the principal product from the reaction of 1,3-cyclopentadiene with hydrogen bromide?

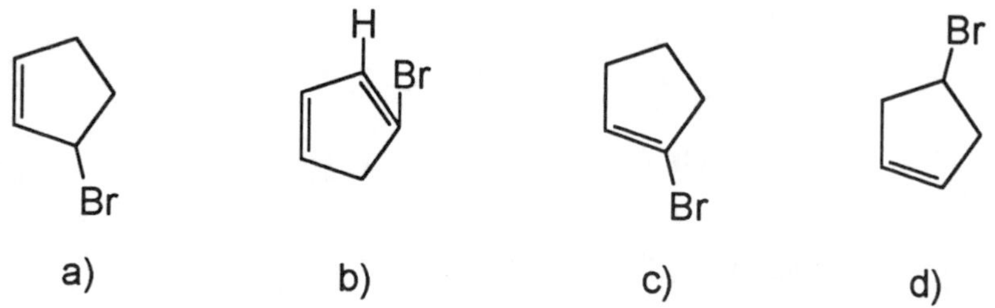

265. What is the order of increasing rate of reaction with bromine for the following compounds (slowest first)?

$CH_2{=}CH{-}CH{=}CH_2$ (I)

$CH_2{=}CH{-}CH{=}CH_2{-}CH_3$ (II)

$CH_2{=}CH{-}CH_2{-}CH{=}CH_2$ (III)

$CH_3{-}CH_2{=}CH{-}CH{=}CH_2{-}CH_3$ (IV)

a) IV, III, II, I
b) I, II, IV, III
c) III, I, II, IV
d) III, II, IV, I

266. What is the product from the reaction of 1,3-butadiene with HOCl under conditions of kinetic control?

a) $HOCH_2{-}CH(Cl){-}CH{=}CH_2$

b) $ClCH_2{-}CH(OH){-}CH{=}CH_2$

c) $HOCH_2{-}CH{=}CH{-}CH_2Cl$

d) $HOCH_2{-}C(Cl){=}CH{-}CH_3$

TIP There are two concerns here. First, the direction of addition is determined by oxygen being more electronegative than chlorine, so the bond is polarized and Cl adds as the electrophile. Second, since this is under conditions of kinetic control, the addition is 1,2. Thus Cl adds first to carbon 1 to give the allylic carbocation followed by OH addition at the 2 position and b) is the correct product.

C. DIELS - ALDER REACTIONS

267. Which of the following substituents activate a diene in a Diels -Alder reaction?

I. $-CH_3$ II. $-CHO$ III. $-C{\equiv}N$ IV. $-C(CH_3)_3$

a) I, III b) II, III c) III, IV d) I, IV

268. Which of the following substituents activate a dienophile in a Diels -Alder reaction?

I. $-CH_3$ II. $-COOCH_3$ III. $-CH{=}CH_2$ IV. $-OCH_3$

a) I, III b) II, III c) II, IV d) I, IV

269. Which compounds will react with maleic anhydride in a Diels-Alder reaction?

I II III IV

a) I, II b) II, IV c) II, III, IV d) I, II, IV

270. Which compounds react readily with cyclopentadiene in a Diels-Alder reaction?

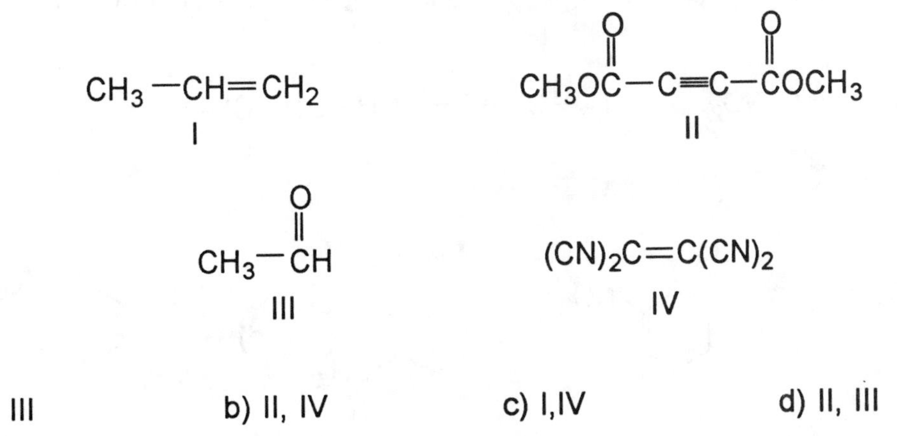

a) I, III b) II, IV c) I,IV d) II, III

271. What is the major product from the thermal reaction of two moles of 1,3-butadiene?

a) 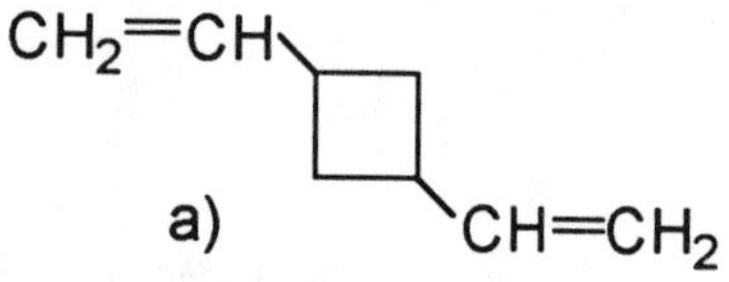 $CH_2{=}CH$ (cyclobutane) $CH{=}CH_2$

b) 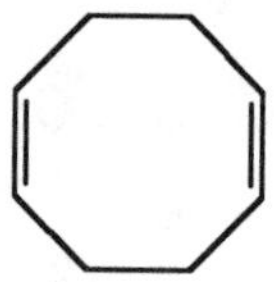 (cyclooctadiene)

c) 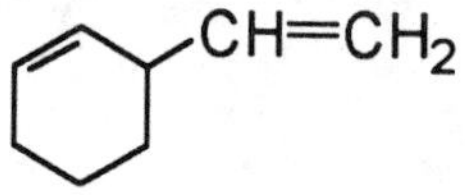 (cyclohexene) $CH{=}CH_2$

d) 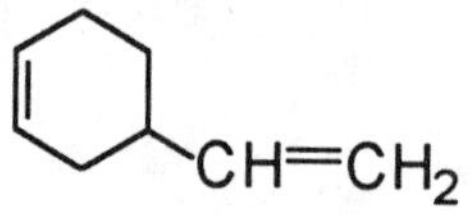 (cyclohexene) $CH{=}CH_2$

272. What are the reactants that produce the following compound in a Diels-Alder reaction?

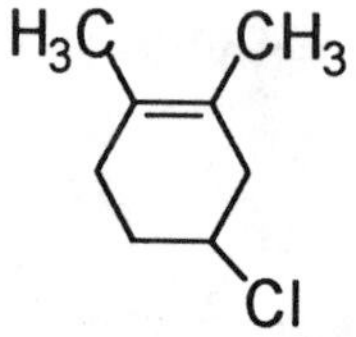

H_3C CH_3 Cl

a) 2,3-dimethyl-1,3-butadiene and vinyl chloride
b) 2-methyl-1,3-butadiene and 1-chloropropane
c) 2-chloro-1,3-butadiene and cis-2-butene
d) 1,3-pentadiene and cis-2-chloropropane

273. What is the major product from the following Diels-Alder reaction?

$$CH_2{=}CH{-}NO_2 \ + \ CH_3{-}CH{=}CH{-}CH{=}CH{-}CH_3 \longrightarrow$$

a) 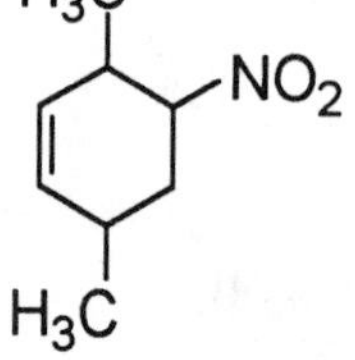 H_3C NO_2 H_3C

b) 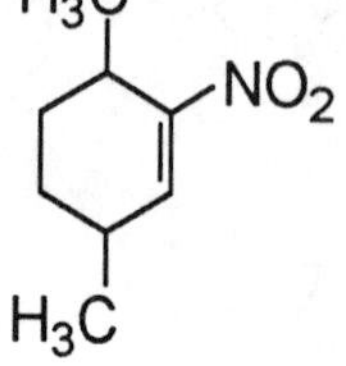 H_3C NO_2 H_3C

c) 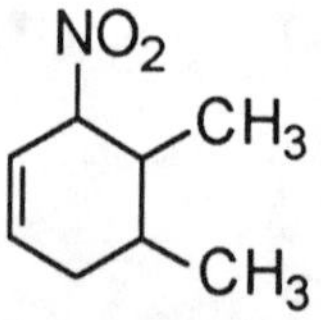 NO_2 CH_3 CH_3

d) 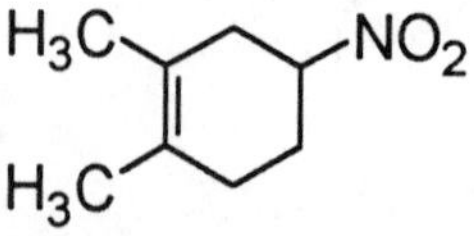 H_3C H_3C NO_2

274. What is the major product from the following Diels-Alder reaction?

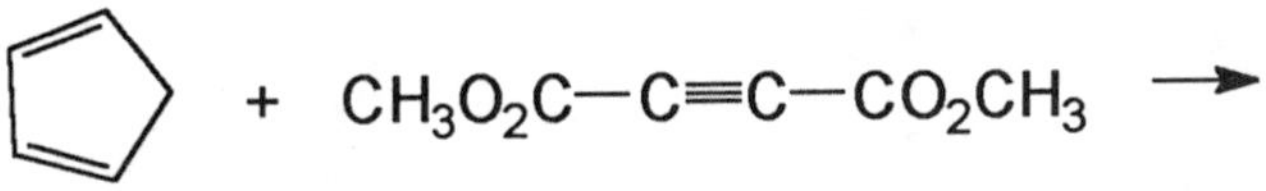

+ $CH_3O_2C-C\equiv C-CO_2CH_3$ →

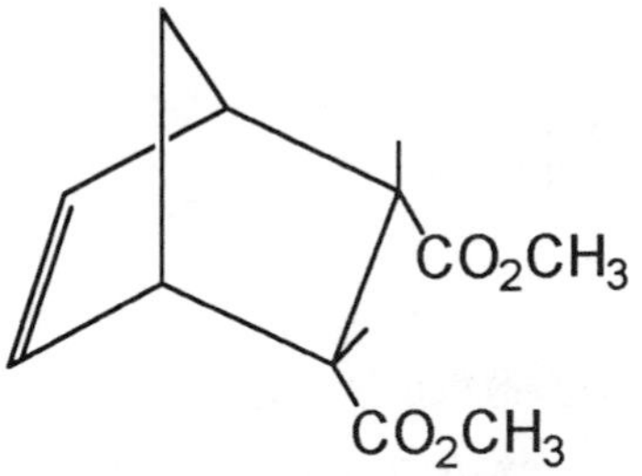

a) CO_2CH_3 CO_2CH_3

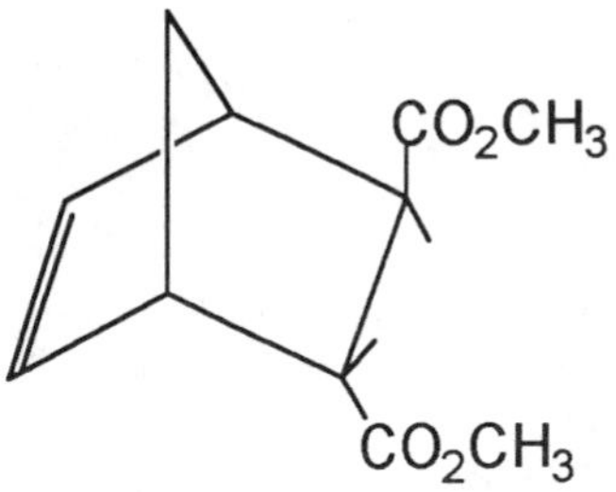

b) CO_2CH_3 CO_2CH_3

c) CO_2CH_3 CO_2CH_3

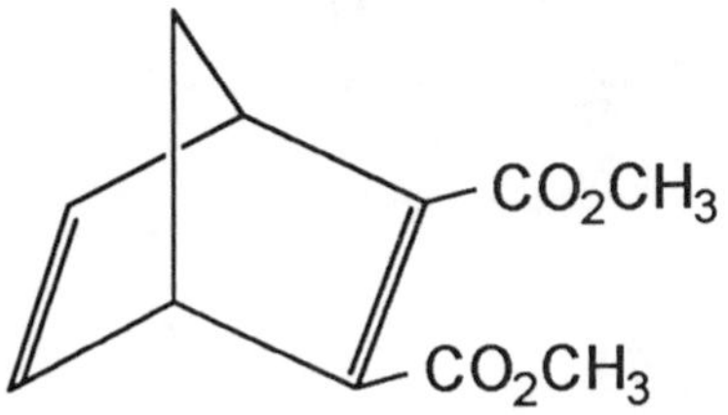

d) 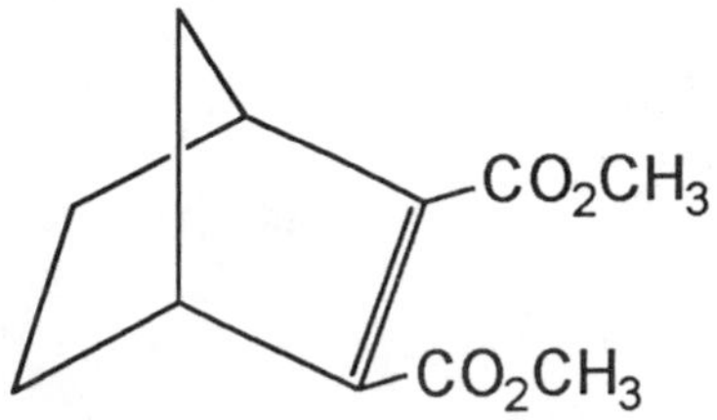 CO_2CH_3 CO_2CH_3

TIP Remember that the Diels-Alder reaction makes 2 carbon-carbon bonds and generates a cyclohexene. The groups attached to the diene and dienophile will end up on the cyclohexene in exactly the way they are substituted originally. Here think of the CH_2 in cyclopentadiene as a group attached to the diene which forms a 1-carbon bridge to the resultant cyclohexene. Next, since the dienophile has two pi bonds (a triple bond), the product will have one double bond in the original position and the groups attached to the double bond will be in the same positions as in the original dienophile. Answers a) and b) are wrong because the double bond from the dienophile is missing and answer d) is wrong because the double bond formed in generating the cyclohexene is missing.

275. What is a likely product from the following Diels-Alder reaction?

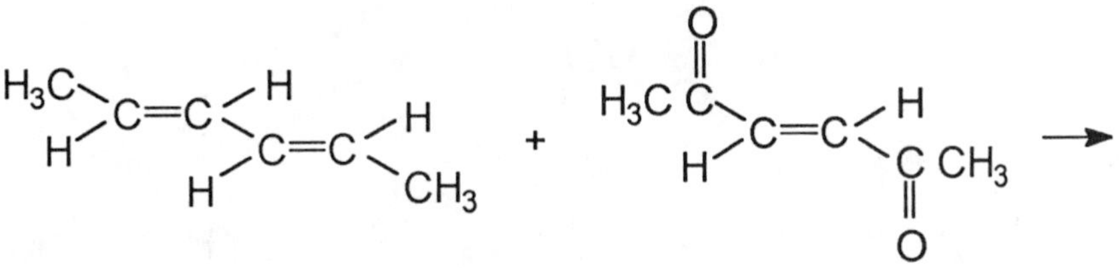

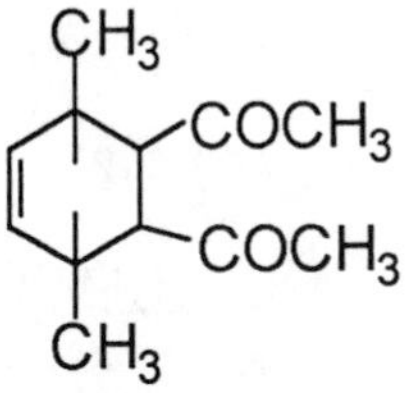

H_3C, H, C=C, H; H, C=C, H, CH_3 + H_3C C(=O), H, C=C, H, C(=O) CH_3 →

a) CH_3 $COCH_3$ $COCH_3$ CH_3

b) CH_3 $COCH_3$ H_3C $COCH_3$

c) H_3C $COCH_3$ $COCH_3$ H_3C

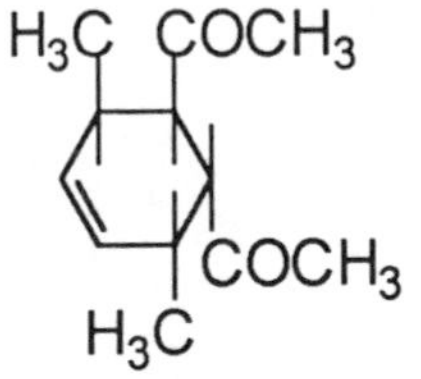

d) H_3C $COCH_3$ $COCH_3$ H_3C

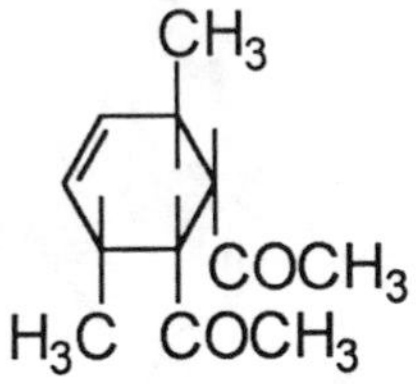

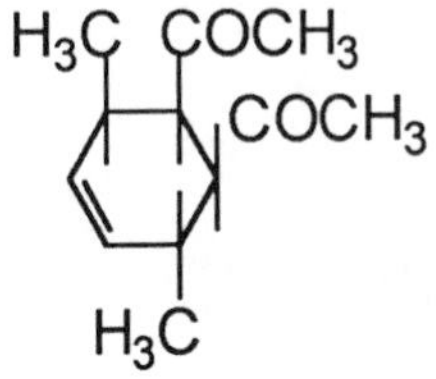

276. What is the major product from the following Diels-Alder reaction?

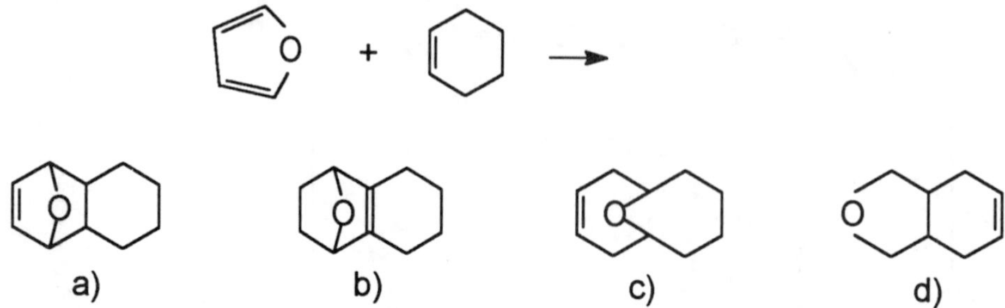

277. What is the major product from the following Diels-Alder reaction?

a) b) c) d)

278. What are the reactants that produce the following compound in a Diels-Alder reaction?

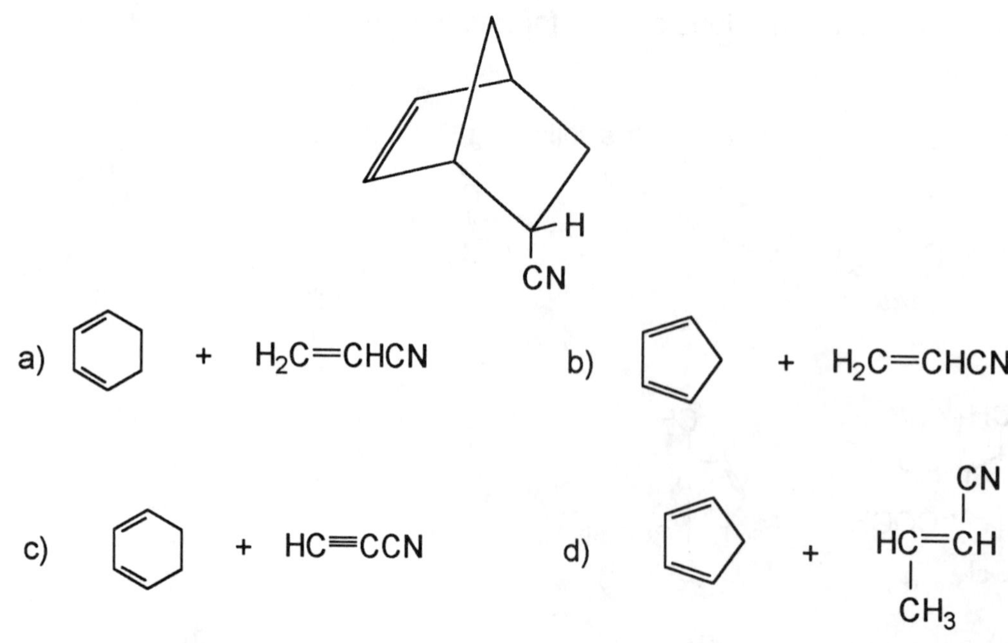

279. 1,3-Butadiene reacted with cis-1,2-dichloroethylene to give compound A. Compound A reacted with chlorine in carbon tetrachloride to form compound B. Which of the following structures is most likely compound B?

a)

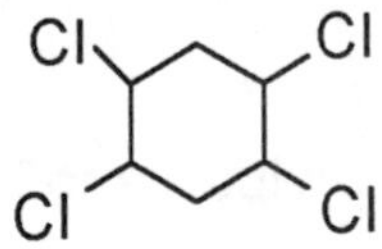

b)

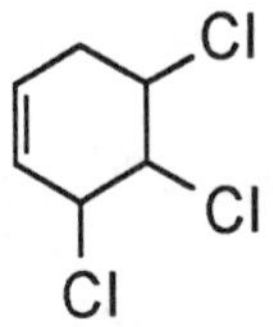

c)

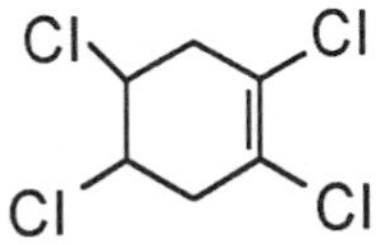

d)

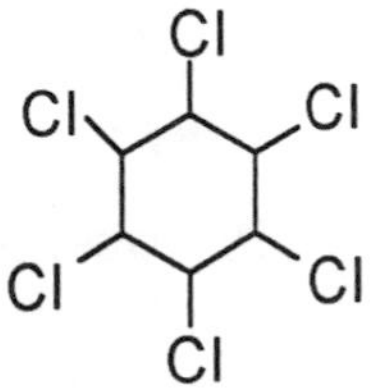

TIP This is a "roadmap" question, or in other words, how do you get from here to there? Working backwards from compound B to compound A we note that addition of chlorine in carbon tetrachloride to a double bond would suggest that A is an alkene and a cyclohexene at that. Since all Diels-Alder reactions give a cyclohexene, we deduce that the reaction to give A is such an example. Setting up the first reaction as a Diels-Alder reaction with 1,3-butadiene as the diene and dichloroethylene as the dienophile, the product is 4,5-dichlorocyclohexene, compound A. So far so good. The second reaction involving chlorine in carbon tetrachloride is an example of electrophilic addition to a double bond. So adding Cl-Cl to a cyclohexene forms 1,2-dichloro-cyclohexane, and in this case, forms 1,2,4,5-hexachlorocyclohexane,answer a).

CHAPTER 8 CHIRALITY

A. STEREOCENTERS

280. How many stereocenters are there in 2,2,4-tribromo-3-methyl-1-pentanol?

a) 2 b) 3 c) 4 d) 5

281. How many stereocenters are there in 2,3-dibromo-4-methylpentane?

a) 1 b) 2 c) 3 d) 4

282. How many stereocenters are there in cyclohex-3-en-1,2-diol?

a) 1 b) 2 c) 3 d) 4

283. How many stereocenters are present in the following structure?

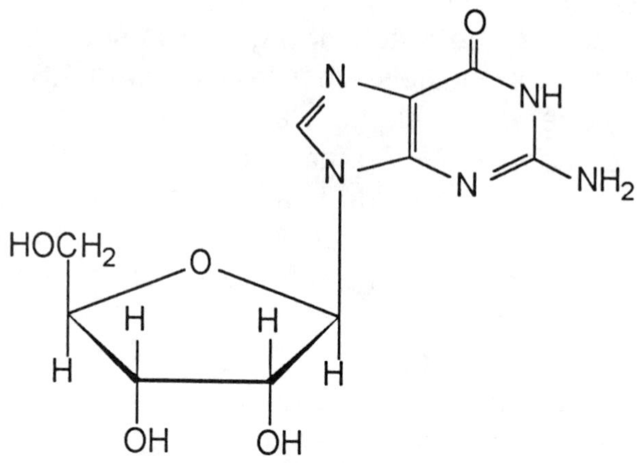

O
N
NH
N
N
NH_2
$HOCH_2$
O
H H
H H
OH OH

a) 1 b) 2 c) 3 d) 4

284. How many stereocenters are present in the following structure?

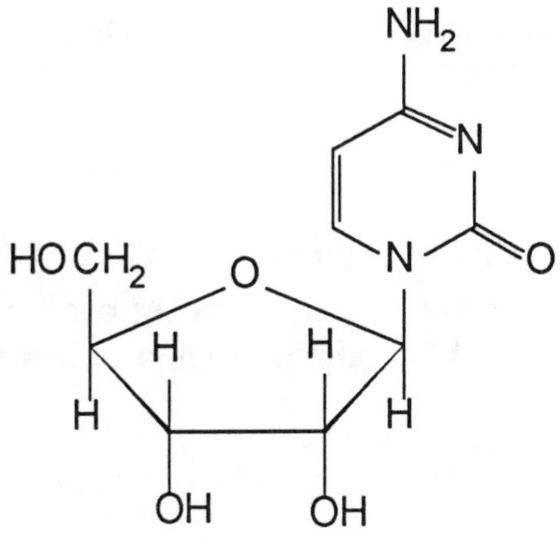

a) 1 b) 2 c) 3 d) 4

TIP At a stereocenter, interchange of any two groups leads to a stereoisomer. Thus a carbon attached to four different groups is a stereocenter. In this problem, only the carbons in the five membered ring should be considered. We note that there are four carbon atoms in the ring, and all are attached to four different groups.

B. CHIRALITY

285. Which of the following molecules are chiral?

I. trans-1-chloro-2-methylcyclopropane
II. cis-1-chloro-2-methylcyclopropane
III. 1-chloro-1-methylcyclopropane
IV. cis-1,2-dichlorocyclopropane

a) I, IV b) II, III c) I, II d) III, IV

286. Which of the following molecules are chiral?

I. 1-chloro-2-methylpropane
II. 2-bromobutane
III. 3-chloropentane
IV. 2-chloro-3-methylbutane

a) I, IV b) II, III c) I, III d) II, IV

287. Which of the following molecules are achiral?

I. trans-1,2-cyclohexanediol
II. cis-1,2-cyclohexanediol
III. trans-1,4-cyclohexanediol
IV. trans-1,3-cyclohexanediol

a) I, IV b) II, III c) I, II d) III, IV

TIP Note that there is no plane of symmetry for I and IV even in a planar cyclohexane ring. So only II and III contain a plane of symmetry. Structures I and IV are chiral.

288. Which of the following molecules are chiral?

I. 4-chlorocyclopentene
II. 2-chloro-3,4-dimethylpentane
III. trans-2-chlorocyclohexanol
IV. cis-3,5-dichlorocyclopentene

a) I, IV b) II, III c) I, II d) III, IV

289. Which of the following molecules are chiral?

I. cis-2-chlorocyclohexanol
II. 2-chloropentane
III. 2-chloro-1-pentene
IV. 2-chloropropane
V. 2-chlorobutane
VI. cis-3,5-dichlorocyclopentene

a) I, II, V b) II, IV, VI c) I, II, VI d) IV, V, VI

C. STEREOISOMERS

290. If tetravalent carbon formed square planar compounds, how many isomers are possible for CH_2Cl_2?

a) 1 b) 2 c) 3 d) 4

291. If tetravalent carbon formed square planar compounds, how many isomers are possible for $CHFCl_2$?

a) 1 b) 2 c) 3 d) 4

292. If tetravalent carbon formed square planar compounds, how many isomers are possible for CHFClBr?

a) 1 b) 2 c) 3 d) 4

TIP The key to this problem is how many ways can you arrange four different objects at the corners of a square? If we start counting by considering objects on the diagonal, we can rule out answers a) and d). Further, we note that for a given object at one point on the diameter, each of the remaining four can be placed at the diagonal. Thus the answer is c).

293. What is the maximum number of stereoisomers possible for the following structure?

```
              CH2OH
              |
HOCH------CH
  |          \
  |           O
  |          /
HOCH------CH
              |
              OH
```

a) 2 b) 4 c) 8 d) 16

294. How many total stereoisomers are possible for 1,2,3,4-pentanetetraol, and of these, how many are chiral?

a) 8 and 4 b) 8 and 8 c) 6 and 4 d) 4 and 4

295. There are X isomers of dimethylcyclopropane of which Y are chiral. Select the right numbers for X and Y.

	x	y
a)	4	4
b)	2	2
c)	2	0
d)	4	2

TIP Since both methyl groups can be on the same carbon atom, and this structure is clearly achiral, answer a) can be ruled out (Y cannot be 4). Moreover the trans-1,2-compound is chiral, so answer c) is incorrect (Y cannot be 0). The cis compound, which is achiral, will add another to the X column but not to the Y column, so answer b) is incorrect.

296. Which of the following are stereoisomers?

I	II	III	IV
(H)(H)C=C(H)(CH_2CH_3)	(H)(CH_3)C=C(CH_3)(H)	(H_3C)(H)C=C(CH_3)(H)	(CH_3CH_2)(H)C=C(H)(H)

a) I, II b) I, III c) I, IV d) II, III

297. Which of the following structures have the meso form?

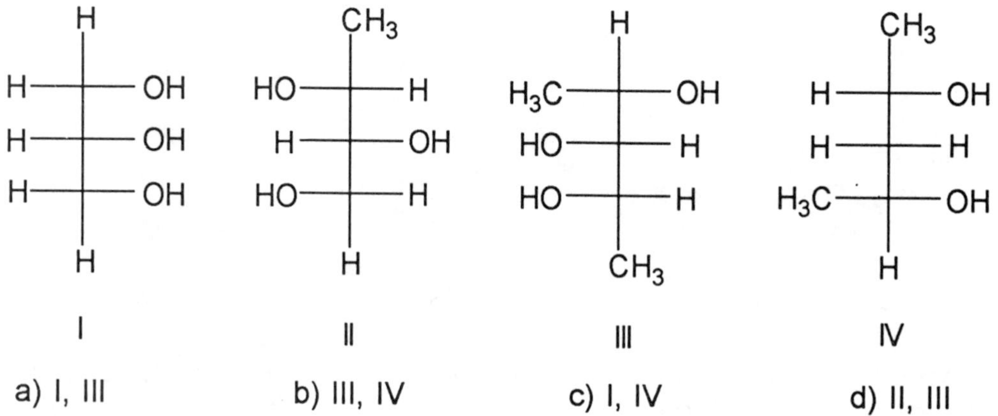

a) I, III b) III, IV c) I, IV d) II, III

298. The following structures are stereoisomers. Which are enantiomers?

	I	II	III	IV
top	CH_3	CH_3	CH_3	CH_3
C2	H — OH	H — OH	H — OH	HO — H
C3	H — OH	HO — H	H — OH	H — OH
C4	H — OH	H — OH	HO — H	H — OH
bottom	CH_3	CH_3	CH_3	CH_3

a) I, II b) III, IV c) II, III d) I, IV

TIP Enantiomers have non-superimposable mirror images. In this Fischer notation, we note that the mirror image for compound I is superimposable on I, thus answers a) and d) are incorrect. A 180° rotation of structure IV reveals that it is a non-superimposable mirror image of structure III.

299. For the following sets of structures, which pairs are enantiomers?

I. 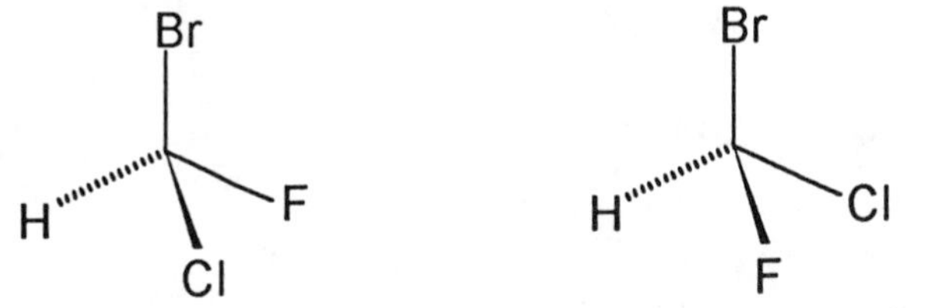

II. 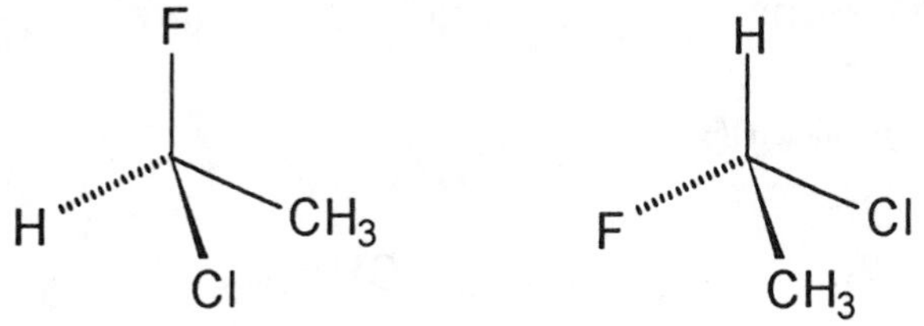

III. 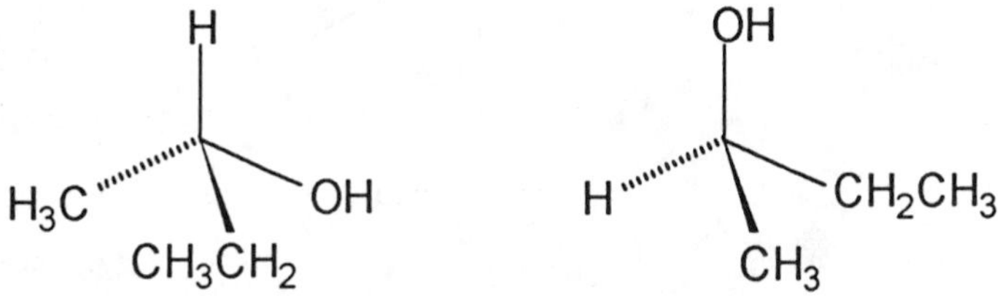

a) I, II b) I, III c) II, III d) none

300. What relationship do the following structures have?

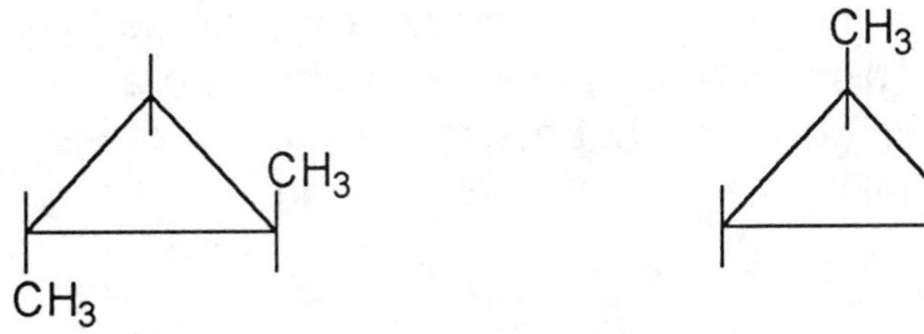

a) enantiomers b) diastereomers c) structural isomers d) identical

301. What relationship do the following structures have?

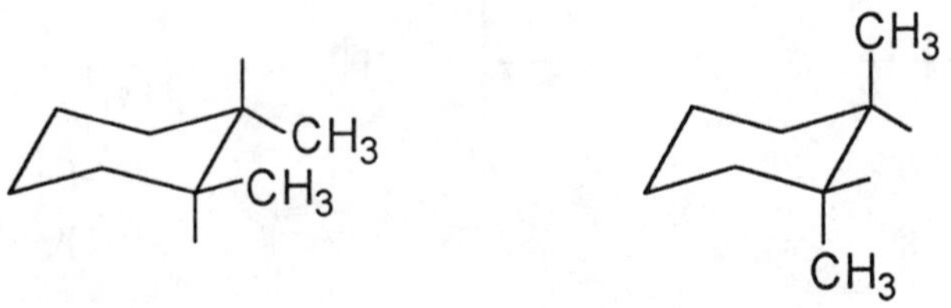

a) enantiomers
b) diastereomers
c) structural isomers
d) conformational isomers

302. What relationship do the following structures have?

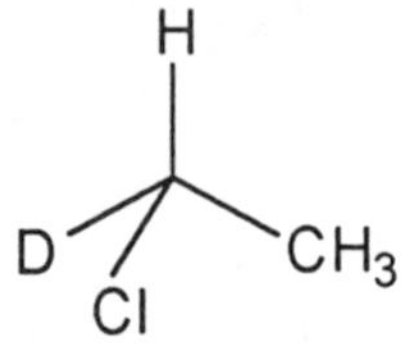

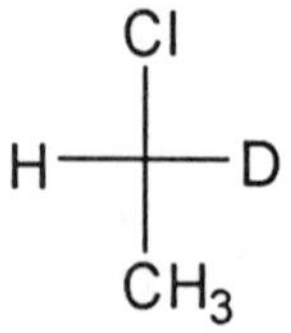

a) enantiomers b) diastereomers c) structural isomers d) identical

303. What relationship do the following structures have?

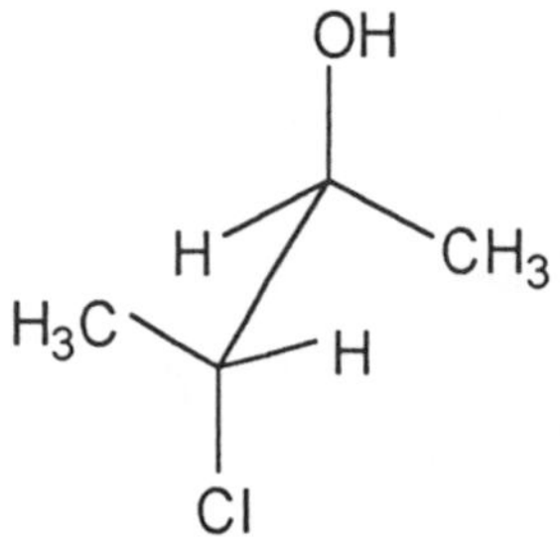

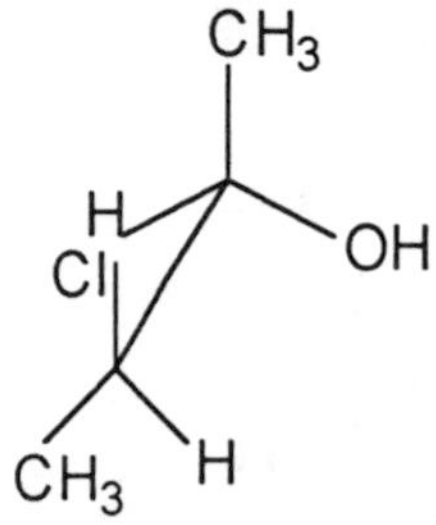

a) enantiomers b) diastereomers c) structural isomers d) identical

TIP Of the several ways of attacking this problem, one sure method is to place the methyl groups in eclipsed conformations and then consider the other groups. For the first structure, when the methyl groups are eclipsed, the hydrogen atoms are also eclipsed. This is true for the second structure as well. The remaining groups (Cl) are in positions that are mirror images of each other in the two structures. This makes the structures enantiomers.

304. Which of the following structures are pairs of enantiomers?

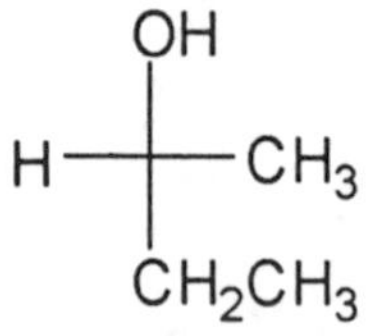

I 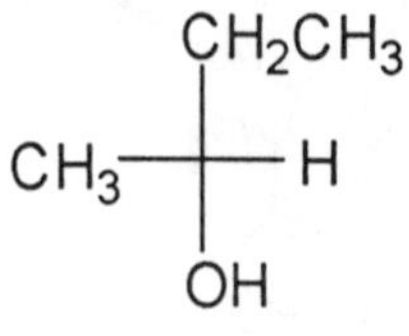 II 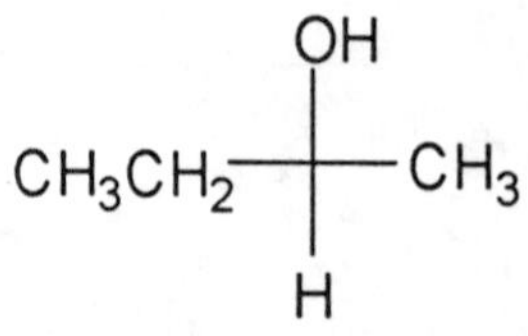 III 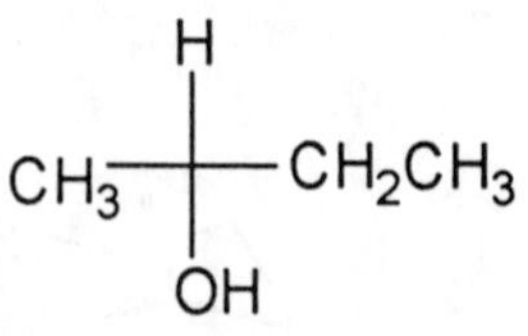 IV

a) I, II b) III, IV c) II, IV d) none

305. What is the total number of pairs of enantiomers for the following molecules?

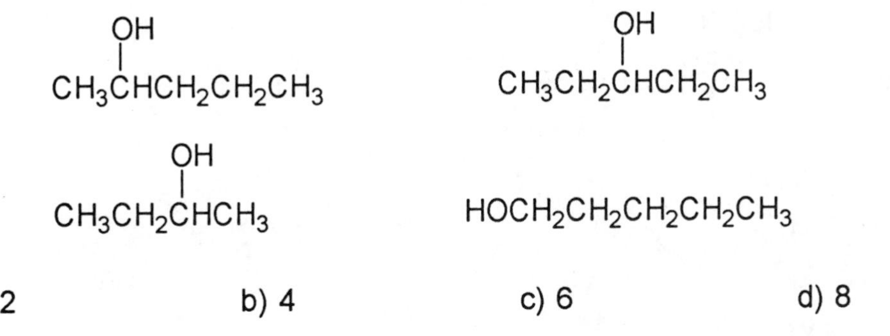

a) 2 b) 4 c) 6 d) 8

306. How many pairs of diastereomers does 1,2-dimethylcyclopentane have?

a) 1 b) 2 c) 3 d) 4

307. Which of the following molecules can have diastereomers but not enantiomers?

I: $CH_3CH(OH)CH(OH)CH_3$

II: $CH_3C{\equiv}CCH_2CH_3$

III: $CH_2{=}CHCH_2CH_3$

IV: $CH_3CH{=}CHCH_2CH_3$

V: $CH_3CH(OH)CH_3$

VI: H_3C–(cyclohexane-1,4-diyl)–CH_3

a) I, III b) III, IV c) II, V d) IV,VI

TIP Diastereomers are non-identical stereoisomers that are not related as mirror images, whereas enantiomers are non-identical stereoisomers that are related as mirror images. Structure I can have chiral forms, so answer a) is wrong. Structure V has no stereoisomers, so answer c) is wrong. The remaining answers both contain structure IV which is correct, and so we look at III and VI. Structure III has no stereoisomers but VI does have cis and trans forms.

308. Which of the following molecules can have both diastereomers and enantiomers?

a) $CH_3CH(OH)CH(OH)CH_3$

b) $CH_3CH(OH)CH_2CH_3$

c)  H_3C–(cyclohexane)–CH_3

d) $CH_3CH{=}CHCH_3$

D. THE R / S SYSTEM

309. What is the R,S configuration for the following structure?

Fischer projection: top $CH_2CH_2CH_3$; HO—C—H; HO—C—H; bottom CH_2CH_3

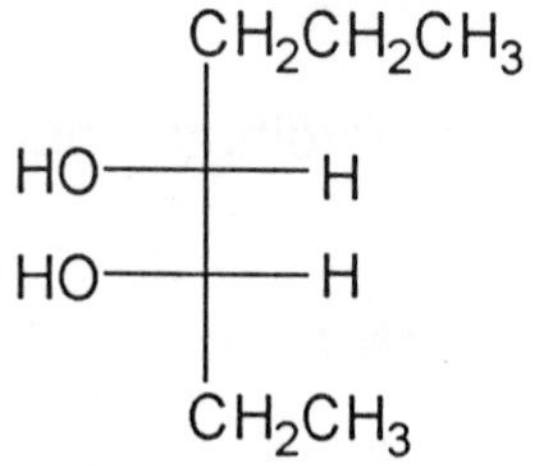

a) 3R, 4R   b) 3S, 4S   c) 3R, 4S   d) 3S, 4R

310. What is the R,S configuration for the following structure?

CO_2H; H; CH_2CO_2H; HO_2C; H; Cl

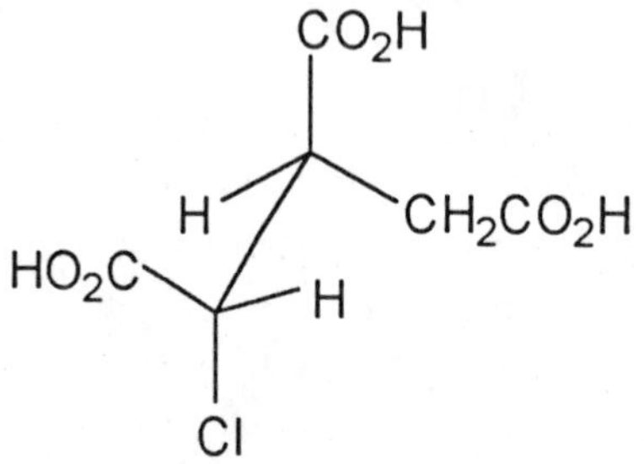

a) 2R, 3R   b) 2R, 3S   c) 2S, 3R   d) 2S, 3S

311. Which of the following structures are designated S?

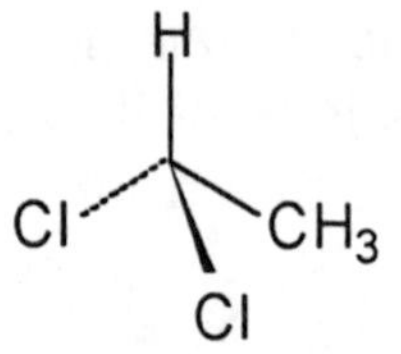

I: H, Cl, CH_3, Cl

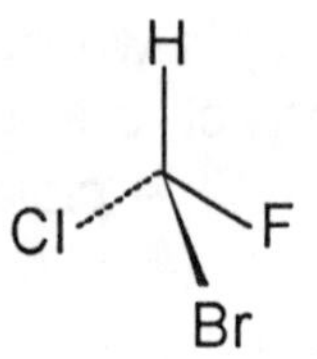

II: H, Cl, F, Br

III: OH, H_2N, H, CH_3

IV: H, CH_3CH_2, $C(CH_3)_3$, CH_3

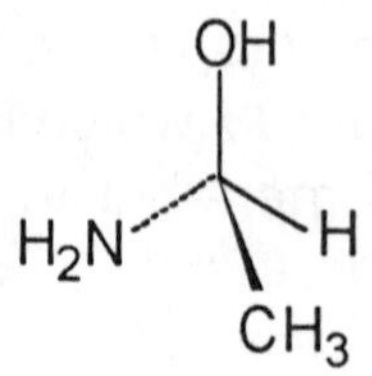

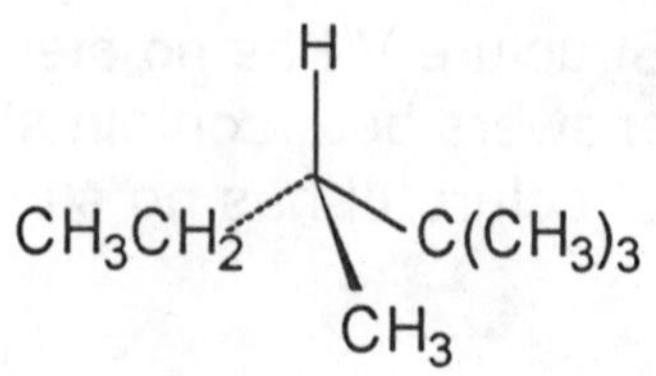

a) I, II   b) II, III   c) II, IV   d) I, III

312. Which of the following structures are designated S?

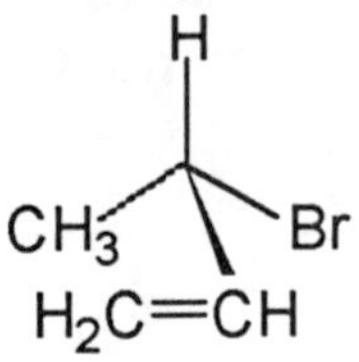

H; CH_3; Br; $H_2C{=}CH$ — I

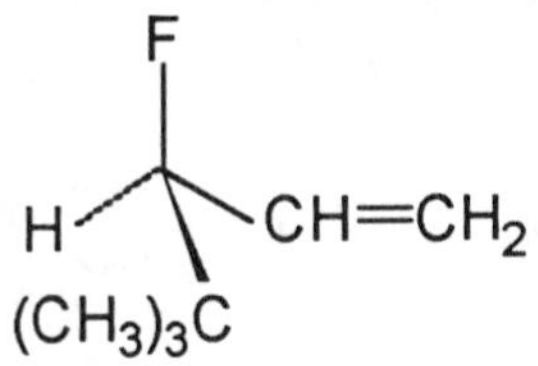

F; H; $CH{=}CH_2$; $(CH_3)_3C$ — II

H; CH_3CH_2; $CH(CH_3)_2$; CH_3 — III

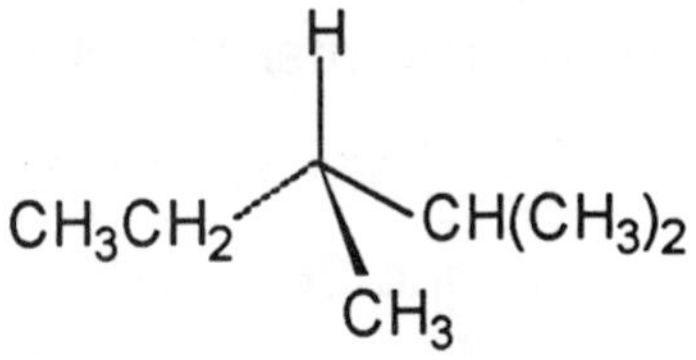

a) I, II b) I, III c) II, III d) all

313. Which of the following structures have an R configuration?

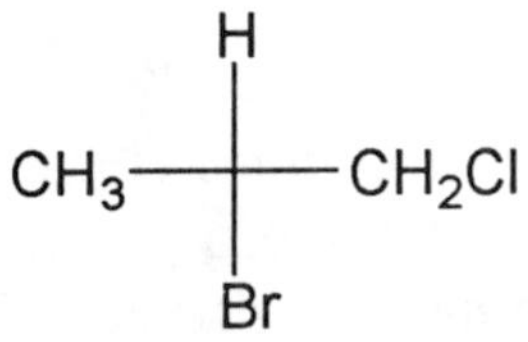

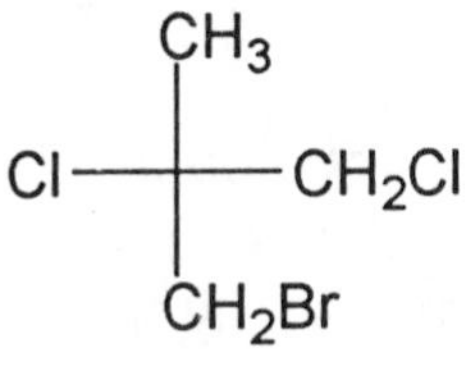

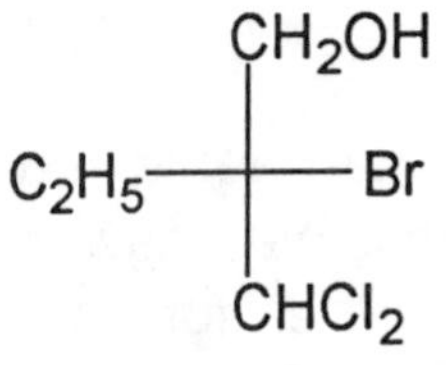

H; CH_3; CH_2Cl; Br — I

CH_3; Cl; CH_2Cl; CH_2Br — II

CH_2OH; C_2H_5; Br; $CHCl_2$ — III

CH_3; H; CH_3; CH_3 — IV

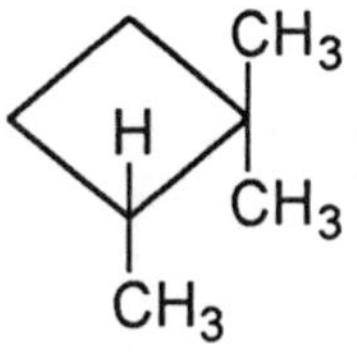

a) I, II b) I, III c) II, IV d) III, IV

TIP In the Fischer notation, when the group of lowest priority is on the vertical line (as it is for structures I and II) it is behind the plane and one simply numbers the remaining groups in priority. When the group of lowest priority is on the horizontal as in structure III, the situation is more complex since that group is coming out of the plane. One sure way to deal with this is to assign the R and S forms as if the group were behind the plane and then reverse the assignment. So structures III and IV are S forms.

314. What is the enantiomeric excess for the reaction that gives 66% of the S-form and 33% of the R-form?

a) 25% b) 33% c) 50% 75%

315. The specific rotation of dextrarotatory tartaric acid is +12.7 degrees. A mixture of dextrarotatory and levorotatory tartaric acid has a specific rotation of -6.35 degrees. What is the optical purity of the mixture?

a) 25% b) 33% c) 50% d) 75%

E. REACTIONS INVOLVING STEREOCENTERS

316. Trans-2-pentene was reacted with bromine in carbon tetrachloride to give two 2,3-dibromopentanes. What are the configurations of the two compounds?

a) 2R, 3R and 2S, 3S
b) 2S, 3S and 2R, 3S
c) 2S, 3R and 2R, 3S
d) 2R, 3S and 2S, 3S

317. A chiral compound with the molecular formula of C_5H_8 undergoes catalytic hydrogenation to form an achiral compound with the molecular fromula C_5H_{10}. What is the most reasonable structure for the original chiral compound?

a) 1-methylcyclobutene
b) 3-methylcyclobutene
c) 1,2-dimethylcyclopropene
d) cyclopentene

TIP This problem can be solved easily if we concentrate on the word chiral and look at the possible answers. Structures a), c) and d) all have a plane of symmetry and therefore are achiral. Finally, hydrogenation of structure b) leads to methyl cyclobutane which has a plane of symmetry.

318. A terminal alkyne with a molecular formula of C_6H_{10} loses its chirality when catalytically hydrogenated to a compound with the molecular formula of C_6H_{14}. What is the alkyne?

a) 1-hexyne
b) 3-methyl-1-pentyne
c) 4-methyl-1-pentyne
c) 3,3-dimethyl-1-butyne

319. When 1-chloro-2-methylbutane is chlorinated, how many pairs of enantiomers will be in the dichloride products?

a) 2 b) 3 c) 4 d) 5

320. What is the product of the reaction of S-3-bromo-1-pentene with hydrogen and platinum?

a) S-3-bromopentane
b) R-3-bromopentane
c) pentane
d) 3-bromopentane

TIP Upon hydrogenation, the carbon-carbon double bond is converted to an ethyl group in this instance. With the bromine atom on carbon atom number 3 such a conversion means that carbon atom number 3 is no longer a stereocenter! Thus answers a) and b) are wrong because they designate a stereocenter, and answer c) is wrong because the chemistry is wrong.

CHAPTER 9 ALCOHOLS and THIOLS

A. STRUCTURE and NOMENCLATURE

321. Which of the following structures have the correct IUPAC name?

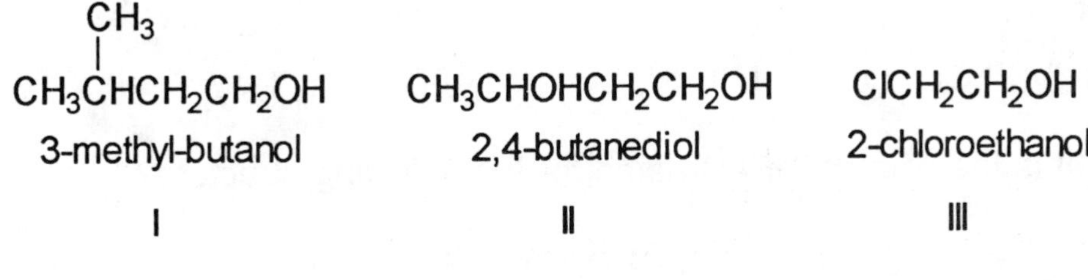

$CH_3CHOHCH_2Cl$

2-chloroethyl-1-propyl alcohol

IV

H
CH_2OH
OH
H

trans-2-(hydroxymethyl) cyclohexanol

V

a) III, IV, V   b) I, IV, V   c) I, III, V   d) I, II, IV

322. Which of the following structures have the correct common name?

OH
isopropyl alcohol
I

OH
sec butyl alcohol
II

OH
tert butyl alcohol
III

Br OH
bromobutyl alcohol
IV

OH
neopentyl alcohol
V

OH
methyl alcohol
VI

a) III, IV, VI   b) I, III, VI   c) I, II, III   d) II, III, V

323. Place the following functional groups in the order of increasing priority according to IUPAC nomenclature (lowest first).

halide	alkene	alcohol	alkyne
I	II	III	IV

a) I, IV, II, III b) I, III, IV, II c) IV, II, III, I d) II, IV, I, III

324. What is the correct IUPAC name for the following structure?

$$\begin{array}{l} ClCH_2 \\ \quad\quad\;\; \rangle C{=}C\langle \\ H \end{array}\begin{array}{l} CHOHC{\equiv}CH \\ \\ CH_3 \end{array}$$

a) E-6-chloro-4-methyl-4-hexen-1-yn-3-ol
b) E-1-chloro-3-methyl-hex-2-en-5-yn-4-ol
c) Z-6-chloro-4-methyl-4-hexen-1-yn-3-ol
d) Z-1-chloro-3-methyl-hex-2-en-5-yn-4-ol

TIP The priority for the groups on the double bond are $ClCH_2$ and CHOH, and since they are on the same side, (Z), answers a) and b) are wrong. The priority for numbering the chain comes from the triple bond not the chloro group, so answer d) is wrong because the substituent numbers are wrong.

325. Arrange the following substances in the order of increasing boiling point (lowest first).

I. ethanol II. 95% ethanol in water III. diethyl ether IV. ethylene glycol

a) II, III, IV, I b) III, I, II, IV c) III, II, I, IV d) IV, III, II, I

326. Which of the following compounds are readily soluble in water?

$CH_3CH_2CH_2CH_2OH$ I

$(CH_3)_3COH$ II

$HOCH_2CH_2CH_2CH_2OH$ III

$CH_3CH_2CH_2CH_2SH$ IV

a) II, III b) II, IV c) III, IV d) I, III

327. Arrange the following solvents in the order of decreasing polarity (greatest first).

CH_3CH_2OH	$CH_3CH_2OCH_2CH_3$	H_2O
I	II	III

a) III, II, I b) I, II, III c) II, I, III d) III, I, II

328. What is the angle for the oxygen-hydrogen-oxygen atoms in the hydrogen bond?

a) 90° b) 104 ° c) 108 ° d) 180 °

329. How many different kinds of hydrogen bonds are possible for a solution of ethanol and diethyl ether in water?

a) 3 b) 5 c) 6 d) 8

330. How many different kinds of hydrogen bonds are possible for a solution of 3-methoxy propanol in diethyl ether?

a) 2 b) 4 c) 5 d) 6

TIP Hydrogen bonds require an electronegative atom such as O, N or F with a lone pair of electrons and one bonded to a hydrogen atom. Diethyl ether cannot form a hydrogen bond with itself since it lacks a hydrogen atom on the oxygen atom. However 3-methoxy propanol can form 3 different kinds of hydrogen bonds with itself ($OH\cdots OCH_3$, $OH\cdots OH$, and internally to form a 6 membered ring). Also diethyl ether can form a hydrogen bond with 3-methoxy propanol, so the total is 4.

331. Place the following in the order of increasing strength of the hydrogen bonds (weakest first).

I. $HNH{-}{-}{-}{-}{-}OH_2$ II. $H_3N{-}{-}{-}{-}{-}HNH_2$

III. $H_2O{-}{-}{-}{-}{-}HOH$ IV. $H_3N{-}{-}{-}{-}{-}HOH$

a) I, II, III, IV b) III, I, IV, II c) II, IV, I, III d) I, IV, II, III

332. What is the metabolite that is responsible for the toxicity of methanol?

a) acetaldehyde b) acetic acid c) formaldehyde d) formic acid

333. Arrange the following ions in the order of decreasing basicity (greatest first).

$(CH_3)_2CHO^{\ominus}$	$NH_2^{\ominus}$	$OH^{\ominus}$	$CN^{\ominus}$
I	II	III	IV

a) II, I, III, IV b) III, I, II, IV c) IV, II, III, I d) I, III, II, IV

334. Arrange the following compounds in the order of decreasing acidity (greatest first).

CH_3CH_2OH	$ClCH_2CH_2OH$	$CH_3CH_2CH_2OH$	CF_3CH_2OH
I	II	III	IV

a) IV, I, II, III b) II, I, III, IV c) IV, II, I, III d) IV, II, III, I

335. What is the order of increasing acidity for the following compounds (least first)?

$CH_3CH_2CH_2OH$	$CH_3CH_2CH_3$	$CH_3C{\equiv}CH$	H_2O	NH_3
I	II	III	IV	V

a) II, III, I, IV, V b) V, II, I, III, IV c) II, V, III, I, IV d) V, II, I, IV, III

TIP Electronegativity of the central atom is an important contributor to acidity, as is hybridization and inductive effects. Oxygen is the most electronegative of the above atoms (followed by nitrogen) and one of the oxygen acids (I or IV) is the most acidic. I is less acidic than IV because of the inductive effect of the alkyl group. The least acidic is II, and because of the hybridization of the carbon atom in the triple bond, III is more acidic than V.

336. Which of the following reactions are favored to the right?

$$\text{I. } C_4H_9OH + \overset{\oplus}{H} \rightleftharpoons C_4H_9\overset{\oplus}{O}H_2$$

$$\text{II. } C_4H_9OH + \overset{\ominus}{CN} \rightleftharpoons C_4H_9CN + \overset{\ominus}{OH}$$

$$\text{III. } C_4H_9OH + HBr \rightleftharpoons C_4H_9Br + H_2O$$

$$\text{IV. } C_4H_9OH + \overset{\ominus}{Br} \rightleftharpoons C_4H_9Br + \overset{\ominus}{OH}$$

a) I, III b) II, IV c) II, III d) I, IV

337. Which of the following reactions are favored to the right?

$$\text{I. } CH_3CH_2OH + \overset{\ominus}{NH_2} \rightleftharpoons CH_3CH_2\overset{\ominus}{O} + NH_3$$

$$\text{II. } (CH_3)_2CHOH + \overset{\ominus}{OH} \rightleftharpoons (CH_3)_2CH\overset{\ominus}{O} + H_2O$$

$$\text{III. } CH_3CH_2OH + \overset{\ominus}{Br} \rightleftharpoons CH_3CH_2Br + \overset{\ominus}{OH}$$

$$\text{IV. } C_6H_{11}\text{–}OH + \overset{\ominus}{OH} \rightleftharpoons C_6H_{11}\text{–}\overset{\ominus}{O} + H_2O$$

a) I, III b) II, IV c) II, III d) I, IV

338. What is the order of decreasing reactivity of the following alcohols with hydrogen bromide (greatest first)?

$CH_2{=}CHCH_2OH$	CH_3CH_2OH	$(CH_3)_2CHOH$	$(CH_3)_3COH$
I	II	III	IV

a) II, III, I, IV b) I, II, III, IV c) IV, I, III, II d) I, IV, III, II

TIP The formation of a stable carbocation is the heart of the answer, that is the more stable the carbocation formed, the greater the reactivity. Answers a) and d) are incorrect because the tertiary cation (from IV) is more stable than the allylic carbocation (from I). Answer b) is incorrect because the carbocation stability is tertiary > secondary > primary.

339. For the following alcohols, what is the order of increasing reactivity in an acid-catalyzed dehydration (least first)?

I. butanol II. sec butanol III. isobutanol

a) I, II, III b) III, II, I c) I, III, II d) II, III, I

340. What is the major product when 2,2-dimethylcyclohexanol is heated in acid?

a) 3,3-dimethylcyclohexene b) 2,3-dimethylcyclohexene
c) 2,2-dimethylcyclohexene d) 1,2-dimethylcyclohexene

TIP Carbocation reactions often have rearrangements and two factors are of note here. First, the rearrangement always proceeds to give a more stable carbocation, and second, the presence of two methyl groups next to the incipient carbocation is a signal for rearrangement. Answer c) is incorrect because the double bond cannot be formed with a trisubstituted carbon atom. Answer a) is incorrect because it does not involve rearrangement. Answer b) is incorrect because the rearrangement is not to the more stable carbocation.

341. When 1-butanol is heated with an acid, which of the following is NOT likely to be in the product mixture?

a) 1-butene b) cis-2-butene
c) 2-methylpropene d) trans-2-butene

342. Which of the following chemical tests could distinguish between cyclohexanol and cyclohexene?

Br_2 in CCl_4	cold conc H_2SO_4	chromic acid	$AgNO_3$
I	II	III	IV

a) I, II b) I, III c) II, IV d) III, IV

B. PREPARATION

343. Which of the following synthetic methods for preparing alcohols will not yield 2-pentanol from 1-pentene?

a) acid-catalyzed hydration
b) hydroboration-oxidation
c) oxymercuration-demercuration
d) hydroboration-base hydrolysis

344. What is the major product from the acid-catalyzed hydration of 3,3-dimethyl-1-butene?

a) 2,3-dimethyl-2-butanol
b) 3,3-dimethyl-2-butanol
c) 3,3-dimethyl-1-butanol
d) 2,3-dimethyl-1-butanol

345. Alcohols are inexpensive and readily available from commercial preparations not usually done in the laboratory. Which alcohols can be made by these commercial methods?

A. hydration of alkenes
B. oxo process
C. fermentation

I. CH_3OH
II. CH_3CH_2OH
III. $(CH_3)_2CHOH$
IV. $CH_3CH_2CH_2OH$

a) A and II, B and I, C and I
b) A and III, B and IV, C and II
c) A and IV, B and III, C and II
d) A and III, B and II, C and IV

346. What is the major product from the acid-catalyzed hydration of 3,3-dimethylcyclopentene?

a) 1,2-dimethylcyclopentanol
b) 2,2-dimethylcyclopentanol
c) 1,3-dimethylcyclopentanol
d) 3,3-dimethylcyclopentanol

347. Which of the following reaction sequences is best for converting 1-propanol to 2-propanol?

a) $\xrightarrow[\text{EtOH, heat}]{\text{NaOEt}}$ $\xrightarrow{\text{dil } H_2SO_4}$ b) $\xrightarrow[\text{heat}]{H_2SO_4}$ $\xrightarrow{B_2H_6}$ $\xrightarrow[\text{NaOH}]{H_2O_2}$

c) $\xrightarrow[\text{heat}]{H_2SO_4}$ $\xrightarrow{Hg(OAc)_2}$ $\xrightarrow{NaBH_4}$ d) $\xrightarrow[\text{EtOH, heat}]{\text{NaOEt}}$ $\xrightarrow{KMnO_4}$

TIP This scheme requires a dehydration to give an alkene followed by a Markovnikov addition of water to give the alcohol in the 2 position. Answer a) and d) are wrong conditions for the dehydration of 1-propanol. Answer b) adds water in an anti-Markovnikov orientation and would actually give back 1-propanol.

C. REACTIONS

348. Which of the following compounds cannot be prepared from 1-butanol in one step?

a) 1-butene b) butane c) butyl chloride d) 1-butanal

349. What is the product from the reaction of HCl with 2,2-dimethylcyclohexanol?

a) 2,2-dimethylcyclohexyl chloride b) 1,2-dimethylcyclohexyl chloride
c) 2,3-dimethyl-2-chlorocyclohexene d) 3,3-dimethyl cyclohexyl chloride

350. The carbon-oxygen bond in alcohols does not break in which of the following reactions?

I. oxidation with H_2CrO_7
II. reaction with NaOH
III. formation of inorganic esters
IV. dehydration
V. hydrobromination

a) I, II, III b) II, IV, V c) II, III, V d) I, IV, V

351. Which of the following reagents will cleave the oxygen-hydrogen bond in alcohols?

Na	$K_2Cr_2O_7$, acid	NaH	HBr	conc. H_2SO_4, heat
I	II	III	IV	V

a) I, II, III b) II, IV, V c) I, III, V d) III, IV, V

352. Which of the following oxidizing reagents can be used to prepare CH_3CH_2CHO from 1-propanol?

$KMnO_4$	$CrO_3 \cdot 2$ pyridine	PCC	H_2O_2
I	II	III	IV

a) I, II b) II, III c) I, III d) III, IV

353. Which of the following reactions will proceed with retention of configuration?

a) $CH_3CHBrCH_2CH_3$ + NaOH ⟶ $CH_3CH(OH)CH_2CH_3$

b) $CH_3CHBrCH_2CH_3$ + H_2O (in acetone) ⟶ $CH_3CH(OH)CH_2CH_3$

c) $CH_3CH(OH)CH_2CH_3$ + $SOCl_2$(in diethyl ether) ⟶ $CH_3CHClCH_2CH_3$

d) $CH_3CH(OH)CH_2CH_3$ + HCl ⟶ $CH_3CHClCH_2CH_3$

TIP In order to have retention of configuration, the simple S_N1 and S_N2 reactions have to be avoided because S_N1 will often give racemized products and S_N2 reactions give inverted products. Answers a) and b) are wrong because both processes are involved. Answer d) is wrong because it is an S_N1 process. Answer c) involves a cyclic mechanism that proceeds with retention.

354. What are the best conditions for converting 3-methyl-2-pentanol to 2-chloro-3-methyl pentane?

a) HCl　　b) $HCl + ZnCl_2$ + heat　　c) PCl_3　　d) $SOCl_2$

355. What are the best conditions for converting neopentyl alcohol to neopentyl chloride?

a) HCl　　b) $HCl + ZnCl_2$ + heat　　c) PCl_3　　d) $SOCl_2$

356. Cyclopentanol and phosphoric acid were heated and the product was then reacted with cold dilute potassium permanganate. Which of the following possibilities is in the final product?

a) $HO—(CH_2)_5—OH$

b) (cyclopentane ring with OH and HO on adjacent carbons, trans) OH, HO

c) (cyclopentane ring with OH and HO on adjacent carbons, cis) OH, HO

d) $HCCH_2CH_2CH_2COCH_2CH_3$ (with C=O at each marked C: $HC(=O)CH_2CH_2CH_2C(=O)OCH_2CH_3$)

357. Isobutyl bromide was dissolved in ethanol and heated with potassium hydroxide. The product of this reaction was further heated with ethanethiol and peroxide. Which of the following is the most likely final product?

a) $(CH_3)_2CHCH_2SCH_2CH_3$

b) $CH_3CH_2SC(CH_3)_2CH_3$

c) $(CH_3)_2CHCH_2SH$

d) $(CH_3)_2CHCH_2CH_2SCH_3$

358. Cyclopentene was treated with hydrogen bromide, then with lithium in ether, then with cuprous iodide, and finally with methyl iodide to give product A. Product A was brominated and then heated with potassium hydroxide to give product B. Hydroboration of product B, followed by treatment with hydrogen peroxide and sodium hydroxide gave product C. Which of the following structures is the most likely for product C?

a) 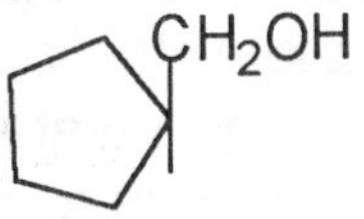 CH_2OH b) 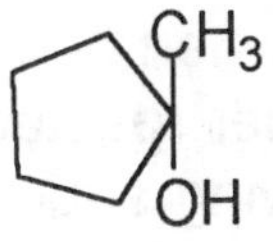 CH_3 OH c) 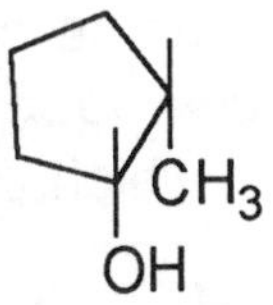 CH_3 OH d) 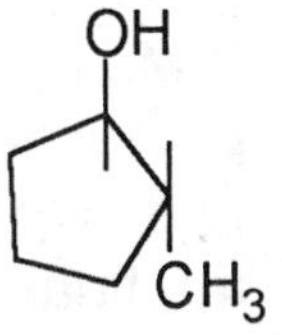 OH CH_3

TIP The important parts of this scheme are: (1) cyclopentene is alkylated with a methyl group (product A); (2) free radical bromination will give the tertiary bromide that is dehydrohalogenated to methyl cyclopentene (product B); (3) a syn, anti-Markovnikov addition of water gives product C. Answer a) is incorrect because step (1) did not occur. Answer b) is incorrect because it is the Markovnikov product, and answer c) is incorrect because it is not a syn product.

359. Hydroboration of 2-methyl-1-pentene, followed by treatment with aqueous hydrogen peroxide, gave an organic compound A. This compound was reacted with hydrogen bromide to give the final product. Which of the following is the most likely final product?

a) $(CH_3)_2C(Br)(CH_2)_2CH_3$

b) $CH_3CH_2CH_2CH(CH_3)CH_2Br$

c) $CH_3CH_2CH{=}C(CH_3)_2$

d) $CH_3CH_2CH_2CH(Br)CH_2OH$

360. What is the final product of the reaction of propene with hydrogen bromide and peroxides, followed by reaction with sodium methylsulfide?

a) 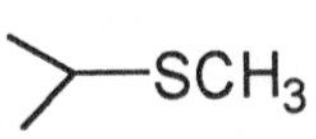 $-SCH_3$ b) 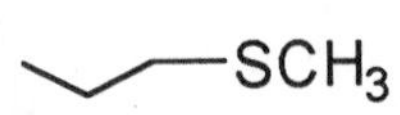 $-SCH_3$ c) 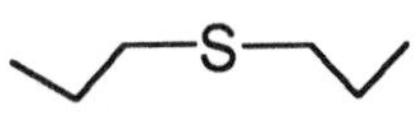 $-S-$ d) 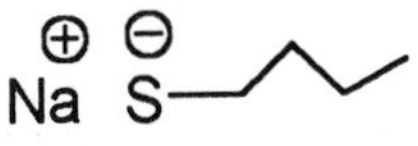 $Na^{\oplus}$ $S^{\ominus}-$

361. Methylene cyclohexane was dissolved in dilute acid and then heated with concentrated sulfuric acid. What is the final product?

a) 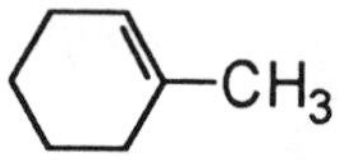 $-CH_3$ b) 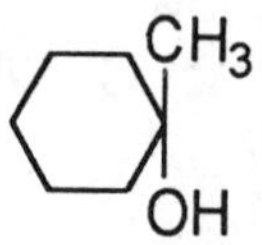 CH_3, OH c) 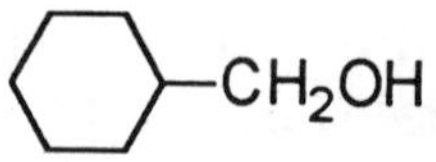 $-CH_2OH$ d) 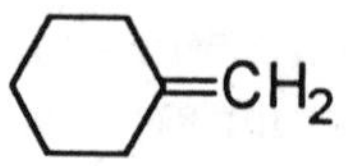 $=CH_2$

362. 2-Methylpentane gave two products when heated with bromine, however both products gave the same compound when heated in ethanol with potassium hydroxide. This compound was further reacted with chlorine in water to give the final product. Which of the following structures is most likely to be the final product?

a) 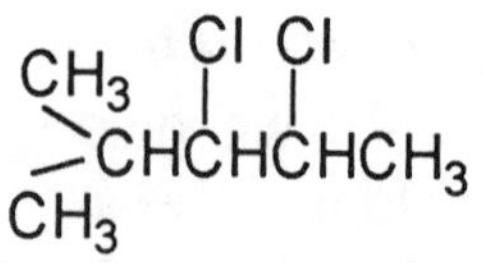 $(CH_3)_2CHCH(Cl)CH(Cl)CH_3$

b) 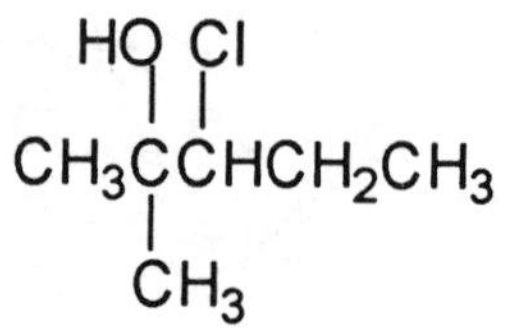 $CH_3C(OH)(CH_3)CH(Cl)CH_2CH_3$

c) 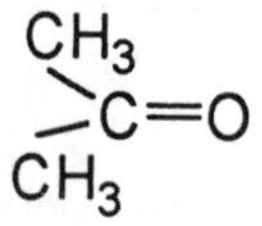 $(CH_3)_2C{=}O$

d) 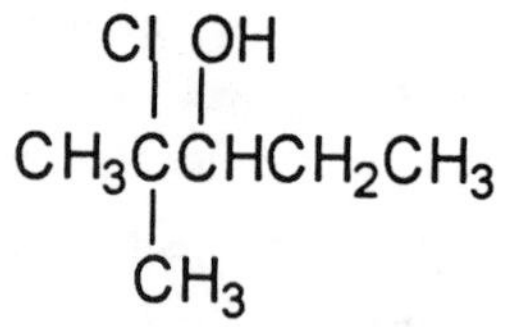 $CH_3C(Cl)(CH_3)CH(OH)CH_2CH_3$

TIP This scheme involves a Markovnikov-like addition of HO-Cl to the 2-methyl-2-pentene generated in the first steps. Answer a) is incorrect because the addition is to the wrong alkene. Answer d) results from an anti-Markovnikov addition of HO-Cl. Answer c) would result from completely different reaction conditions?

CHAPTER 10 ALKYL HALIDES

A. NOMENCLATURE and STRUCTURE

363. Which of the following structures have the correct IUPAC name?

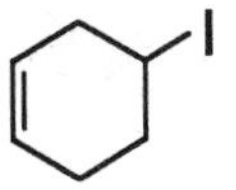

3-iodocyclohexene
I

H_3C Cl
$CH_3CH_2CHCHCH_2CH_3$
3-chloro-4-methylhexane
II

Cl OH
$CH_3CHCHCH_3$
2-chloro-3-butanol
III

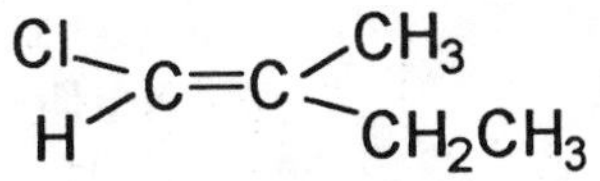

Cl(H)C=C(CH_3)(CH_2CH_3)
E-1-chloro-2-methylbutane
IV

a) I, III   b) II, IV   c) I, IV   d) II, III

364. Which of the following structures have the correct common name?

$CH_2{=}CHCl$	$(CH_3)_3CCH_2Br$	CHF_3	$(CH_3)_2CHCH_2I$
vinyl chloride	neopentyl bromide	methylene fluoride	sec-butyl iodide
I	II	III	IV

a) I, II   b) II, IV   c) I, IV   d) II, III

365. Which of the following matches are correct?

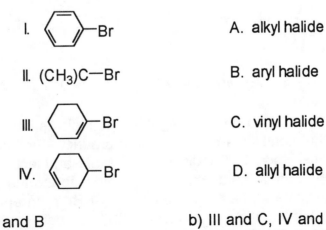

I. 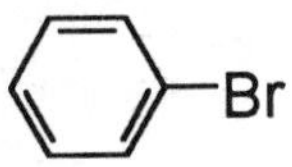 (benzene ring)–Br   A. alkyl halide

II. $(CH_3)C{-}Br$ B. aryl halide

III. 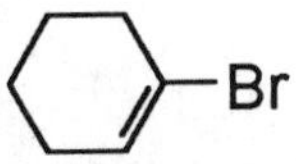 (cyclohexene ring)–Br   C. vinyl halide

IV. 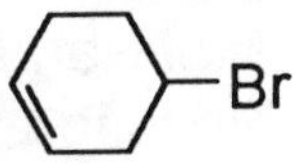 (cyclohexene ring)–Br   D. allyl halide

a) I and A, II and B   b) III and C, IV and A
c) I and B, IV and C   d) II and A, III and D

366. Place the following halides in the order of increasing polarity (least first).
I. methylene bromide II. chloroform III. iodoform IV. carbon tetrachloride

a) II, I, III, IV b) IV, I, III, II c) III, I, IV, II d) IV, II, I, III

B. PREPARATIONS

367. What are the best conditions for preparing the following halide?

(methylenecyclohexane, =CH_2) → (cyclohexyl–CH_2Br)

a) Br_2 in CCl_4 b) HBr and peroxides c) HBr d) NBS and light

368. What are the best conditions for preparing the following halide?

(methylenecyclohexane, =CH_2) → (1-bromo-1-methylcyclohexane: CH_3, Br)

a) Br_2 in CCl_4 b) HBr and peroxides c) HBr d) NBS and light

369. What are the best conditions for preparing the following halide?

(methylenecyclohexane, =CH_2) → (cyclohexane with CH_2Br and Br on same carbon)

a) Br_2 in CCl_4 b) HBr and peroxides c) HBr d) NBS and light

370. What are the best conditions for preparing the following halide?

(methylenecyclohexane, =CH_2) → (2-bromo-1-methylenecyclohexane: =CH_2, Br)

a) Br_2 in CCl_4 b) HBr and peroxides c) HBr d) NBS and light

TIP For questions 367-370, understanding the mechanisms involved will lead to choosing the right conditions for the desired product. First look at the starting compound vs. the product. Conditions a) indicate an electrophilic addition of bromine at both ends of the double bond. This leads to the product in 369. Conditions b) indicate a free radical addition of HBr in an anti-Markovnikov configuration. This leads to the product in 367. Conditions c) indicate a carbocation mechanism for electrophilic addition. This leads to the product in 368. Conditions d) use NBS which is a special reagent for allylic substitution and leads to the product in 370 and another isomer not shown.

371. Which alkyl halide is prepared from the following reaction?

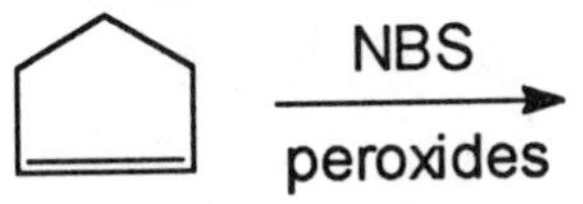

cyclopentene $\xrightarrow[\text{peroxides}]{\text{NBS}}$

a) 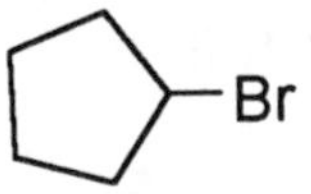 bromocyclopentane (cyclopentyl–Br) b) 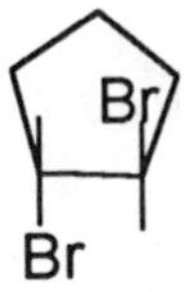 1,2-dibromocyclopentane (Br, Br) c) 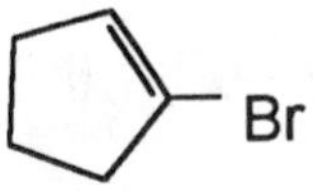 3-bromocyclopentene (ring C=C, –Br) d) 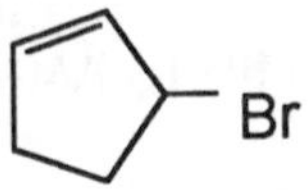 3-bromocyclopentene (ring C=C, –Br)

372. What is the product from the following reaction?

$$CH_3CH_2\overset{CH_3}{\overset{|}{C}}HCH_2OH + HCl \xrightarrow[\text{heat}]{ZnCl_2}$$

a) $CH_3CH_2\underset{CH_3}{\underset{|}{\overset{CH_3}{\overset{|}{C}}}}Cl$ b) $CH_3CH_2\overset{CH_3}{\overset{|}{C}}HCH_2Cl$

c) $(CH_3)_2\overset{Cl}{\overset{|}{C}}CH_2CH_3$ and $CH_3CH_2\overset{CH_3}{\overset{|}{C}}HCH_2Cl$ d) $CH_3CH_2\overset{Cl}{\overset{|}{C}}{=}CH_2$

373. What are the best conditions for preparing 1-bromo-2-butene from 2-butene?

a) HBr, $ZnBr_2$ and heat b) HBr c) PBr_3 d) NBS and peroxides

C. NUCLEOPHILIC ALIPHATIC SUBSTITUTION

374. What is the order of increasing reactivity for the following nucleophiles (least first)?

$CH_3O^{\ominus}$	H_2O	$CH_3NH^{\ominus}$	$CH_3COO^{\ominus}$
I	II	III	IV

a) IV, II, I, III b) II, IV, I, III c) IV, III, I, II d) I, II, III, IV

375. What is the order of decreasing nucleophilicity for the following compounds and ions (strongest first)?

CH_3OH	$CH_3O^{\ominus}$	H_2O	$^{\ominus}OH$
I	II	III	IV

a) I, III, II, IV b) IV, III, II, I c) II, IV, III, I d) II, IV, I, III

376. Which of the following are examples of strong nucleophiles but weak bases?

$CH_3S^{\ominus}$	$CH_3O^{\ominus}$	$I^{\ominus}$	H_2O	$Cl^{\ominus}$
I	II	III	IV	V

a) I, IV b) IV, V c) I, III d) III, IV

TIP Nucleophilicity includes basicity, charge and polarizability. In questions 374 and 375, polarizability is not a factor. The oxygen species are stronger bases than nitrogen, but since methanol is a weaker acid than water, the conjugate base is a stronger base. The charged species are stronger nucleophiles than the neutral species. So the order is II, IV, I, III in 374 and II, IV, I, III in 375. For larger central atoms, polarizability becomes a major factor and the polarizability of a group inverts the basicity argument order so that the highly polarizable groups in question 376 are strong nucleophiles but weak bases (I and III).

377. Arrange the following leaving groups in the order of increasing reactivity in an S_N2 reaction (least first).

$-Br$	$-Cl$	$-OH$	$-\overset{\oplus}{O}H_2$
I	II	III	IV

a) III, II, IV, I b) I, II, III, IV c) III, IV, II, I d) II, I, IV, III

378. What is the order of increasing reactivity for the following halides in an S_N2 reaction (least first)?

$CH_3CHClCH_3$	$CH_3CH_2CH_2Cl$	$(CH_3)_3CCl$	CH_3Cl
I	II	III	IV

a) IV, III, I, II b) III, I, II, IV c) III, I, IV, II d) I, II, IV, III

379. Place the following alkyl halides in the order of increasing reactivity in an S_N2 reaction (least first).

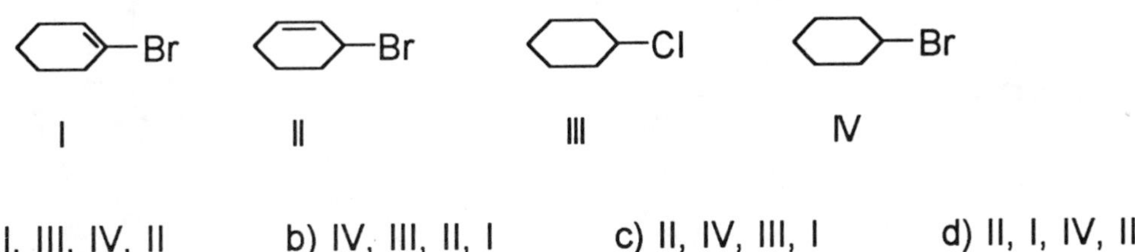

a) I, III, IV, II   b) IV, III, II, I   c) II, IV, III, I   d) II, I, IV, III

TIP The leaving group activity depends solely on the basicity of the atom or group. The weaker the base, the stronger the conjugate acid, the better the leaving group. In question 377, the order of acid strengths is HBr > HCl > H_3O^+ > H_2O so the leaving group order of reactivity is I, II, IV, III.
In question 378, steric considerations are the primary contributions to halide structure differences, and the order for reactivity is methyl, 1°, 2°, 3° .
Since the bond to the leaving group is broken in the rate determining step, the bond strength contributions determine the reaction rate, and therefore, the strong bond in a vinyl halide is very unreactive but the weak bond in an allyl halide is very reactive.

380. For an S_N1 reaction, what is the order of increasing reactivity for the following alkyl halides (least first)?

$CH_3CHClCH_3$	$CH_3CH_2CH_2Cl$	$(CH_3)_3CCl$	CH_3Cl
I	II	III	IV

a) II, IV, I, III   b) I, III, IV, II   c) III, I, IV, II   d) IV, II, I, III

381. Place the following alkyl halides in the order of increasing reactivity in an S_N1 reaction.

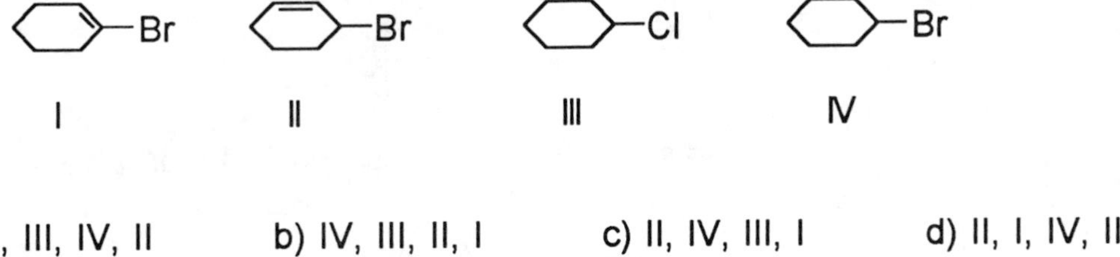

a) I, III, IV, II   b) IV, III, II, I   c) II, IV, III, I   d) II, I, IV, III

382. What is the order of increasing stability for the following carbocations (least first)?

$CH_3CH_2\overset{\oplus}{C}H_2$	$(CH_3)_2\overset{\oplus}{C}H$	$(CH_3)_3\overset{\oplus}{C}$	(bicyclic cation, bridgehead ⊕)
I	II	III	IV

a) IV, III, II, I   b) IV, I, II, III   c) I, II, III, IV   d) I, II, IV, III

383. What is the order of increasing rate of solvolysis for the following halides (slowest first).

$(CH_3)_3CCl$	$(CH_3CH_2)_2CHCl$	$(CH_3)_2CHCl$	$CH_3CH_2CH_2Cl$
I	II	III	IV

a) IV, III, II, I   b) I, II, III, IV   c) III, II, I, IV   d) IV, II, III, I

TIP For a solvolysis reaction, carbocation stability is critical. In general, carbocation stability is 3°, 2°, 1°, methyl. A second feature is a steric concern. Although ion IV in question 382 is tertiary, it cannot be planar as is required for the sp^2 hybridization and is therefore highly disfavored. Steric concerns also render the secondary carbocation derived from compound II in question 383 less stable than the secondary carbocation from compound III.

384. In the following reaction, if the concentration of t-butyl bromide and ethanol are doubled, what effect does this have on the rate of reaction?

$$CH_3\underset{CH_3}{\overset{CH_3}{\underset{|}{\overset{|}{C}}}}Br + CH_3CH_2OH \longrightarrow CH_3CH_2O\underset{CH_3}{\overset{CH_3}{\underset{|}{\overset{|}{C}}}}CH_3$$

a) no change   b) doubles   c) triples   d) quadruples

385. What are the products from the following reaction?

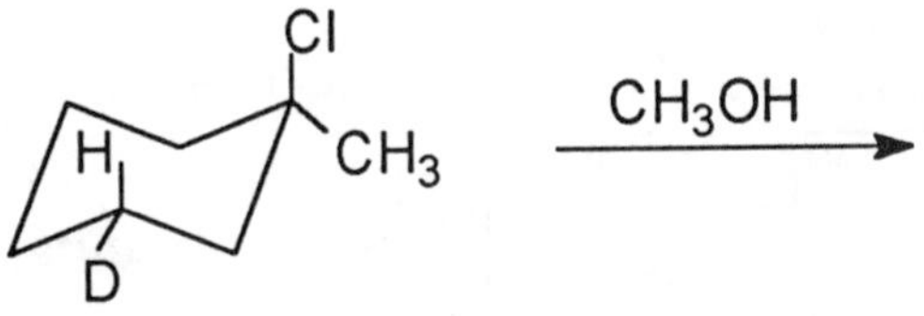

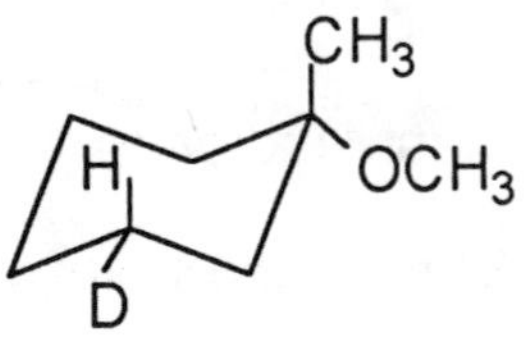

$$\text{(cyclohexane with Cl, } CH_3 \text{, H, D)} \xrightarrow{CH_3OH}$$

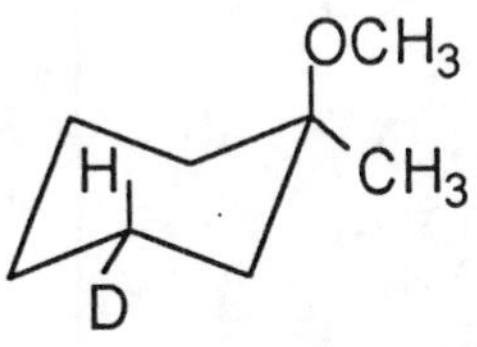

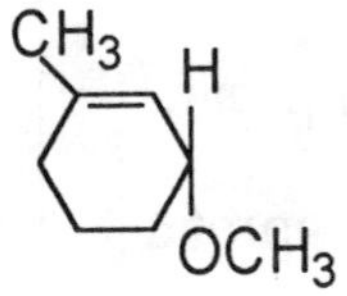

I: CH_3, OCH_3, H, D

II: OCH_3, CH_3, H, D

III: CH_3, H, OCH_3

a) I, II b) II, III c) I, III d) I, II, III

386. Which of the following is an optimum set of conditions for an S_N1 reaction?

	solvent	nucleophile	substrate
a)	polar aprotic	$F^{\ominus}$	methyl iodide
b)	polar protic	CH_3OH	t-butyl bromide
c)	polar protic	$CH_3O^{\ominus}$	neopentyl iodide
d)	polar aprotic	$t\text{-BuO}^{\ominus}K^{\oplus}$	isopropyl chloride

TIP Tertiary halides cannot undergo S_N2 reactions. Under solvolysis conditions (ethanol in 384, methanol in 385 and 386), an S_N1 reaction is expected. The S_N1 reaction is first order in halide and zero order in nucleophile or base. Thus doubling the concentration of ethanol in question 384 has no effect on the rate, while doubling the concentration of halide doubles the reaction rate. Additionally, the reaction of the planar carbocation leads to a racemized product so that both compounds I and II are formed in question 385.

387. Which of the following is an optimum set of conditions for an S_N2 reaction?

	solvent	nucleophile	substrate
a)	polar aprotic	$F^{\ominus}$	methyl iodide
b)	polar protic	CH_3OH	t-butyl bromide
c)	polar protic	$CH_3O^{\ominus}$	neopentyl iodide
d)	polar aprotic	t-$BuO^{\ominus}K^{\oplus}$	isopropyl chloride

388. Which of the following reactions involve inversion of configuration?

I: R-2-bromohexane $\xrightarrow[CH_3CH_2OH]{^{\ominus}OH}$

II: S-2-iodobutane $\xrightarrow{NaSCH_3}$

III: R-3-bromo-3-methylhexane $\xrightarrow[CH_3CH_2OH]{^{\ominus}OH}$

IV: R-2-bromohexane $\xrightarrow[heat]{CH_3OH}$

a) I, II b) III, IV c) II, III d) I, IV

389. Which of the following statements characterize the reaction of R-3-bromocyclopentene with sodium iodide in acetone?

I. The reaction involves a carbocation intermediate.
II. The reaction involves inversion of configuration.
III. The reaction involves retention of configuration.
IV The reaction gives predominately a racemic product.

a) I b) II c) I, IV d) I, III

390. Which of the following statements apply to an S_N1 reaction?

I. The reaction is first order in alkyl halide and first order in the nucleophile.
II. The order of reactivity is methyl > 1°> 2° > 3°.
III. The reaction is first order in alkyl halide and zero order in the nucleophile.
IV. Rearrangements are common.

a) I, II b) III, IV c) I, IV d) III

391. Which of the following statements apply to an S_N2 reaction?

I. The reaction is first order in alkyl halide and first order in the nucleophile.
II. The order of reactivity is methyl > 1°> 2° > 3°.
III. The reaction is first order in alkyl halide and zero order in the nucleophile.
IV. Rearrangements are common.

a) I, II b) III, IV c) I, IV d) II, IV

TIP S_N2 and S_N1 reactions are distinguished by three criteria: kinetics, structure, and stereochemistry. S_N2 reactions are bimolecular with a rate dependency on both nucleophile and halide. S_N1 reactions are unimolecular with a rate dependency only on halide. Reactivity orders for structure of the halide are opposite with S_N2 decreasing and S_N1 increasing in the order of methyl >1°> 2° > 3°. Finally, all S_N2 reactions involve inversion of configuration whereas S_N1 reactions lead to racemization.

D. ELIMINATION

392. Which of the following halides will give propene in the most efficient and rapid reaction with base.

a) 1-bromopropane
b) 1-chloropropane
c) 2-bromopropane
d) 2-chloropropane

393. Which of the following halides give the highest yield of isobutylene by dehydrohalogenation?

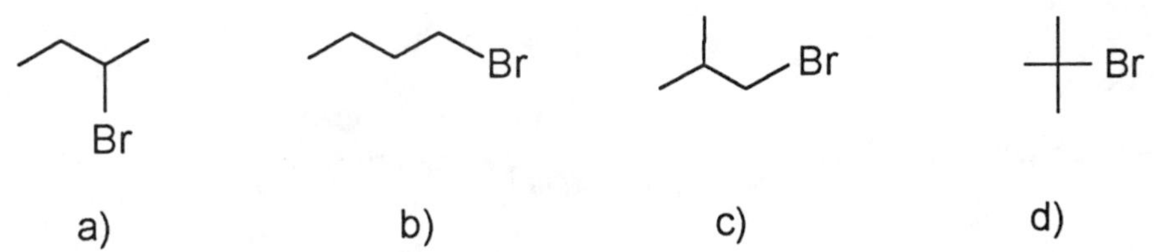

a) b) c) d)

394. Which halide gives the highest yield of 2-methyl-2-butene by dehydrohalogenation?

a) 1-chloro-2-methylbutane
b) 1-bromo-2-methylbutane
c) 2-chloro-2-methylbutane
d) 2-bromo-2-methylbutane

395. Which of the following is an optimum set of conditions for an E2 reaction?

	solvent	nucleophile	substrate
a)	polar aprotic	$F^{\ominus}$	methyl iodide
b)	polar protic	CH_3OH	neopentyl iodide
c)	polar protic	$CH_3O^{\ominus}$	neopentyl iodide
d)	polar aprotic	t-$BuO^{\ominus}$ $K^{\oplus}$	isopropyl chloride

TIP An E2 reaction requires a hydrogen atom beta to the leaving group. Note that only isopropyl chloride has a beta hydrogen atom.

396. Which halide gives only one elimination product from reaction with sodium ethoxide in ethanol?

a) 2-bromo-4,4-dimethylpentane
b) 2-bromo-3,3-dimethylpentane
c) 3-bromo-2-methylhexane
d) 3-bromo-2,2-dimethylpentane

397. When a mixture of equal amounts of 2-bromopentane and 2-bromo-3-methylbutane react with sodium ethoxide in ethanol, what is the order of increasing amounts of the following alkenes produced (least first)?

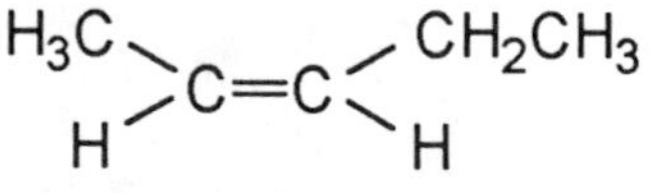

I: $(H_3C)(H)C{=}C(CH_2CH_3)(H)$ (H₃C and CH₂CH₃ on the same side)

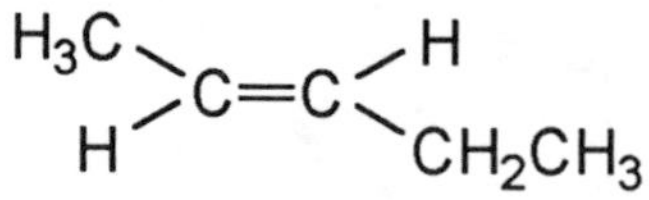

II: $(H_3C)(H)C{=}C(H)(CH_2CH_3)$ (H₃C and CH₂CH₃ on opposite sides)

III: $CH_2{=}CHCH_2CH_2CH_3$

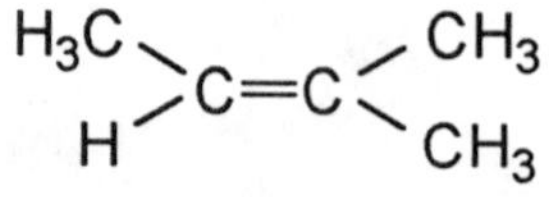

IV: $(H_3C)(H)C{=}C(CH_3)(CH_3)$

a) IV, I, III, II b) III, I, II, IV c) IV, I, II, III d) I, III, II, IV

398. How many products are formed and which isomer reacts faster when a 1:1 mixture of cis and trans-1-bromo-2-methylcyclohexane is dehydrohalogenated with 0.5 equivalents of sodium ethoxide in ethanol?

a) trans and 1 b) trans and 2 c) cis and 1 d) cis and 2

399. What is the product of the reaction of trans-1-bromo-2-methylcyclohexane with strong base?

a) methylcyclohexene
b) 3-methylcyclohexene
c) 4-methylcyclohexene
d) cis-1-bromo-2-methylcyclohexene

TIP The preferred geometry for the E2 reaction is anti-coplanar. In a cyclohexane ring, this is achieved when the leaving group and the beta hydrogen are diaxial. Additionally, when possible, elimination is preferred to give the more highly substituted alkene. For the trans compound in questions 398 and 399, the preferred geometry and the preferred alkene product are in conflict. Thus reaction is slower and the less stable alkene in 399 is formed (answer b).

400. Which of the following is an optimum set of conditions for an E1 reaction of t-butyl bromide?

	solvent, nucleophile	temperature ^{0} C
a)	CH_3OH	25^0
b)	CH_3CH_2OH	80^0
c)	$CH_3CH_2O^{\ominus}$ CH_3CH_2OH	25^0
d)	$CH_3CH_2O^{\ominus}$ DMSO	25^0

401. In which of the following reactions will the stability of a carbocation NOT determine the structure of the product?

I. hydration II. hydroboration III. E1 IV. E2

a) I, II b) I, III c) II, III d) II, IV

402. In which of the following reactions will the stability of a carbocation intermediate determine the structure of the product?

I. acid-catalyzed dimerization of alkenes II. S_N1
III. catalytic hydrogenation IV. S_N2

a) I, II b) III, IV c) I, III d) II, IV

403. What is the best procedure to prepare isopropyl methyl ether?

a) isopropyl iodide and sodium methoxide
b) isopropyl iodide and methanol
c) methyl iodide and potassium isopropoxide
d) methyl iodide and isopropanol

404. In each of the following reactions, products other than those shown will predominate. What is the most likely reason for this?

$$(CH_3)_3COH + HBr \xrightarrow{heat} (CH_3)_3CBr$$

$$(CH_3)_2CBrCH_2CH_3 + Na^{\oplus}\,{}^{\ominus}OCH_3 \longrightarrow (CH_3)_2C(OCH_3)CH_2CH_3$$

1-chloro-1-methylcyclohexane + CH_3CH_2OH → 1-ethoxy-1-methylcyclohexane

a) rearrangements
b) elimination dominates substitution
c) radical processes give mixtures
d) competing solvolytic reactions

E. REACTIONS INVOLVING SUBSTITUTION and ELIMINATION

405. Cyclopentane was reacted with bromine and heat to give compound A, which was then reacted with sodium ethoxide in ethanol to give compound B. Compound B was reacted with NBS and peroxides to form compound C. Which of the following is a reasonable structure for compound C?

a) cyclopentane with OCH_2CH_3 and Br on adjacent carbons
b) cyclopentene with Br
c) cyclopentene with Br
d) cyclopentene with Br

TIP Key mechanistic steps are E2 followed by allylic bromination.

406. What is the final product when 2-butanol is heated with concentrated sulfuric acid, and then this product is reacted with hydrogen and platinum?

a) butane b) cis-2-butene c) 1-butene d) cyclobutane

TIP Key mechanistic step is E1.

407. An achiral compound, $C_4H_{10}O$, reacts with sulfuric acid and heat to form compound B. Compound B gives compound C when reacted first with mercuric acetate and then with sodium borohydride. Compound C is chiral. What is a reasonable structure for compound A?

a) $CH_3CH_2CH(OH)CH_3$

b) cyclopropyl–CH_2OH

c) $CH_3C(CH_3)_2OH$

d) $CH_3CH_2CH_2CH_2OH$

TIP Key mechanistic steps are dehydration (E1) followed by Markovnikov addition of water.

408. Which of the following compounds can be prepared by reacting 1-methylcyclohexanol with aqueous sulfuric acid, and treating the product with bromine in carbon tetrachloride?

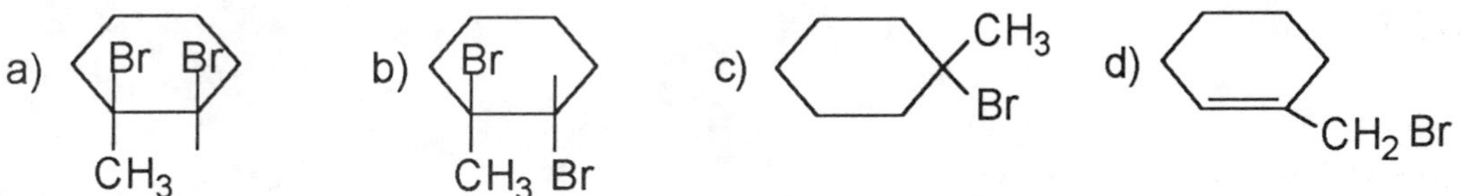

TIP Key mechanistic steps are dehydration (E1) followed by electrophilic addition of bromine.

409. Which of the following is a reasonable product formed when R-2-bromobutane is treated with lithium methyl acetylide and then with hydrogen and Lindlar's catalyst?

a) R-4-methyl-2-hexene
b) S-4-methylhexane
c) S-4-methyl-2-hexene
d) 3-methylhexane

TIP Key mechanistic step is S_N2 with inversion, but look at 'chirality' of final product.

410. Dehydration of cis-1,2-dimethylcyclohexanol yields a product which is then reacted with a peracetic acid solution. What is the most likely description of the final product?

a) a mixture of acids
b) a dicarboxylic acid
c) a chiral diol
d) a meso diol

TIP Key mechanistic step is E1 followed by trans hydroxylation.

411. Which of the following compounds is the most likely product formed when 2R,3R-2-bromo-3-methylhexane is treated with potassium tert-butoxide?

a) Z-3-methyl-3-hexene
b) E-3-methyl-3-hexene
c) E-3-methyl-2-hexene
d) Z-3-methyl-2-hexene

TIP Key mechanistic step is E2. Anti-coplanar elimination is preferred.

CHAPTER 11 ETHERS and EPOXIDES

A. NOMENCLATURE and STRUCTURE

412. Which of the following structures have the correct names?

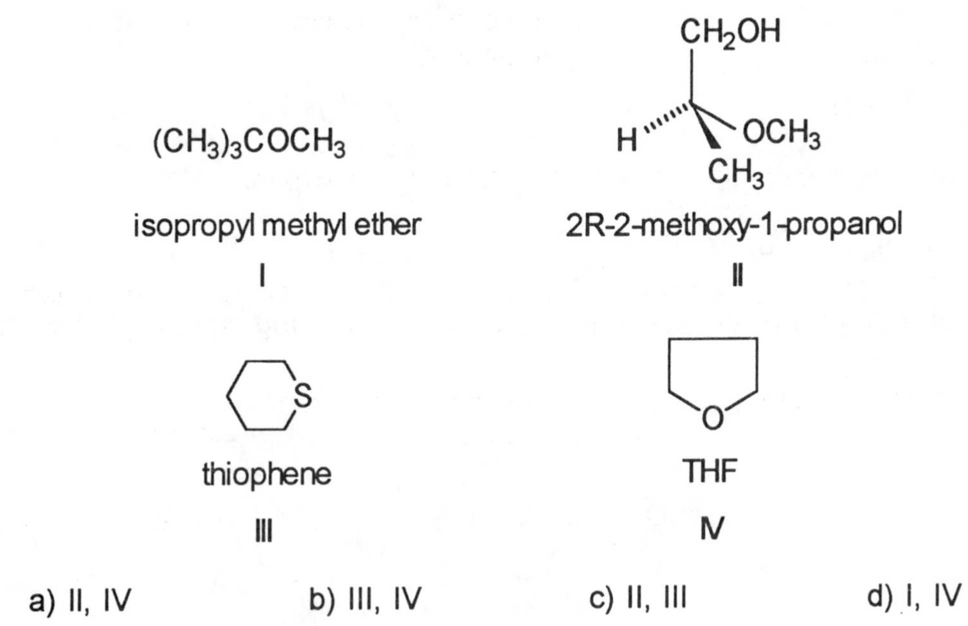

a) II, IV   b) III, IV   c) II, III   d) I, IV

413. Which of the following structures have the correct names?

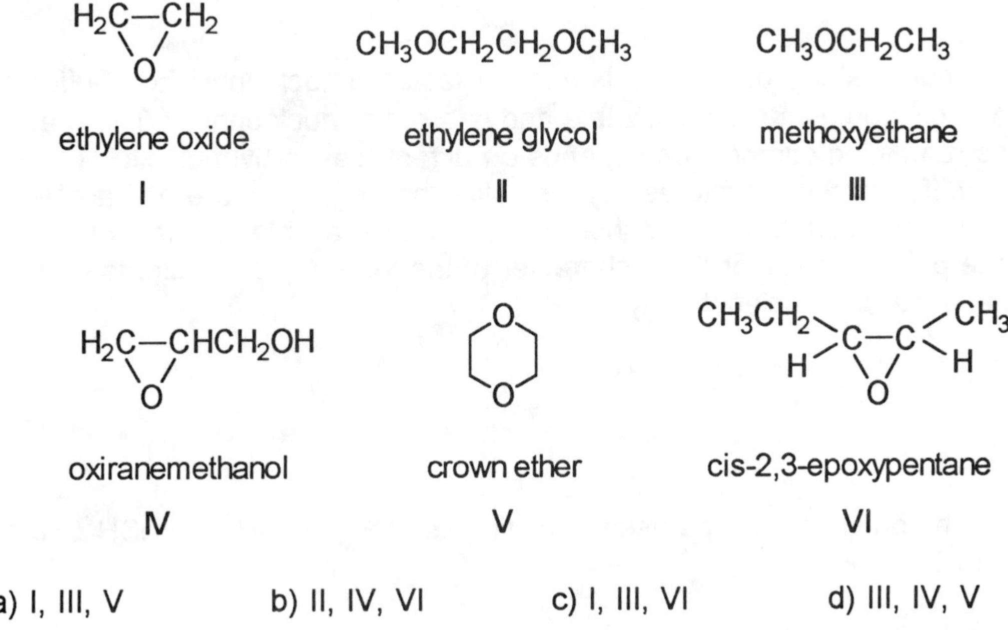

a) I, III, V   b) II, IV, VI   c) I, III, VI   d) III, IV, V

414. What is the order of increasing solubility in water for the following compounds (least first)?

a) II, I, IV, III b) IV, III, II, I c) III, IV, I, II d) I, II, III, IV

415. What is the order of increasing reactivity for the following nucleophiles (least first)?

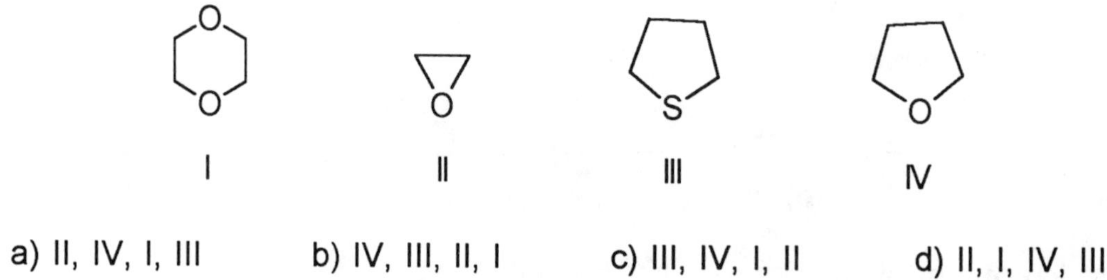

a) II, IV, I, III b) IV, III, II, I c) III, IV, I, II d) II, I, IV, III

TIP The polarizability of an atom is a major factor in nucleophilicity. Sulfur is larger than oxygen, is more polarizable and is a better nucleophile. The reactivity of the remaining compounds depends on differences in hybridization because of the different ring structures. The smaller the ring, the more p character there is in the ring-atom bonds and therefore the less available for the oxygen atom lone pair. The lesser the p character of the lone pair of electrons on oxygen, the lesser the nucleophilicity.

416. Which conditions can be used to distinguish $CH_3CH_2CH_2OCH_2CH_2CH_3$ from $CH_3(CH_2)_4Cl$?

a) solubility in water b) solubility in acetone
b) solubility in concentrated H_2SO_4 d) solubility in dilute HCl

B. PREPARATIONS

417. Which of the following statements is not true for the Williamson synthesis of ethers?

a) The rate of reaction depends on the leaving group.
b) The rate of reaction is proportional to the concentration of the nucleophile.
c) The mechanism involves one step.
d) Rearrangements occur in some situations.

418. What is the ether product from the following series of reactions, starting with 1,5-hexadiene?

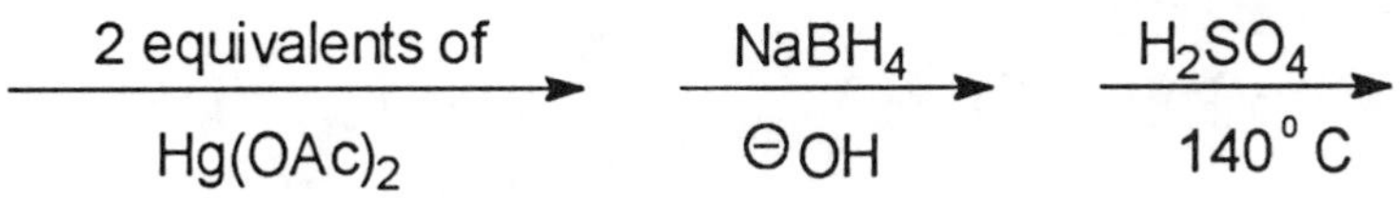

$\xrightarrow[Hg(OAc)_2]{\text{2 equivalents of}}$ $\xrightarrow[^{\ominus}OH]{NaBH_4}$ $\xrightarrow[140^\circ C]{H_2SO_4}$

a) O (oxepane)

b) O, CH_3

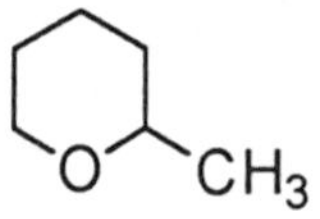

c) H_3C, O, CH_3

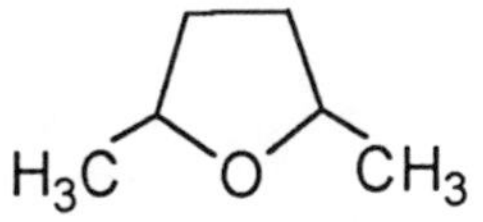

d) 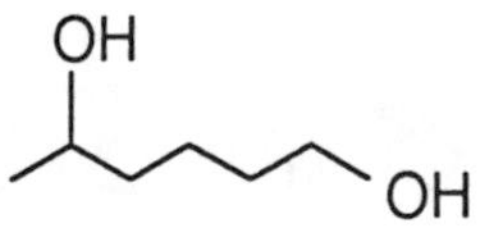 OH, OH

419. What is the ether that is formed from the following series of reactions starting with 1,4-divinylcyclohexane?

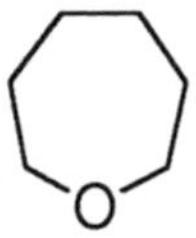

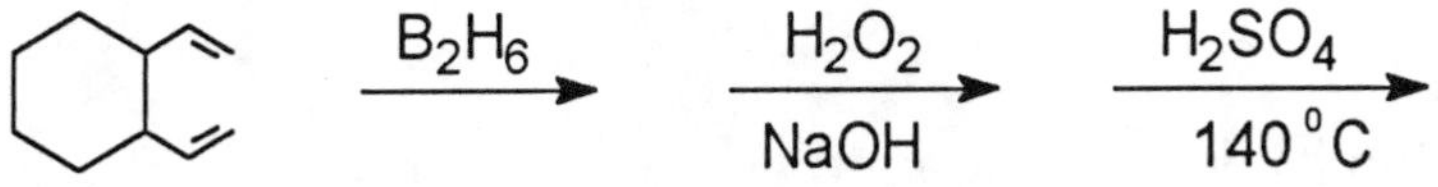

$\xrightarrow{B_2H_6}$ $\xrightarrow[NaOH]{H_2O_2}$ $\xrightarrow[140^\circ C]{H_2SO_4}$

a) 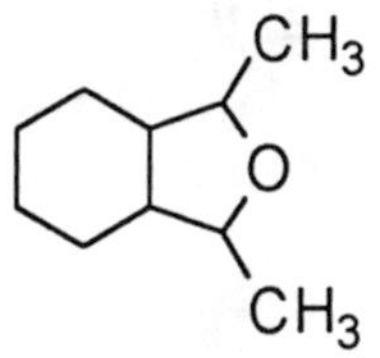 CH_3, O, CH_3

b) 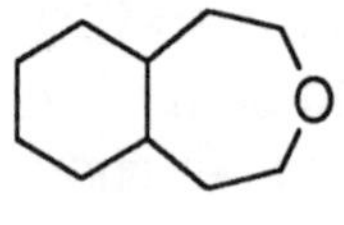 O

c) O, CH_3

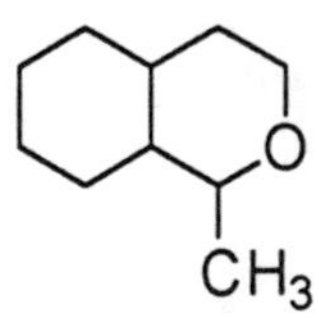

d) 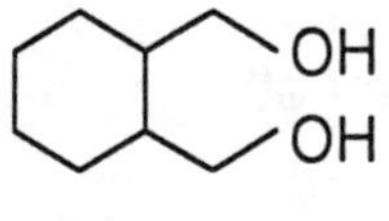 OH, OH

420. What is the major product from the reaction of tert-butanol with sodium hydride, followed by reaction with 2-chloropropane?

a) $CH_2{=}CHCH_3$

b) $(CH_3)_2C{=}CH_2$

c) $(CH_3)_3COCH(CH_3)_2$

d) $(CH_3)_3CO(CH_3)_3$

421. What is the product from the reaction of 4-bromo-1-butanol with NaOH?

a) $(CH_3CH_2CH_2CH_2)_2O$

b) $(BrCH_2CH_2CH_2CH_2)_2O$

c) $(HOCH_2CH_2CH_2CH_2)_2O$

d) 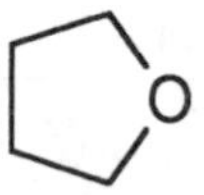 O

422. What is the best laboratory procedure for making tert-butyl methyl ether?

a) $CH_3OH \xrightarrow{Na} \xrightarrow{(CH_3)_3CBr}$

b) $(CH_3)_3CBr \xrightarrow[H_2O\ 100\ ^{\circ}C]{NaOH}$

c) $(CH_3)_3COH \xrightarrow{K} \xrightarrow{CH_3I}$

d) $(CH_3)_3COH + CH_3OH \xrightarrow[100\ ^{\circ}C]{H_2SO_4}$

TIP Reaction b) does not lead to an ether and reaction d) will give complex mixtures, so this is best accomplished with a Williamson ether synthesis (a and c). Note, the competing reaction is an elimination reaction. In a) the halide is tertiary which makes it fastest for an E2 reaction and slowest for an S_N2 reaction. The halide in c) is primary and so it is fastest for an S_N2 reaction, and best yet, not a candidate for the competing reaction.

423. The Williamson ether synthesis proceeds by which of the following mechanisms?

a) E1

b) E2

c) S_N1

d) S_N2

424. What is the major product from the following reaction?

$$Cl(CH_2)_4Cl \xrightarrow{Na_2S} \xrightarrow{H_2O_2} \xrightarrow{H_2O_2}$$

a) 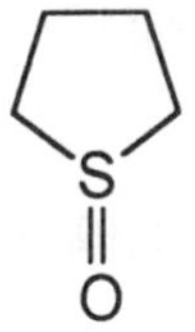 S=O (ring)

b) 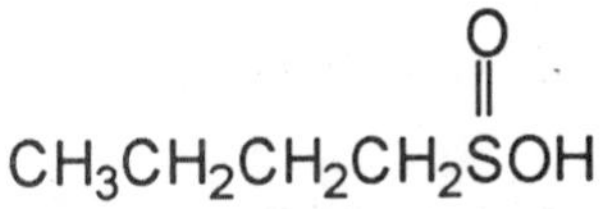 $CH_3CH_2CH_2CH_2S(=O)OH$

c) 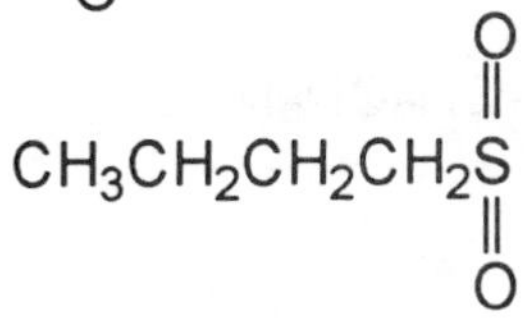 $CH_3CH_2CH_2CH_2S(=O)_2$

d) 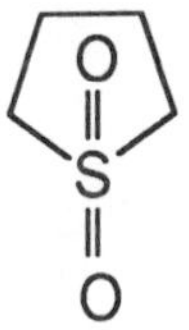 O=S=O (ring)

425. What are the best conditions for preparing diethyl sulfide?

a) $2\ CH_3CH_2Cl + Na_2S \longrightarrow$

b) 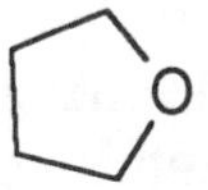 O (ring) $+ Na_2S \longrightarrow$

c) $CH_3CH_2S\ Na + CH_3CH_2I \longrightarrow$

d) $ClCH_2CH_2CH_2Cl + Na_2S \longrightarrow$

C. REACTIONS of ETHERS and EPOXIDES

426. Which of the following ethers react with concentrated sodium ethoxide?

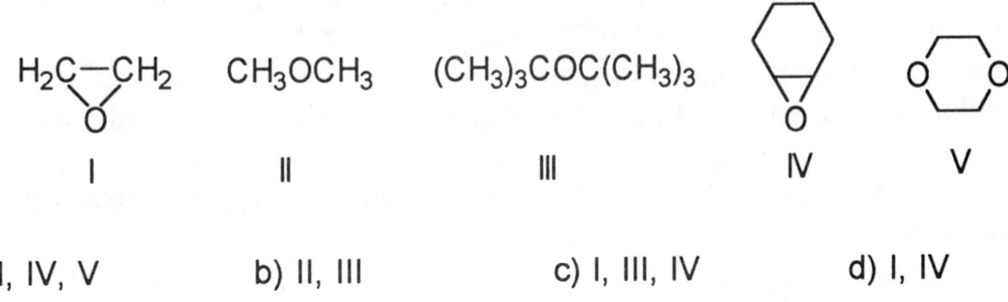

a) I, IV, V b) II, III c) I, III, IV d) I, IV

427. Which of the following ethers react with HI but not with $NaOCH_2CH_3$?

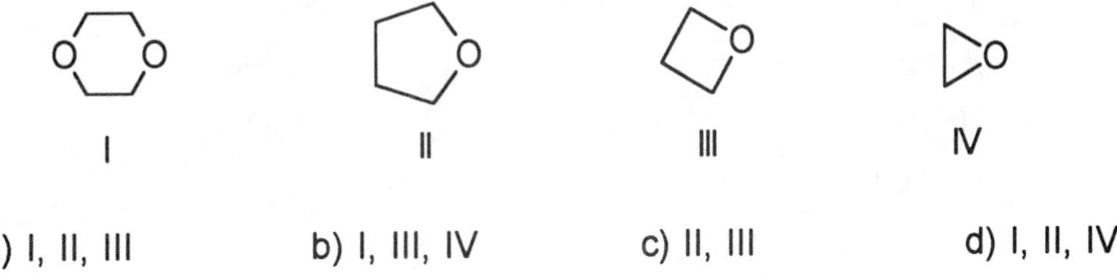

a) I, II, III b) I, III, IV c) II, III d) I, II, IV

428. What is the major product from the reaction of excess, concentrated HBr with dioxane?

a) CH_3CH_2Br b) $BrCH_2CH_2OH$

c) $BrCH_2CH_2Br$ d) $HOCH_2CH_2OCH_2CH_2OH$

429 What is the product from the reaction of tetrahydrofuran with HI?

a) $CH_3CH_2CH_2CH_2I$ b) $ICH_2CH_2CH_2CH_2I$

c) $ICH_2CH_2CH_2CH_2CH_2I$ d) ICH_2CH_2I

TIP The mechanistic scheme involves protonation of the ether oxygen atom, followed by an S_N2 reaction with iodide. This is followed by a second protonation on the alcohol that was formed in the first step, followed by a second S_N2 reaction. So the key is knowing the ether structure. Compound a) comes from dibutyl ether. Compound c) comes from tetrahydropyran. Compound d) comes from dioxane.

430. Which reaction conditions are best for the following conversion?

(cyclopentane)=CH_2 → (cyclopentane with CH_2OH and OCH_3 on the same carbon)

a) chlorine and water, followed by aqueous sodium hydroxide
b) perbenzoic acid, followed by methanol and acid
c) perbenzoic acid, followed by methanol and sodium methoxide
d) chlorine and water, followed by sodium hydroxide, then methanol and acid

431. What is the major product from the reaction of propylene oxide with sodium methoxide?

a) $CH_3CH(OH)CH_2OH$ b) $CH_3CH(OH)CH_2OCH_3$

c) $CH_3CH(OCH_3)CH_2OH$ d) $CH_3OCH(CH_3)OCH_3$

432. What is the major product from the reaction of cyclohexene oxide with aqueous acid?

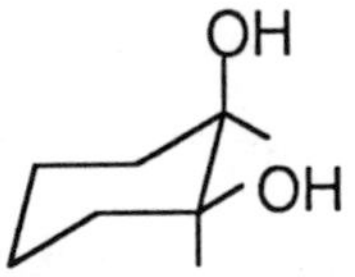

OH, OH a) OH, OH b) OH, OC_2H_5 c)  d)

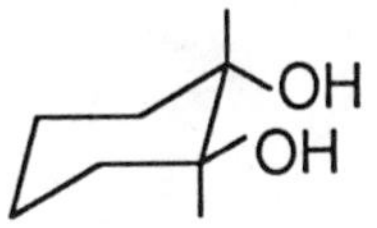

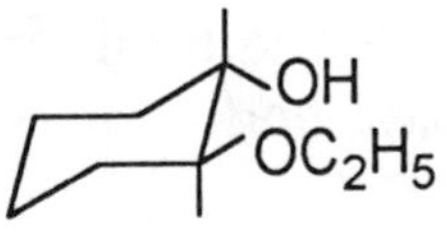

TIP Understanding the mechanism is critical. Step one involves protonation of the epoxide oxygen atom which makes it a good leaving group. The next step is nucleophilic attack by water, most likely an S_N2 process with inversion of configuration. So compounds in c) and d) come from different reagents. Answer a) is wrong because it is not the product from inversion.

433. What is the final product when cyclopentene is treated with aqueous chlorine, then by concentrated aqueous sodium hydroxide?

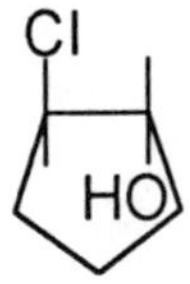

Cl, HO a) O b) O c) O d)

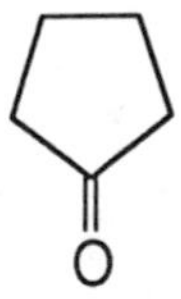

434. What is the major product from the following reaction?

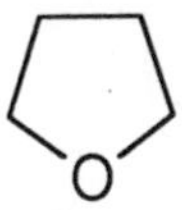

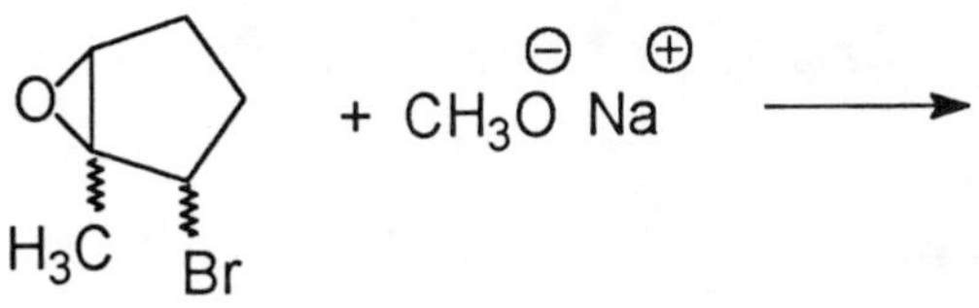

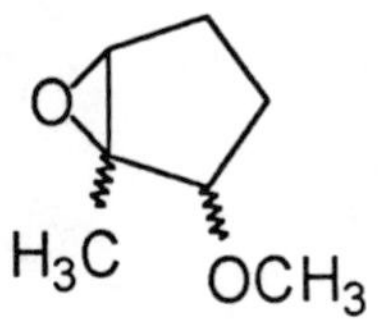

O, H_3C, Br + $CH_3\overset{\ominus}{O}\ \overset{\oplus}{Na}$ ⟶

O, H_3C, OCH_3 a) OCH_3, CH_3, OH b) OCH_3, CH_3, OCH_3 c) CH_3O, H_3C, O d)

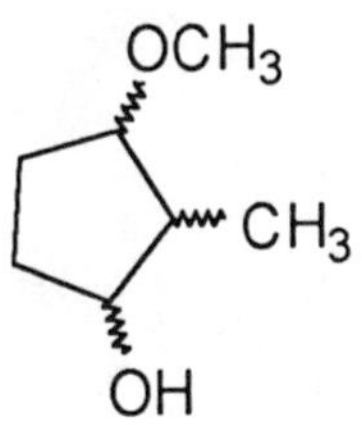

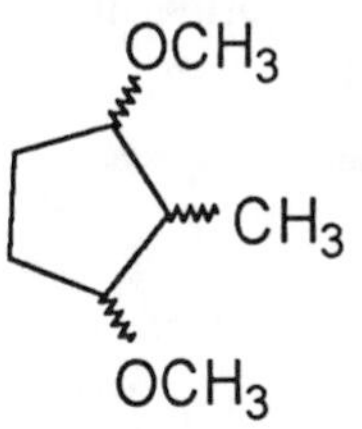

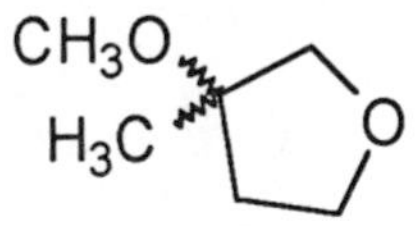

435. What is the product of refluxing methyl phenyl ether with HBr?

a) phenol and methyl bromide
b) bromobenzene and methyl bromide
c) bromobenzene and methanol
d) benzyl bromide and methyl bromide

436. What is the product of the following reaction sequence?

$$(CH_3)_3COH \xrightarrow[\text{heat}]{H_2SO_4} \xrightarrow{ArCO_3H} \xrightarrow[(CH_3)_2CHOH]{(CH_3)_2CHO^{\ominus}K^{\oplus}} \xrightarrow[CH_3CH_2CH_2Cl]{NaH}$$

a) $(CH_3)_2C(OCH(CH_3)_2)CH_2OCH_2CH_2CH_3$ b) $(CH_3)_2C(OCH_2CH_2CH_3)CH_2OCH(CH_3)_2$

c) $(CH_3)_2C(OCH(CH_3)_2)CH_2OCH(CH_3)_2$ d) $(CH_3)_2C(OCH_2CH_2CH_3)CH_2OCH_2CH_2CH_3$

437. What are the reactants needed to prepare diethanolamine?

$$HOCH_2CH_2NHCH_2CH_2OH$$

a) tetrahydrofuran and ammonia b) ethylene oxide and ethyl amine
c) dioxane and ethyl amine d) ethylene oxide and ammonia

438. Cleavage of $CH_3CH_2OCH_2CH_3$ by HI is much faster than by HCl. What is the best reason for this?

a) HI is a stronger acid than HCl.
b) I is a better leaving group than Cl.
c) I is a better nucleophile than Cl.
d) both answers a) and c)

439. What is the product from the following reaction?

HO Cl + NaOH →

HO OH a) O b) c) HO HO d)

440. What is the product from the following reaction?

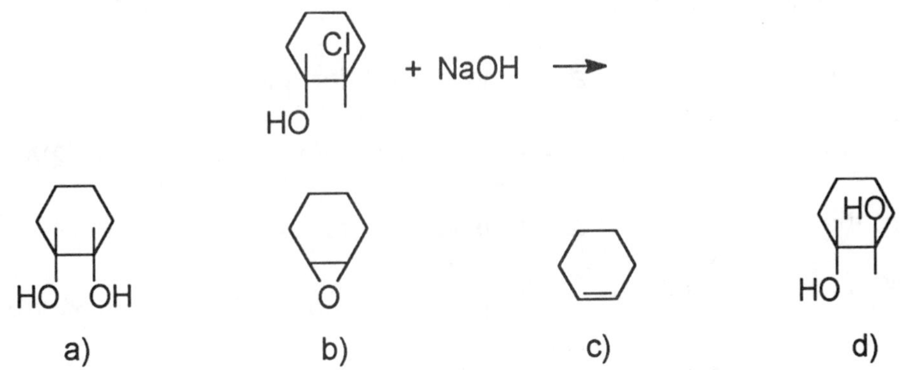

TIP Questions 439 and 440 involve a nucleophilic displacement of chloride with inversion of configuration. In the next step, the nucleophile must be able to attack from the rear. In question 440, the attack comes from the oxygen anion (alkoxide), after loss of the proton from the alcohol, to give an intramolecular reaction and form the epoxide. In question 439, although the alkoxide is formed, it cannot react by an intramolecular mechanism because of the rigid ring, and instead an attack by the hydroxide nucleophile will lead to answer d).

CHAPTER 12 MASS SPECTROMETRY

441. Which of the following species can a mass spectrometer detect?

electrons	water	cations	alkenes	radical cations
I	II	III	IV	V

a) I, V b) II, IV c) III, V d) I, III

442. Which of the following are the most likely fragmentation ions formed in a mass spectrometer?

$\overset{\oplus}{CH_3}$ (I) $C_6H_5-CH_2\overset{\oplus}{C}H_2$ (II) $CH_2{=}CH\overset{\oplus}{C}HCH_3$ (III) $CH_3\overset{\oplus}{C}{=}O$ (IV)

a) I, II b) III, IV c) II, III d) I, IV

443. Which of the following is a characteristic peak for sulfides?

a) (m+2)/m = 1 b) (m+2)/m = 0.3 c) (m+2)/m = .002 d) (m+2)/m = .045

444. Which of the following is a characteristic peak for chlorides?

a) (m+2)/m = 1 b) (m+2)/m = 0.3 c) (m+2)/m = .002 d) (m+2)/m = .045

445. Which of the following is a characteristic peak for bromides?

a) (m+2)/m = 1 b) (m+2)/m = 0.3 c) (m+2)/m = .002 d) (m+2)/m = .045

446. Which of the following is a characteristic peak for ethers?

a) (m+2)/m = 1 b) (m+2)/m = 0.3 c) (m+2)/m = .002 d) (m+2)/m = .045

TIP The relative abundance of naturally occuring isotopes gives rise to the characteristic intensities for the ratio of peaks surrounding the molecular ion. For example, the ratio of $^{32}S/^{34}S$ is 22/1, and if a compound contains sulfur, the ratio of the (m+2)/m peaks is 1/22 or 0.045. Similarly the ratios of $^{35}Cl/^{37}Cl$ is 3/1, and of $^{79}Br/^{81}Br$ is 1/1, and therefore the correct answer for 444 is b) and for 445 is a). For compounds containing oxygen, the (m+2)/m ratio is .002, which makes identification by this method far less useful because of the small number.

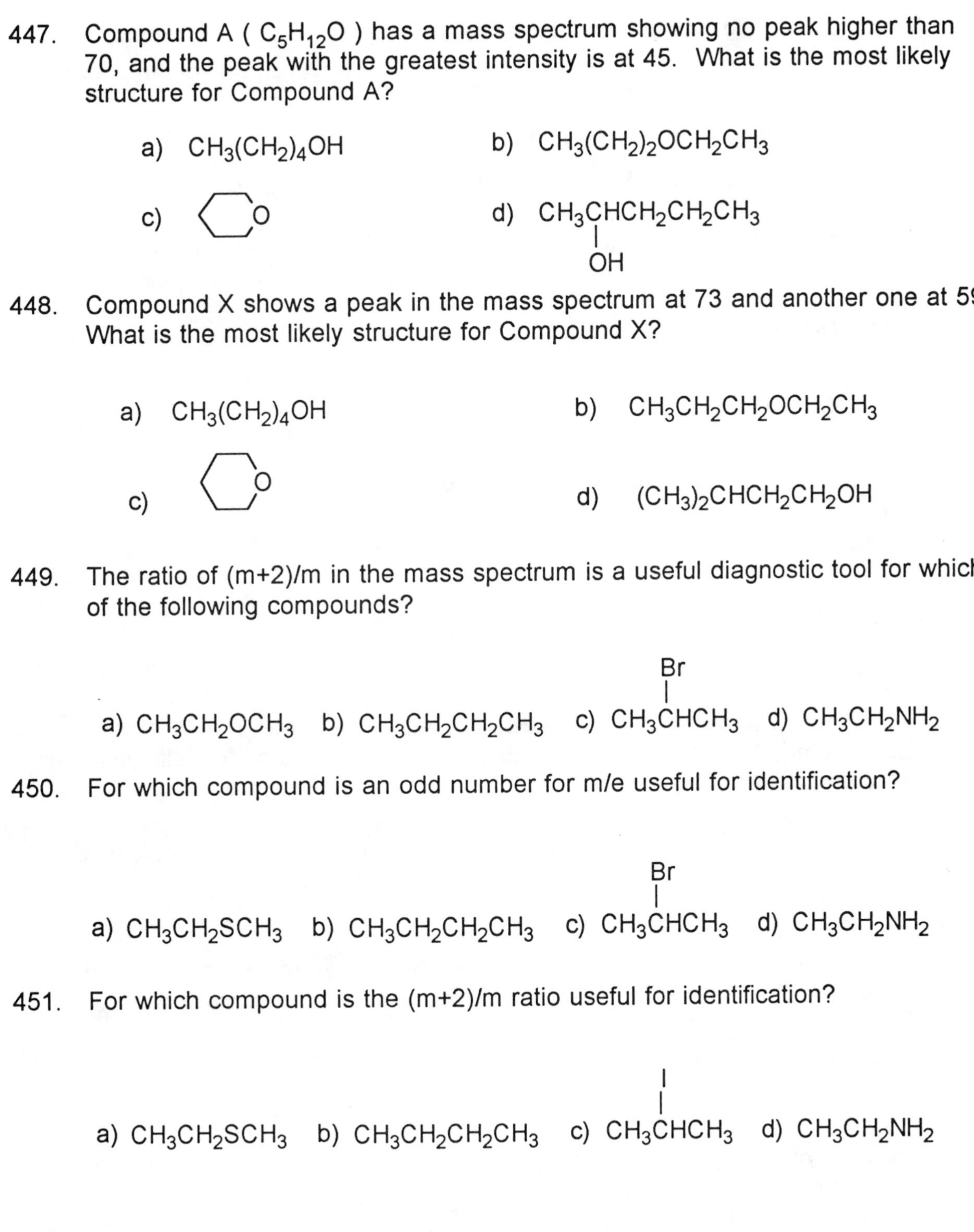

447. Compound A ($C_5H_{12}O$) has a mass spectrum showing no peak higher than 70, and the peak with the greatest intensity is at 45. What is the most likely structure for Compound A?

a) $CH_3(CH_2)_4OH$

b) $CH_3(CH_2)_2OCH_2CH_3$

c) (ring structure with O)

d) $CH_3\underset{OH}{C}HCH_2CH_2CH_3$

448. Compound X shows a peak in the mass spectrum at 73 and another one at 59. What is the most likely structure for Compound X?

a) $CH_3(CH_2)_4OH$

b) $CH_3CH_2CH_2OCH_2CH_3$

c) (ring structure with O)

d) $(CH_3)_2CHCH_2CH_2OH$

449. The ratio of (m+2)/m in the mass spectrum is a useful diagnostic tool for which of the following compounds?

a) $CH_3CH_2OCH_3$ b) $CH_3CH_2CH_2CH_3$ c) $CH_3\overset{Br}{C}HCH_3$ d) $CH_3CH_2NH_2$

450. For which compound is an odd number for m/e useful for identification?

a) $CH_3CH_2SCH_3$ b) $CH_3CH_2CH_2CH_3$ c) $CH_3\overset{Br}{C}HCH_3$ d) $CH_3CH_2NH_2$

451. For which compound is the (m+2)/m ratio useful for identification?

a) $CH_3CH_2SCH_3$ b) $CH_3CH_2CH_2CH_3$ c) $CH_3\overset{I}{C}HCH_3$ d) $CH_3CH_2NH_2$

TIP The only (m+2)/m ratio that is useful in this problem is for sulfur (see the TIP for question 446). No such straightforward relationship exists for carbon, iodine or nitrogen.

452. Which of the following compounds has an m/e peak at 127 in the mass spectrum?

a) $CH_3CH_2SCH_3$ b) $CH_3CH_2CH_2CH_3$ c) $CH_3CH(I)CH_3$ d) $CH_3CH_2NH_2$

453. Which compound has the greatest intensity at 43 in the mass spectrum?

a) $CH_3CH_2CH_2CH_2CH_2CH_3$

b) $(CH_3)_2CHCH_2CH_2CH_3$

c) [tetrahydropyran ring with O]

d) $(CH_3)_2CHCH(CH_3)_2$

454. Which of the following compounds has the greatest intensity for the ratio of the (m+2)/m peaks ?

a) [cyclohexane ring]–Cl b) [ring with O] c) [ring with S] d) [cyclohexane ring]–Br

455. The following compounds each show a molecular ion peak in their mass spectrum at 102 m/e. Where in each spectrum do you look to distinguish between the two compounds?

$CH_3CH_2CH_2OCH_2CH_2CH_3$ [ring with S]

a) m - 34 b) m -18 c) m+2 d) m+1

456. The mass spectrum for a hydrocarbon shows a series of peaks separated by 14 m/e units. Which of the following is most likely the hydrocarbon?

a) straight chain b) cyclic c) branched chain d) unsaturated

TIP The methylene fragmentation ion (14 m/e units) is characteristic of only the simple straight chain compounds.

457. Which heteroatom is responsible for a molecular ion peak and an m+2 peak having the intensity ratio of 3:1?

a) N b) O c) Cl d) Br

458. Which peak corresponds to a common fragmentation ion for 2,2,4-trimethylpentane?

a) 28　　b) 43　　c) 57　　d) 84

459. Which of the following compounds could show a McLafferty rearrangement?

a) $(CH_3)_2CHCH_2CH_2OH$　　b) $(CH_3)_2CHCH_2OH$

c) cyclohexyl–OH　　d) $CH_3CH_2CH_2OH$

460. What is the fragment that is observed in a mass spectrum from a McLafferty rearrangement?

a) $CH_3C{\equiv}\overset{\oplus}{O}$　　b) $CH_3\overset{\oplus}{C}HCH{=}CH_2$　　c) $CH_2{=}\overset{\oplus\cdot}{C}H_2$　　d) $\overset{\oplus}{C}H_3$

TIP The McLafferty rearrangement of alcohols involves transfer of a hydrogen atom five atoms removed from the oxygen by way of a six membered ring transition state to give water, and alkene and a new radical cation. For butanol the radical cation will be the ethylene radical cation answer c).

461. Which of the following compounds gives a molecular ion peak at 162 (100%), a second peak at 163 (6.6%), a third peak at 164 (96%), and a fourth peak at 165 (6.6%)?

a) cyclohexyl–Br　　b) $CH_3CH_2CH_2CH_2CH_2CH_2Br$

c) $ClCH_2CH_2CH_2CH_2CH_2Cl$　　d) dioxin ring (O, O) with CH_2Br substituent

462. Which of the following compounds gives a molecular ion peak at 176 (100%), a second peak at 177 (5.6%), a third peak at 178 (96%), and a fourth peak at 179 (5.6%)?

a) 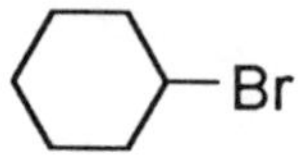 Br

b) $CH_3CH_2CH_2CH_2CH_2CH_2Br$

c) $ClCH_2CH_2CH_2CH_2CH_2Cl$

d) 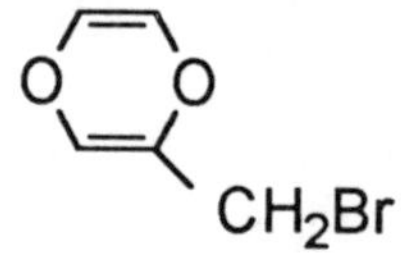 O O CH_2Br

TIP In this problem we note the m+2/m ratio of approximately 1 which denotes the presence of bromine and thus rules out answer c. Answers a and b are incompatible with the m+1/m intensity ratio of 6.2 and the molecular ion of 176. Answer d which is $C_5H_5O_2Br$ has a molecular ion at 176. The m+1/m intensity ratio of 6.2 comes from the natural abundance of 5 X ^{13} carbon (5x 1.1) plus the natural abundance of 2 X ^{17}oxygen (2X .04). Similar considerations apply to the peak at 179.

463. What information is given in a mass spectrum showing a peak at 86 (100%), another at 87 (5.6%), and another at 88 (4.0%)?

I. The compound contains a nitrogen atom.
II. The compound contains 5 carbon atoms.
III. The compound contains a sulfur atom.
IV. The compound contains an oxygen atom.

a) I, II b) III, IV c) I, III d) II, III

464. The fragmentation peaks in a mass spectrum depend on the stability of which of the following species?

I. carbocations II. radicals III. radical cations IV. neutral compounds

a) II, III b) II, III, IV c) I, II, III d) I, II, III, IV

465. Which of the following compounds has a peak at 73 (100%), another at 74 (3.74%), and another at 75 (0.23%)?

a) C_3H_7NO b) $C_3H_9N_2$ c) $C_4H_{11}N$ d) C_4H_9O

466. A sample of methanol, which is enriched with ^{13}C, shows a (m+1)/m peak ratio of 0.043. What is the % enrichment of the sample?

a) 4.3 b) 3.2 c) 5.4 d) 1.6

467. The $LiAlH_4$ reduction of a sample of acetaldehyde, which is enriched with ^{13}C in the C-1 position, yields ethanol. The mass spectrum of this sample of ethanol shows a 47/46 ratio of 0.142. What is the % enrichment of the sample of acetaldehyde?

a) 14.2 b) 13.1 c) 7.0 d) 12.0

TIP The natural abundance of ^{13}C in ethanol has a value of 0.022 for the ratio of 47/46 (2 carbon atoms = 2 x 0.011). This value is subtracted from the observed value of 0.142 to give 0.120 or 12%, which is the contribution from the ^{13}C enrichment. This enrichment is the same for ethanol and acetaldehyde since the reduction step does not affect the enrichment at the C-1 position.

CHAPTER 13 NUCLEAR MAGNETIC RESONANCE SPECTROSCOPY

A. PROTON NMR SPECTROSCOPY

468. How many sets of signals are there in the ^{1}H NMR spectrum for the following compound?

$$CH_3CH(OH)CH_3$$

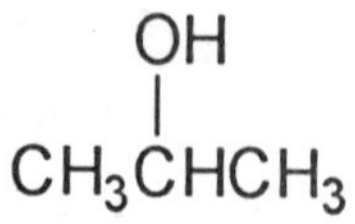

a) 2 b) 3 c) 4 d) 8

469. How many sets of signals are there in the ^{1}H NMR spectrum for 3-chloropropene?

a) 2 b) 3 c) 4 d) 5

470. How many sets of equivalent hydrogen atoms are there for the following compound?

CH3 / O / CH3

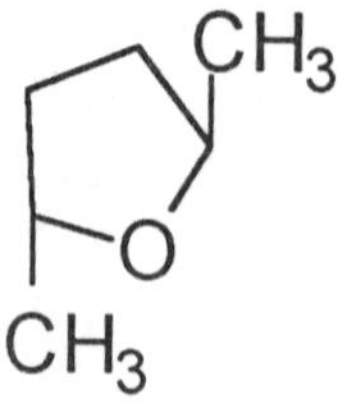

a) 2 b) 3 c) 4 d) 5

471. How many sets of equivalent hydrogen atoms are there for the following compound?

CH3

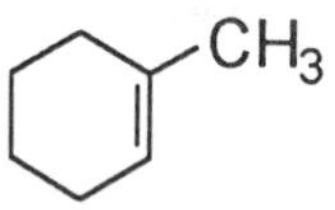

a) 3 b) 4 c) 5 d) 6

TIP Equivalent hydrogen atoms are atoms which if replaced by a substituent, for example chlorine, would lead to identical compounds. Thus inequivalent hydrogen atoms lead to different compounds. For methyl cyclohexene some differences are obvious, the methyl and the vinyl hydrogens. For the remaining methylene hydrogens, replacement by a chlorine at carbon atoms 3,4,5, and 6 lead to 4 different compounds and thus the total number of sets of equivalent hydrogens is 6.

472. How many sets of equivalent protons are there in the following compound?

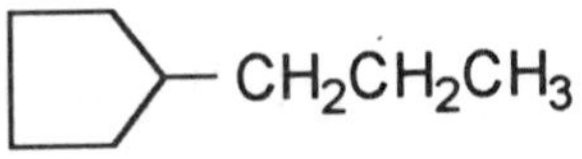

(cyclopentyl)– $CH_2CH_2CH_3$

a) 6 b) 7 c) 8 d) 9

473. How many sets of proton NMR resonances are there for all isomers of dichloropropane?

a) 7 b) 8 c) 9 d) 10

474. How many sets of proton NMR resonances are there for all isomers of dichlorocyclopropane?

a) 4 b) 5 c) 6 d) 10

475. How many sets of equivalent protons are there for CH_3CHDCH_2Br?

a) 2 b) 3 c) 4 d) 5

476. How many sets of equivalent protons are there for $CH_3CHClCH_2Cl$?

a) 2 b) 3 c) 4 d)5

TIP This is an example of a difference that is not obvious. The key to this problem is to note that carbon 3 is chiral and therefore the test of replacement of the methylene hydrogens on carbon 2 leads to diastereomers (i.e different compounds). In all then the answer is 4 sets of equivalent protons.

477. What are the relative areas by integration for the proton NMR spectrum of 3-chloro-2-methylbutane?

a) 3:1 b) 9:2 c) 9:1:1 d) 6:3:1:1

478. In the following compound, what are the total relative areas under the proton NMR signals for each set of signals?

$$CH_3CH_2C(CHCl_2)_2CHCl_2$$

	singlet	triplet	quartet
a)	3	3	2
b)	1	3	4
c)	1	3	2
d)	3	2	3

479. In the proton NMR spectrum for ethyl propyl ether, which of the indicated protons is assigned the lowest field resonance?

$CH_3{-}CH_2{-}O{-}CH_2{-}CH_2{-}CH_3$

a) CH_3 (ethyl) b) CH_2 (ethyl, next to O) c) CH_2 (propyl, middle) d) CH_3 (propyl)

480. Arrange the following compounds in the order of increasing δ values downfield from TMS in the NMR spectrum (lowest first).

$CH_3{-}CH_2{-}CH(Cl){-}CH_3$

I = CH_3, II = CH_2, III = CH, IV = CH_3

a) IV, I, II, III b) I, II, IV, III c) III, II, I, IV d) I, IV, II, III

TIP Deshielding derives from the presence and proximity of an electronegative substituent, and therefore, shielding (lower chemical shift values) is exactly opposite. A good way to do these type of problems is to determine the most shielded (I) and the least shielded (III) in this case. Then since IV is more shielded than II, the overall order is: methyl more shielded than methylene, and methyne more shielded than methylene.

481. What is the order of increasing δ values downfield from TMS in the NMR spectra for the indicated protons (lowest first)?

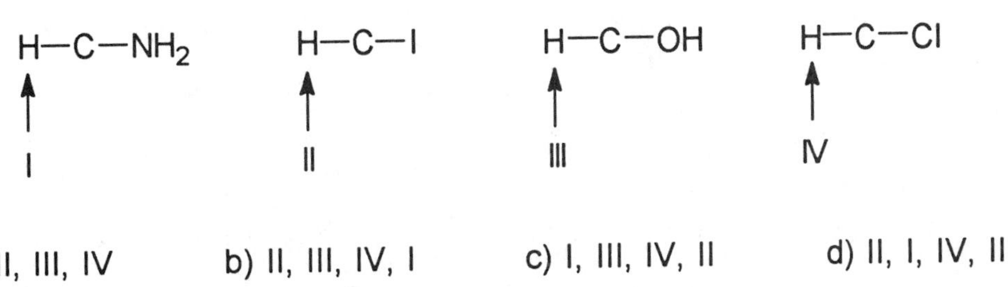

a) I, II, III, IV b) II, III, IV, I c) I, III, IV, II d) II, I, IV, III

482. Arrange the following in the order of increasing frequency downfield from TMS in the NMR spectra (lowest first).

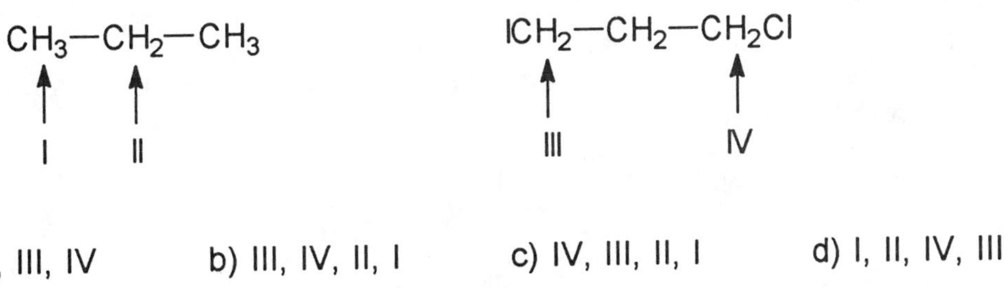

a) I, II, III, IV b) III, IV, II, I c) IV, III, II, I d) I, II, IV, III

483. Which of the following compounds will not show signal splitting in the proton NMR?

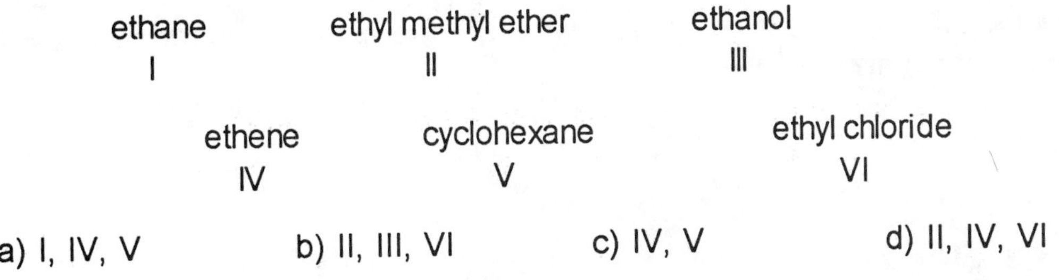

a) I, IV, V b) II, III, VI c) IV, V d) II, IV, VI

484. Which of the following compounds will show signal splitting in the proton NMR?

I. 1,3,5-trichlorocyclohexane
II. cyclohexyl methyl ether
III. 1,2-dichloroethane
IV. 2,3-dimethylbutane

a) I, II b) III, IV c) I, II, IV d) I, II, III, IV

485. What is the multiplicity in the NMR spectrum for the indicated protons in the following compound?

$Cl_2CHCH_2C(CH_3)_3$

(arrows: I → CH, II → CH_2, III → CH_3)

	I	II	III
a)	triplet	quartet	singlet
b)	singlet	triplet	singlet
c)	doublet	singlet	triplet
d)	triplet	doublet	singlet

486. For the five spin system shown below, what are the resonances for H_b in the proton NMR spectrum?

H_a—C(H_a)(H_a)—C(H_b)—H_b

a) 1:3:3:1 quartet
b) 1:2:1 triplet
c) a singlet
d) a nine pattern 1:2:1:2:4:2:1:2:1

487. For the following five spin system, what are the resonances for H_b in the proton NMR spectrum?

J_{ab} = 10 Hz
J_{bc} = 2 Hz

H_a—C(H_a)—C(H_b)—C(H_c)—H_c

a) 1:3:3:1 quartet
b) 1:4:6:4:1 quintet
c) 1:2:3:4:3:2:1 septet
d) a nineline pattern 1:2:1:2:4:2:1:2:1

TIP This is a tough one but yields to careful analysis. Proton b has two kinds of next nearest neighbors. There are two a types and two c types. Depending on the values of the coupling constants, a variety of patterns are possible. In the case given, J_{ab} is large and J_{bc} is small. A reasonable way to proceed is to "turn on" the coupling sequentially. Thus, for the a protons alone, b is expected to be a 1:2:1 triplet with a separation of 10 Hz. Now if we "turn on" the c coupling, each of the lines of the triplet will be turned into a 1:2:1 triplet with now a 2 Hz separation. This leads to a nineline pattern with 1:2:1:2:4:2:1:2:1. What would the answer be if J_{ab} and J_{ac} were equal? Does answer b make sense?

488. Using a 60 MHz instrument, the proton NMR spectrum for 1-bromo-2-chloroethane has the centers of two triplets separated by 48 Hz. What is the separation using a 300 MHz instrument?

a) 48 HZ b) 80 Hz c) 240 Hz d) 300 Hz

489. Using a 60 MHz instrument, the proton NMR spectrum for 1-bromo-2-chloroethane has the centers of two triplets separated by 0.8 ppm. What is the separation using a 100 MHz instrument?

a) 40 Hz b) 48 Hz c) 80 Hz d) 100 Hz

490. If the centers of two triplets are separated by 60 Hz using a 100 MHz NMR spectrometer, what is the separation using a 60 MHz spectrometer?

a) 36 Hz b) 48 Hz c) 60 Hz d) 100 Hz

491. If the two peaks of a doublet appear at 1.1 and 1.2 ppm on a 60 MHz spectrometer, what is the separation on a 100 MHz spectrometer?

a) 10 Hz b) 14 Hz c) 6 Hz d) 0.1 Hz

492. In the proton NMR spectrum for ethyl bromide at 60 MHz, the methyl hydrogens are assigned the peak at 1.7 ppm and the methylene hydrogens at 3.3 ppm. If the coupling constant is 8 Hz, what is the separation of the two closest lines in the two multiplets?

a) 96 Hz b) 88 Hz c) 80 Hz d) 76 Hz

493. If two peaks of approximately equal intensity appear at 1.1 and 1.2 ppm on a 60 MHz spectrometer and they are separated by 6 Hz on a 100 MHz spectrometer, then the peaks are which of the following possibilities?

a) two uncoupled singlets b) two peaks of a doublet
c) two coupled singlets d) two uncoupled doublets

TIP Here we are concerned with differences between the response of chemical shifts and coupling constants to field changes. Chemical shifts in Hz are linearly related to field strength, whereas coupling constants are invariant. If the two lines are separated by 0.1ppm at 60 MHz then they are 6 Hz apart. Since they are also separated by 6 Hz at the higher field, the separation is not field dependent and answer b is correct.

494. What is the structure for a compound, $C_4H_{10}O$, that has a proton NMR spectrum consisting of singlets?

a) $CH_3OCH(CH_3)_2$ b) $(CH_3)_3COH$

c) $CH_3CH_2OCH_2CH_3$ d) $CH_3OC(CH_3)_3$

495. What is the structure for the compound, $C_4H_8Br_2$, that has the following proton NMR spectrum?

doublet δ 1.7 (6H)
quartet δ 4.4 (2H)

a) 1,1-dibromobutane b) 1,2-dibromobutane
c) 1,3-dibromobutane d) 2,3-dibromobutane

496. What is the structure for a compound, $C_3H_6Cl_2$, that has the following proton NMR spectrum?

triplet δ 5.8 (1H)
multiplet δ 3.0 (2H)
triplet δ 1.0 (3H)

a) 1,1-dichloropropane b) 1,2-dichloropropane
c) 1,3-dichloropropane d) 2,2-dichloropropane

497. What is the expected proton NMR spectrum for a compound with the formula, $CH_aBr_2CH_bCl_2$?

	H_a	H_b
a)	low field — singlet	high field — singlet
b)	low field — doublet	high field — doublet
c)	high field — doublet	low field — doublet
d)	high field — singlet	low field — doublet

TIP Since chlorine is more electronegative than bromine, Hb will be more deshielded and at lower field than Ha. The coupling pattern is derived from the next nearest neighbors, one in each case here to give two doublets and answer c is correct.

498. What is the structure for a compound, $C_6H_{13}Br$, that has the following proton NMR spectrum?

triplet	δ	0.9	(6H)
quartet	δ	1.6	(4H)
singlet	δ	1.2	(3H)

a) 1-bromo-2-ethylbutane
b) 3-bromo-3-methylpentane
c) 2-bromo-3-methylpentane
d) 1-bromo-4-methylpentane

499. What is the expected proton NMR spectrum for 2,3-dibromobutane?

	low field signal	high field signal
a)	doublet	triplet
b)	doublet	quartet
c)	quartet	doublet
d)	triplet	doublet

500. What is the expected proton NMR spectrum for 1,1,2-trichloroethane?

	low field signal	high field signal
a)	doublet	triplet
b)	triplet	doublet
c)	quartet	doublet
d)	doublet	quartet

TIP This is similar to 497. Clearly two chlorines are more deshielding than one chlorine so we can predict the position of the signal. The deshielded proton on carbon 1 has two next nearest neighbors and is therefore a triplet. Similarly the more shielded protons on carbon 2 have one next nearest neighbor and will appear as a doublet.

501. What is the expected proton NMR spectrum for 1,3-dichloropropane?

	low field signal	high field signal
a)	triplet	quintet
b)	triplet	triplet
c)	doublet	doublet
d)	triplet	singlet

502. What is the expected proton NMR spectrum for 1,2-dichloropropane on a 500 MHz spectrometer?

$$\underset{\uparrow\ \text{I}}{H_2\overset{Cl}{\overset{|}{C}}}-\underset{\uparrow\ \text{II}}{\overset{Cl}{\overset{|}{C}}H}-\underset{\uparrow\ \text{III}}{CH_3}$$

	I	II	III
a)	doublet	multiplet	doublet
b)	two doublets	multiplet	doublet
c)	doublet	triplet	doublet
d)	singlet	singlet	singlet

503. What is the most likely structure for a compound that has the molecular formula, $C_5H_8O_2$, and the following proton NMR spectral data?

singlet	δ 2.2	(3H)
singlet	δ 3.7	(3H)
singlet	δ 5.5	(1H)
singlet	δ 6.0	(1H)

a) $CH_2{=}CHCO_2CH_2CH_3$

b) (six-membered cyclic lactone: ring O adjacent to C=O)

c) $CH_2{=}\overset{CH_3}{\overset{|}{C}}CO_2CH_3$

d) $CH_2{=}\overset{CH_3}{\overset{|}{C}}OCOCH_3$

504. Compound X has the molecular formula, $C_6H_{12}O_2$, and a proton NMR spectrum with two singlets having relative areas of 3:1. One peak is at 1.2 ppm and the other is at 3.7 ppm. What is the most likely structure for Compound X?

a) CH_3O–(cyclohexane)–OCH_3
b) $(CH_3)_3COC(=O)CH_3$
c) $(CH_3)_3CC(=O)OCH_3$
d) $(CH_3)_2CHOC(=O)CH_2CH_3$

505. Which of the statements concerning the proton NMR spectra for the following compounds is NOT true?

I: $CH_3CH(Cl)CH_3$
II: $CH_3CH_2CH_2Cl$
III: $(CH_3)_2C{=}CH_2$
IV: $CH_2{=}CHCl$
V: cyclopropane with Br, Br on one carbon
VI: cyclopropane with Br, Cl on one carbon
VII: $CH_3CH(Cl)CH_2Br$

a) Two sets of signals are expected from compounds III, V and VI.
b) Three sets of signals are expected from compounds II and IV.
c) Four sets of signals are expected from compound VII.
d) The splitting patterns from compound I are a doublet and a septet.

506. Ignoring the proton-deuterium coupling, what is the expected proton NMR spectrum for 1-bromo-2-deuterioethane?

a) 2 triplets
b) a triplet and a quartet
c) 2 singlets
d) a quartet and a doublet

507. Ignoring the proton-deuterium coupling, what is the expected proton NMR spectrum for 1-bromo-1-deuterioethane?

a) 2 triplets
b) a triplet and a quartet
c) 2 singlets
d) a quartet and a doublet

508. What is the expected proton NMR for 1,1,1-trifluoroethane?

a) a singlet
b) a doublet
c) a triplet
d) a quartet

509. What is the expected proton NMR for tert-butyl fluoride?

a) a singlet   b) a doublet   c) a triplet   d) a quartet

TIP ^{19}F is a spin 1/2 nucleus and therefore couples to other spin 1/2 nuclei, which are protons in these cases. The protons in question 508 have three next nearest neighboring fluorines and will be a quartet whereas the protons in question 509 have one next nearest neighboring fluorine and will be a doublet.

B. ^{13}CARBON NMR SPECTROSCOPY

510. What is the expected off-resonance-decoupled ^{13}C NMR spectrum for propanol?

	low field signal	high field signal
a)	doublet	quartet
b)	triplet	quartet
c)	quartet	doublet
d)	doublet	singlet

511. What is the most likely structure for a compound that has the molecular formula, $C_5H_{11}Cl$, and the following off-resonance-decoupled ^{13}C NMR data?

quartet	δ	21
doublet	δ	25
triplet	δ	41
triplet	δ	43

a) 1-chloropentane
b) 1-chloro-3-methylbutane
c) 2-chloro-2-methylbutane
d) 3-chloropentane

512. What is the most likely structure for a compound that has the molecular formula, $C_3H_5Cl_3$, and the off-resonance-decoupled ^{13}C NMR spectrum that showed a quartet upfield and two doublets downfield?

a) 1,1,1-trichloropropane
b) 1,1,2-trichloropropane
c) 1,2,2-trichloropropane
d) 1,2,3-trichloropropane

513. How many resonances are expected in the proton-decoupled ^{13}C NMR for the following compound?

H_3C CH_3

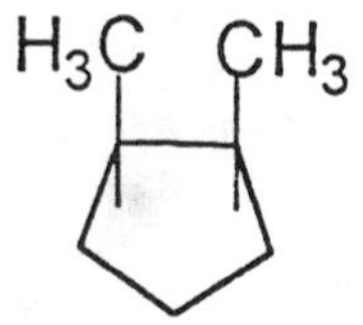

a) 2 b) 3 c) 4 d) 5

514. How many resonances are expected in the proton-decoupled ^{13}C NMR for the following compound?

CH_3

CH_3

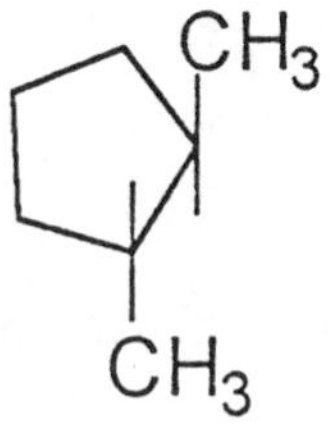

a) 2 b) 3 c) 4 d) 5

515. How many total resonances are expected for all isomers of dibromobenzene in the proton-decoupled ^{13}C NMR?

a) 3 b) 6 c) 9 d) 11

516. Which of the following describes the proton-decoupled ^{13}C NMR for 2,2-difluoropropane?

a) 2 doublets
b) 2 singlets
c) 2 triplets
d) a singlet and a triplet

517. Free radical chlorination of chiral 2-chlorobutane gave as one of the products a chiral dichlorobutane with a ^{13}C NMR spectrum consisting of four resonances. What is the most likely structure for the dichlorobutane?

a) 1,2-dichlorobutane
b) 2,3-dichlorobutane
c) 2,2-dichlorobutane
d) 1,1-dichlorobutane

TIP Both chirality and the number of signals are key here. Note that compounds c) and d) are not chiral. Moreover compound b) will have only two carbon signals.

CHAPTER 14 INFRARED and UV-VISIBLE SPECTROSCOPY

A. IR SPECTROSCOPY

518. Which of the following cannot be distinguished by infrared spectroscopy?

I. resonance hybrids II. tautomers III. enantiomers IV. cis-trans isomers

a) I, II b) II, III c) I, III d) II, IV

519. Which of the following can be readily distinguished by infrared spectroscopy?

I. bond length differences II. bond hybridization changes
III. geometric isomers IV. resonance isomers

a) I, II b) III, IV c) I, II, III d) I, III, IV

520. What is the approximate energy range for absorptions in the IR?

a) 50-100 kcal/mol b) 10-50 kcal/mol
c) 1-10 kcal/mol d) 0.01-0.1 kcal/mol

521. What is the wave number for an IR band at 20 microns?

a) 500 cm^{-1} b) 2000 cm^{-1} c) 5000 cm^{-1} d) 10,000 cm^{-1}

522. What is the wavelength of an IR absorption band at a wave number of 5,000 cm^{-1}?

a) 10 μ b) 2 μ c) 25 μ d) 0.2 μ

523. Infrared radiation has sufficient energy to cause which of the following transitions?

I. electronic II. molecular vibrational III. atomic IV. bonding
V. resonance VI dipole changes VII. symmetrical vibrations

a) I, II, III b) II, IV, VI c) II, VI,VII d) III, VI, VII

524. Which of the following compounds can be distinguished usually by IR spectroscopy?

I. ortho and para substituted benzenes II. secondary and tertiary amines
III. ethers and alcohols IV. bromides and iodides
V. straight chain and branched chain alkanes

a) I, II, III b) II, III, IV c) I, III, IV d) II, III, V

525. Place the following compounds in the order of increasing bond stretching frequencies (lowest first).

CD_4	CH_4	CO_2	CS_2
I	II	III	IV

a) IV, III, I, II b) III, IV, I, II c) I, II, III, IV d) IV, III, II, I

TIP From Hooke's Law, reduced mass and force constants are principal components of estimation of stretching frequencies. Reduced mass considerations lead to C-H bonds having a greater frequency than C-D, and C=O having a greater frequency than C=S. Force constants depend upon whether the bond is single, or double or triple. Thus C-O would absorb at a lower freqency than C=O. In this case the reduced mass contribution from C-H and C-D outweighs the greater force constant of the doubly-bonded compounds.

526. What kind of compound has a very broad IR band in the region of 3300 cm^{-1}?

a) hydrocarbon b) aromatic c) alcohol d) halide

527. Which of the following compounds has a sharp IR absorption band at 1710 cm^{-1} and a broad band at 3300 cm^{-1}?

a) acetic acid b) ethanol c) acetone d) diethyl ether

528. Which of the following compounds has a sharp IR absorption band at 1710 cm^{-1} but no band at 3300 cm^{-1}?

a) acetic acid b) ethanol c) acetone d) diethyl ether

TIP In the three problems (526-528) we are using characteristic bands of oxygen-containing compounds to differentiate among possible structures. The bands are O-H at 3300 cm^{-1} and carbonyl at 1710 cm^{-1}. In question 527, both appear so only the carboxylic acid is correct. In 528 only the band at 1710 cm^{-1} is found so only a carbonyl is present.

529. Compound A, $C_5H_{10}O$, has a strong infrared absorption band at 1745 cm^{-1}. What is the most likely structure for compound A?

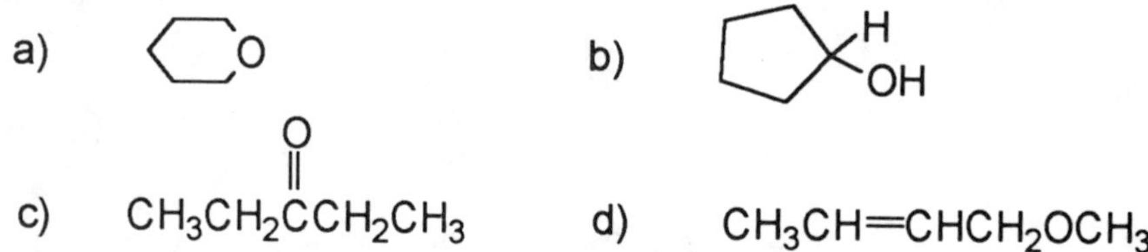

530. What is the most likely structure for a compound, $C_{10}H_{14}$, that has a strong infrared absorption at 1600 cm^{-1}?

a) $H_3C-C_6H_4-\overset{CH_3}{\overset{|}{C}}{=}CH_2$

b) $H_3C-C_6H_2(CH_3)_3$

c)

d) $CH_3CH_2C{\equiv}CC{\equiv}CCH_2CH_2CH_2CH_3$

531. Which of the following has a proton NMR consisting of two singlets of equal intensity and an IR spectrum having a strong band in the 1700 cm^{-1} region?

a) $ClCH_2\overset{O}{\overset{\|}{C}}CH_2Br$

b) $(CH_3)_3COH$

c) $(CH_3)_3\overset{O}{\overset{\|}{C}}CH_3$

d) $H_3CO-C_6H_4-OCH_3$

532. Compound X has a molecular formula of C_5H_8O. The IR spectrum has bands at 1650 cm^{-1} and 1710 cm^{-1}. What is the most likely structure for compound X?

a) =O

b) –OH

c) =O

d) $CH_3CH{=}CHCH_2CHO$

533. What is the most likely structure for a compound that has the molecular formula, $C_{10}H_{14}$, and has a strong infrared absorption at 2200 cm^{-1}?

a) 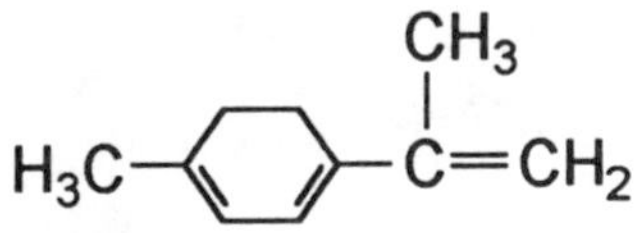 H_3C CH_3 $C{=}CH_2$

b) 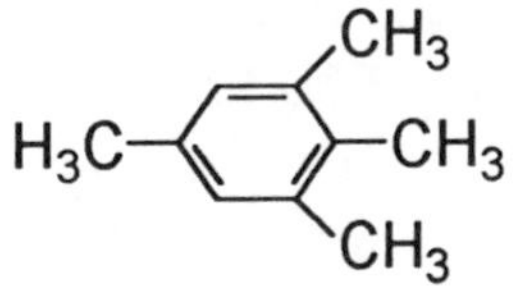 H_3C CH_3 CH_3 CH_3

c)

d) $CH_3CH_2C{\equiv}CC{\equiv}CCH_2CH_2CH_2CH_3$

534. Compound A is a two-carbon compound containing only carbon, hydrogen and chlorine. The IR spectrum has bands at 3125 and 1625 cm^{-1}. The proton NMR has a singlet at 6.3 ppm. What is compound A?

a) 1,2-dichloroethane
b) 1,1-dichloroethene
c) 1,1-dichloroethane
d) chloroethene

TIP The bands at 3125 cm^{-1} and 1625 cm^{-1} denote the presence of C=C, so answers a and c are incorrect. The singlet rules out answer d.

B. UV-VISIBLE SPECTROSCOPY

535. What is the approximate energy range for transitions in the UV?

a) 50-140 kcal/mol b) 10-50 kcal/mol
c) 1,000-1500 kcal/mol d) 500-1000 kcal/mol

536. What is the order of increasing wavelength for the UV absorption of the following species (lowest first)?

I. $CH_3CH_2\overset{\oplus}{C}H_2$ II. $CH_2{=}CH\overset{\oplus}{C}H_2$ III. $CH_2{=}CH\overset{\oplus}{C}HCH{=}CH_2$

a) I, II, III b) I, III, II c) III, II, I d) II, III, I

537. What is the order of increasing energy for the following transitions (lowest first)?

Π -------> Π * (I) n -------> Π * (II) n -------> σ * (III)

a) I, III, II b) II, I, III c) III, II, I d) I, II, III

538. Which of the following compounds can have both an n -------> Π * and a Π -------> Π * transformation?

I: $CH_3\overset{O}{\overset{\|}{C}}CH_3$ II: CH_3CH_2OH III: $CH_2{=}CH_2$ IV: (pyrrole ring, N)

a) I, II b) I, III c) III, IV d) I, IV

TIP For a transition involving n electrons, the molecule must contain a heteroatom such as O or N. For a Π transition, the molecule must contain double bonds. Thus II gives only n-sigma transitions and III gives only Π to Π* transitions. I and IV fulfill the criteria of the problem.

539. Which of the following compounds have an absorption maximum at a wavelength longer than 200 nm?

I. 1,2-pentadiene II. 1,3-pentadiene III. 1,4-pentadiene
IV. 2,3-pentadiene V. 2,4-hexadiene

a) I, II b) II, V c) I, IV d) III, IV

540. Which of the following pairs of compounds can be distinguished by UV spectroscopy?

I. (cyclohexane ring)=O and (cyclohexane ring)–OH

II. (cyclohexene ring) and (cyclohexane ring)=CH_2

III. $CH_2{=}CHCH{=}CH_2$ and $CH_2{=}C{=}CHCH_3$

IV. (cyclohexene ring)=CH_2 and (cyclohexene ring)–CH_3

a) I, IV b) II, IV c) I, III d) II, III

541. Compound A has the molecular formula, C_6H_{10}, and an absorption maximum in the UV above 200 nm. When reacted with acetylene, compound A forms compound B which has 3 sets of signals in the proton NMR spectrum, with 4 equivalent vinyl hydrogens. What is the most likely structure for compound A?

a) $(CH_3)(H)C{=}C(CH_3){-}CH{=}CH_2$

b) $CH_2{=}CH{-}CH_2CH_2{-}CH{=}CH_2$

c) $H_2C{=}C(CH_3){-}C(CH_3){=}CH_2$

d) $(CH_3)(H)C{=}CH{-}CH{=}CH(CH_3)$

542. Compound X is a polyene with an absorption in the UV. The proton NMR spectrum has no peak in the 7-9 ppm region. Compound X is neither aromatic nor nonaromatic, but it can act as either a diene or a dieneophile in a Diels-Alder reaction. Which of the following is most likely Compound X?

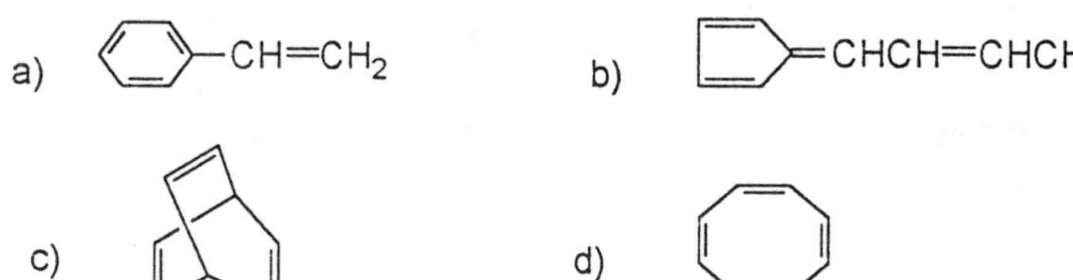

TIP Answer a is aromatic and answer c will not absorb in the uv- visible range above 200 nm. To avoid antiaromaticity, answer d will not be planar and therefore will not fullfill most of the criteria. This leaves answer b as a derivative of cyclopentadiene and meets all the criteria of the problem.

543. A diene has an absorption in the UV and undergoes a Diels-Alder reaction with acetylene to give a product which has 2 triplets in the proton NMR spectrum. What is the most likely structure for the diene?

a) 1,3-butadiene
b) 2,4-hexadiene
c) 1,3-pentadiene
d) 1,2-butadiene

544. Which of the following has 2 nodes in the highest occupied molecular orbital?

a) 1,3-pentadiene
b) 1,3-pentadienyl cation
c) 1,3,5-hexatriene
d) 1,3,5-hexatrienyl anion

545. A compound has the molecular formula, C_6H_8. It decolorizes bromine in carbon tetrachloride and reacts with platinum and 2 moles of hydrogen. The UV spectrum has a maximum at 256 nm. Which of the following is most likely the compound?

a) 1,3-cyclohexadiene
b) 1,4-cyclohexadiene
c) 1,3-hexadiene
d) 1,5-hexadiene

CHAPTER 15 AROMATICS I

A. AROMATICITY

546. Which of the following compounds are aromatic?

I. cyclobutadiene
II. cycloheptatrienyl cation
III. cyclopropenyl cation
IV. cyclopentadienyl cation

a) I, II b) II, III c) III, IV d) II, IV

547. How many pi electrons are there in the cyclooctatetraenyl dianion?

a) 10 b) 8 c) 12 d) 6

548. Which of the following structures are aromatic?

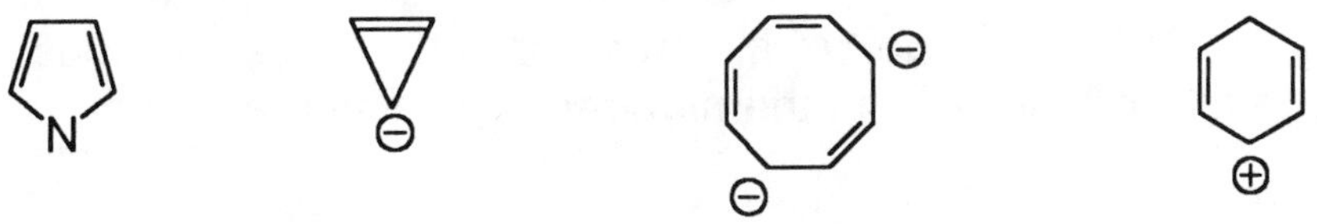

N

I II III IV

a) I, II b) III, IV c) I, III d) II, IV

549. Which of the following compounds or ions are antiaromatic?

I. cyclobutadiene
II. cycloheptatriene
III. cycloheptatrienyl anion
IV. cyclobutadienyl dianion

a) I, II b) III, IV c) II, IV d) I, III

550. Which of the following compounds or ions is neither aromatic nor antiaromatic?

a) cyclobutadiene
b) cycloheptatriene
c) cyclopentadienyl cation
d) cyclooctatetraene

551. Which of the following structures is neither aromatic nor antiaromatic?

N

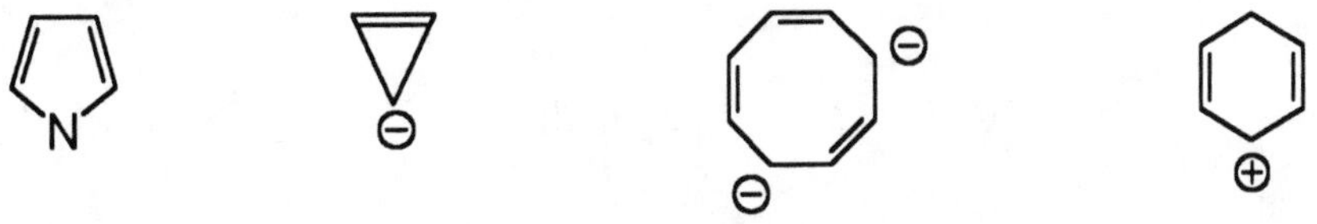

a) b) c) d)

552. Which of the following structures are aromatic?

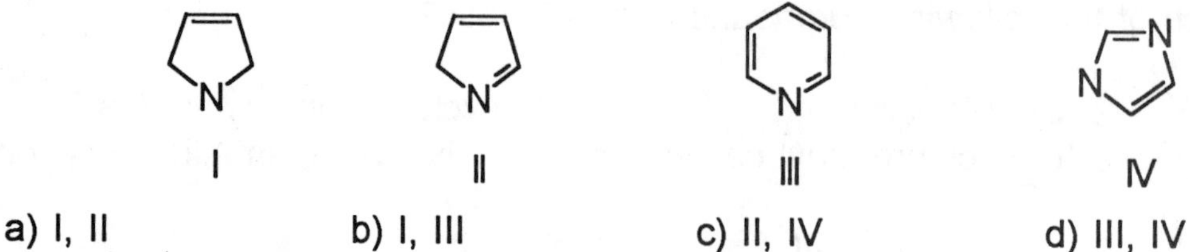

a) I, II b) I, III c) II, IV d) III, IV

TIP In a cyclic compound with p orbitals on continuous atoms, the presence of 2, 6, 10, 14 (or 4n+2) electrons leads to an aromatic compound, while the presence of 4, 8, 12 (or 4n) electrons leads to an antiaromatic compound. Structure I is neither aromatic nor antiaromatic since atoms 2 and 5 do not have p orbitals. Structure II similarly does not have a p orbital on atom 5. Both structures III and IV have the required continuous p orbital requirement and each has 6 pi electrons. In IV the lone pair of electrons on one of the nitrogens is involved, whereas in III the lone pair on the nitrogen is in a sigma orbital.

553. Which of the following structures are aromatic?

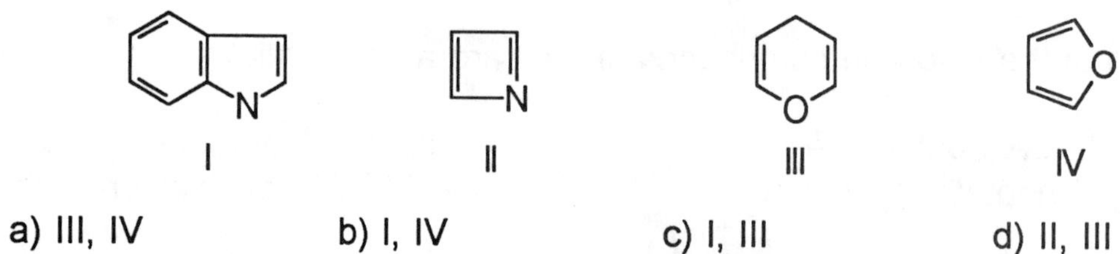

a) III, IV b) I, IV c) I, III d) II, III

554. Which of the following are converted to an aromatic species by the loss of one proton?

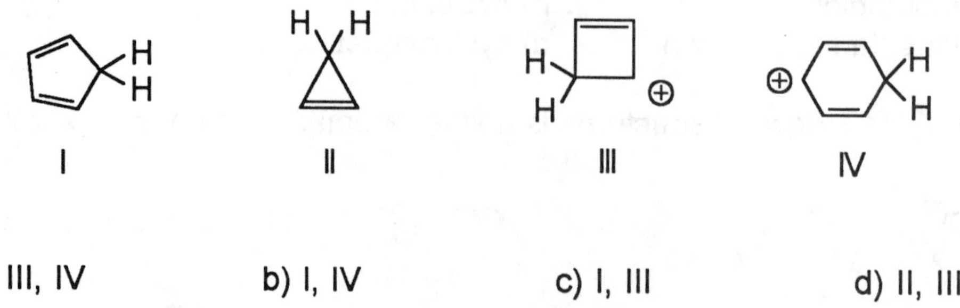

a) III, IV b) I, IV c) I, III d) II, III

555. Which of the following are converted to an aromatic species by the loss of a hydride ion?

I  II 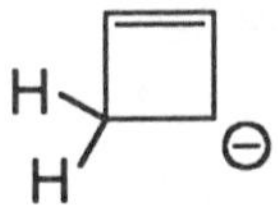 III  IV

a) III, IV b) I, IV c) I, III d) II, III

TIP The key to this problem is to think backwards and ask how many electrons will be in the pi system after the reaction. In 554 loss of a proton will lead to a negatively charged ion for the neutral species, and a neutral species for the charged ions. In 555 loss of a hydride will lead to positively charged ions for I and II and neutral species for III and IV. Then it is a matter of counting electrons. In 554 the number of pi electrons are 6, 4, 4, and 6 for I, II, III and IV. In 555 the number of pi electrons are 2, 4, 4 and 6 for I, II, III and IV.

556. Protonation of a carbonyl compound forms a carbocation according to the following equation.

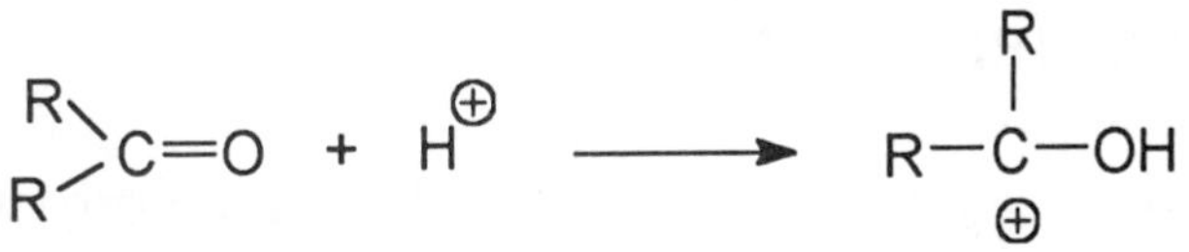

$$R_2C{=}O + H^{\oplus} \longrightarrow R_2C^{\oplus}{-}OH$$

Which of the following carbonyl compounds form an aromatic species upon protonation?

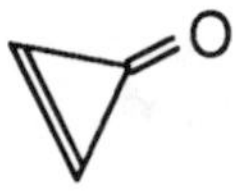

I 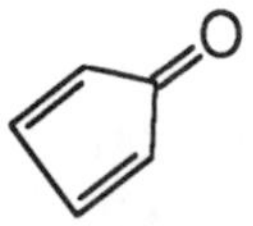 II 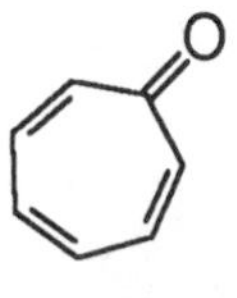 III 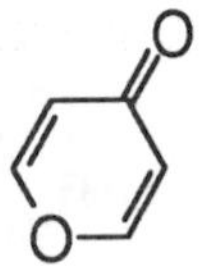 IV 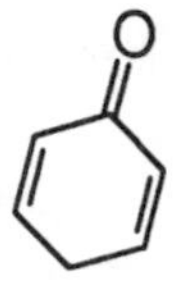 V 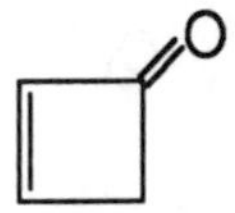 VI

a) I, III, IV b) II, IV, VI c) III, V, VI d) I, II, V

557. How many nodes are there in the highest occupied molecular orbital of benzene?

a) 0 b) 1 c) 2 d) 3

558. Which of the following annulenes is a diradical?

a) [4]annulene b) [6]annulene c) [10]annulene d) [18]annulene

559. Which of the following structures represents the highest occupied molecule orbital of cyclopropenyl cation?

a)

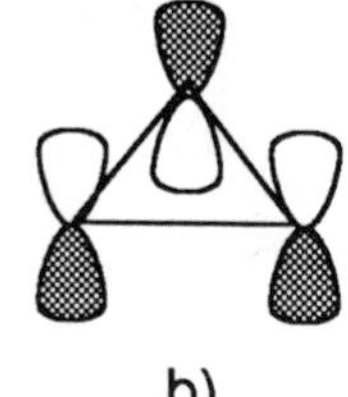
b)

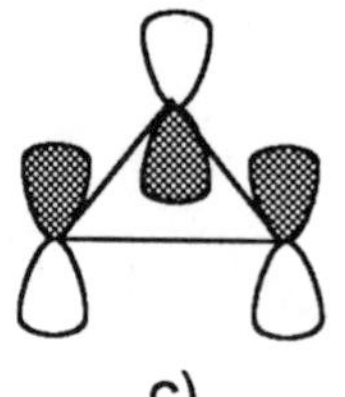
c)

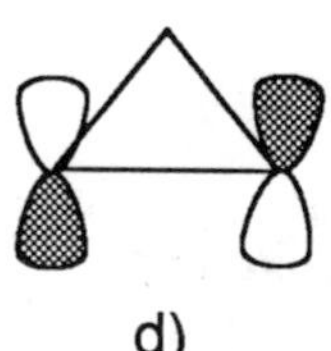
d)

TIP First we must count the number of electrons. In the cyclopropenyl cation with a positive charge there will be two pi electrons. With two electrons only the first molecular orbital with no nodes is occupied. Note that b), c), and d) all contain one node.

B. NOMENCLATURE

560. Which of the following compounds have the correct structure?

I. meta-chlorotoluene

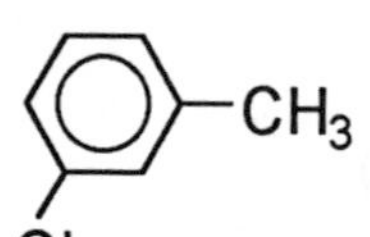

II. cinnamyl bromide

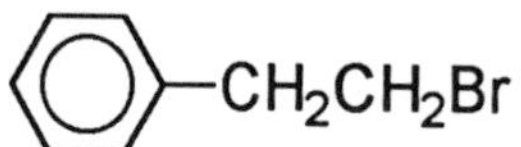

III. para-bromophenol

Br— —COOH

IV. ortho-nitroanisole

—NO_2
OCH_3

V. 2,4,6-trichloroaniline

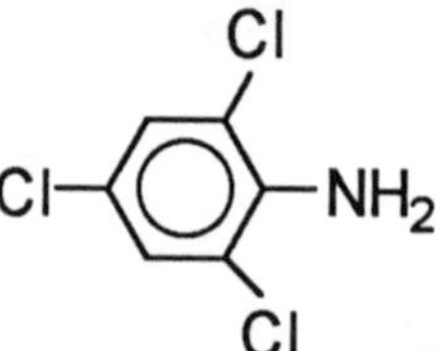

a) I, II, V b) I, III, IV c) I, IV, V d) II, III, V

561. Which of the following structures have the correct names?

I [benzene ring with CH_3, CH_3, CH_3] ortho-meta -dimethyltoluene

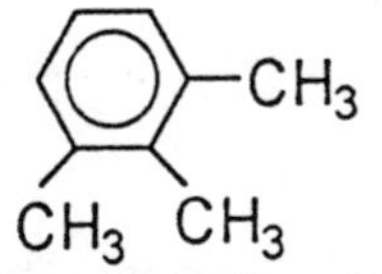

II [benzene ring–C(=O)–benzene ring–Br] para-bromobenzophenone

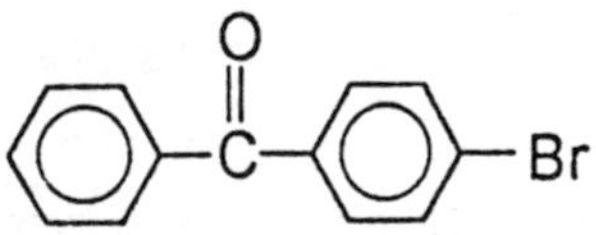

III [benzene ring with $-CH{=}CH_2$ and NO_2] meta-nitroanisole

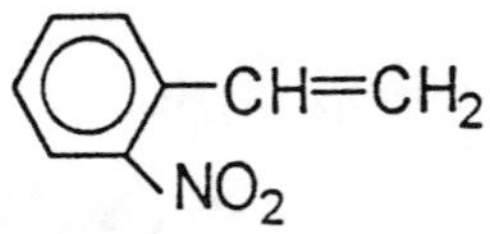

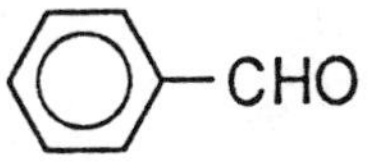

IV [benzene ring–CHO] benzaldehyde

V [benzene ring with O_2N-, NO_2, $-CH_3$, NO_2] 2,4,6-trinitrotoluene (TNT)

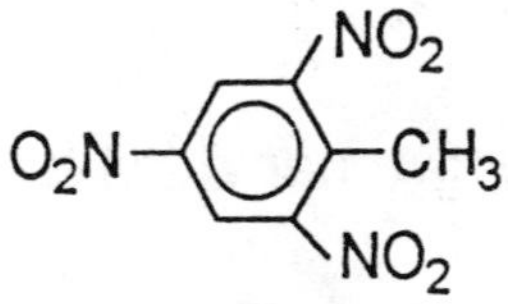

a) I, III, IV b) II, IV, V c) I, II, V d) II, III, V

562. What is the correct name for this structure?

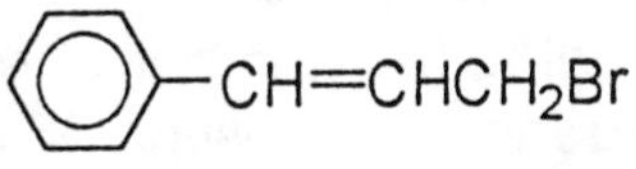

[benzene ring]$-CH{=}CHCH_2Br$

a) α-styryl bromide b) β-styryl bromide
c) cinnamyl bromide d) 1-bromo-3-phenylpropene

563. Which of the following structures match the names?

I.	A philosopher	A.	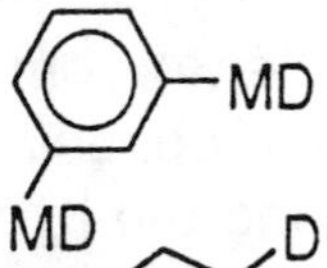 [benzene ring with MD, MD]
II.	An A in organic chemistry	B.	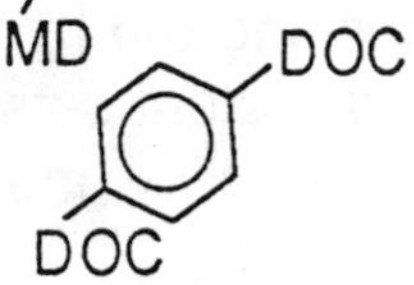 [benzene ring with DOC, DOC]
III.	Apparent contradiction	C.	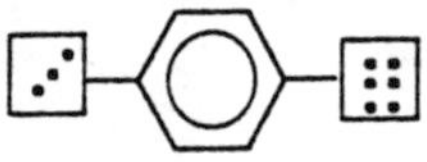 [die face]–[benzene ring]–[die face]
IV.	A carnival structure	D.	COCA—COLA
V.	Aryl—thing	E.	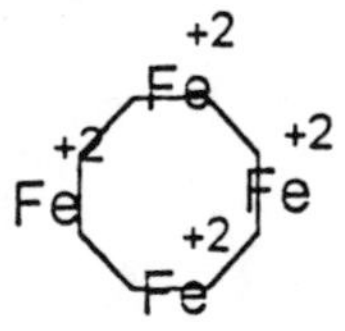 [ring of Fe^{+2} groups]

For most students, C is a match for II. All structures have a matching name.

C. STRUCTURE

564. How many isomers are possible for xylene?

a) 2 b) 3 c) 4 d) 5

565. How many benzylic hydrogen atoms are there in para-diisopropylbenzene?

a) 1 b) 2 c) 4 d) 12

566. How many aryl hydrogen atoms are there in para-diisopropylbenzene?

a) 1 b) 2 c) 4 d) 12

567. How many sets of equivalent protons are there in meta-xylene?

a) 2 b) 3 c) 4 d) 5

TIP For questions 564 to 567. If a xylene is a "methyl substituted" toluene then the number of xylenes will be equal to the number of equivalent sets of aromatic hydrogens (or carbons) in toluene. In question 564, answers a) and d) are clearly incompatible. In questions 565 and 566, benzylic and arylic are often confused. Arylic are ring positions and benzylic are at side chain positions on the carbon atom attached to the ring. There are a total of 18 protons in the molecule in 565, of which 14 are on the side chain. But of these how many are benzylic? And how many does that leave for the answer for 566?
In question 567, a good way to solve this problem is to look for a rotational axis (or mirror plane) from the carbon atom between the two methyl groups to the carbon on the opposite side of the ring. With such a mental guide, the two methyl groups and the hydrogens on the other 'ortho' carbon atoms are easily seen to be equivalent sets. But an answer of two ignores the hydrogens on the symmetry guide. How many are there? Could it be an odd number?

568. What is the order of increasing acid strength for the following compounds (lowest first)?

COOH-benzene	OH-benzene	OH-benzene-NO_2 (para)	CH_2OH-benzene
I	II	III	IV

a) II, IV, III, I b) IV, II, III, I c) IV, II, I, III d) III, II, I, IV

569. How can phenol be distinguished from benzoic acid?

a) solubility in water
b) solubility in hydrochloric acid
c) solubility in sodium bicarbonate
d) solubility in sodium hydroxide

570. Arrange the following compounds in the order of increasing acidity (lowest first).

I: phenol (C_6H_5OH); II: benzyl alcohol ($C_6H_5CH_2OH$); III: p-$HOC_6H_4N^{\oplus}(CH_3)_3$; IV: p-$HOOCC_6H_4CH_3$

a) II, I, III, IV b) II, I, IV, III c) III, II, I, IV d) IV, II, III, I

571. What is the order of increasing stability for the following radicals (least first)?

I: $C_6H_5-CH_2CH(CH_3)\dot{C}H_2$
II: $C_6H_5-CH_2\dot{C}(CH_3)CH_3$
III: $C_6H_5-\dot{C}HCH(CH_3)CH_3$
IV: $C_6H_4^{\bullet}-CH_2CH(CH_3)CH_3$ (radical on a ring carbon)

a) I, III, II, IV b) IV, III, II, I c) IV, I, II, III d) I, II, III, IV

TIP Aryl carbon-hydrogen bonds are particularly strong as a result of hybridization. Benzyl carbon-hydrogen bonds are particularly weak as a result of resonance stabilization of the radical. Thus IV is the least stable and answers b or c are the only possibilities. These are distinguished by alkyl effects on radical stability, primary less than secondary less than tertiary.

572. Compound X has a mass spectrum containing, among other peaks, the molecular ion peak at 134 (100%) and a peak at 135 (11%). This compound forms terephthalic acid on vigorous oxidation. What is the most likely structure for Compound X?

a) 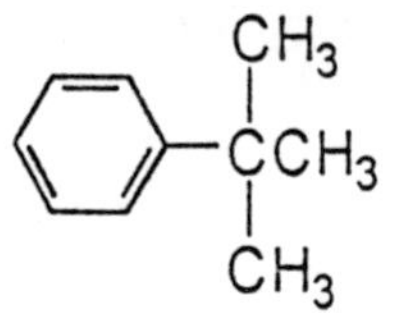 $C_6H_5C(CH_3)_3$

b) $H_3C-C_6H_4-CH(CH_3)CH_3$

c) $C_6H_5CH_2CH_2CH_2CH_3$

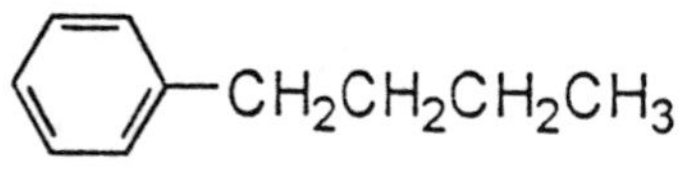

d) 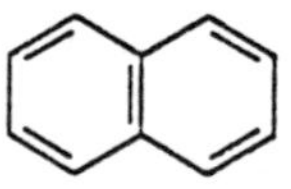 (naphthalene)

573. Compound X has a proton NMR spectrum having 5 aromatic hydrogen atoms in the region of 7.2-7.5 ppm range, a septet of 1 hydrogen atom at 2.2 ppm, and a doublet of 6 hydrogen atoms at 1.3 ppm. What is the most likely structure for Compound X?

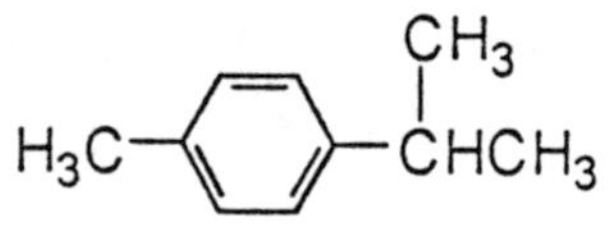

a) 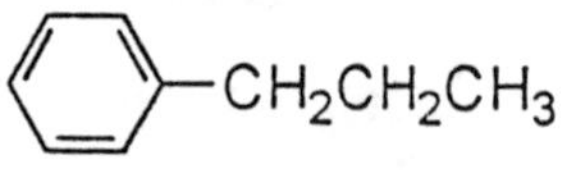 $C_6H_5CH_2CH_2CH_3$

b) $C_6H_5CH(CH_3)CH_3$

c) $C_6H_3(CH_3)_3$ (1,2,3-trimethylbenzene)

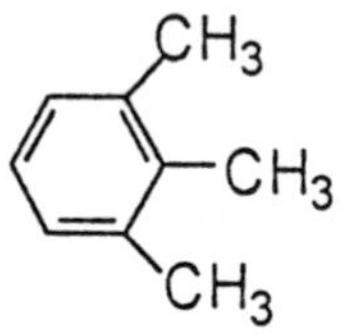

d) 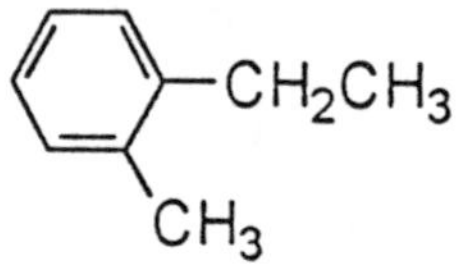 $C_6H_4(CH_2CH_3)(CH_3)$ (ortho)

574. A compound has the molecular formula, C_8H_9Br, and a carbon-13 NMR spectrum having 8 signals. What is the most likely structure for the compound?

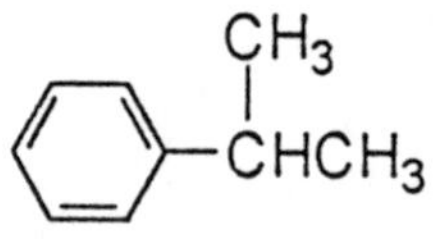

a) 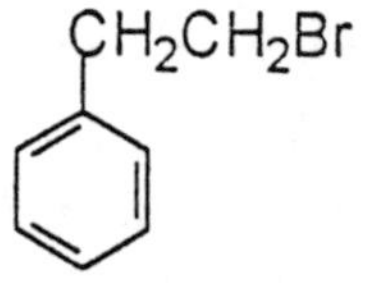 $C_6H_5CH_2CH_2Br$

b) $C_6H_5CH(Br)CH_3$

c) $CH_3CH_2-C_6H_4-Br$ (para)

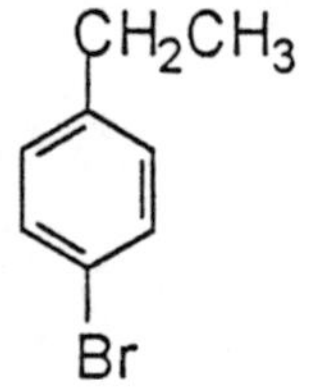

d) 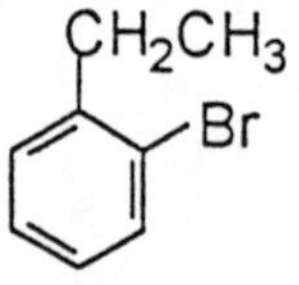 $CH_3CH_2-C_6H_4-Br$ (ortho)

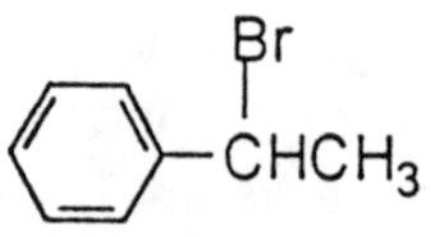

D. REACTIONS at the BENZYLIC POSITION

575. What is the most likely product from the following series of reactions?

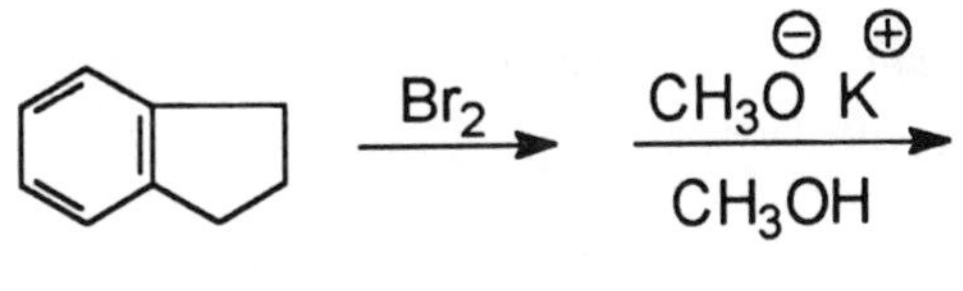

Br_2 → $CH_3O^{\ominus}K^{\oplus}$ / CH_3OH →

a) 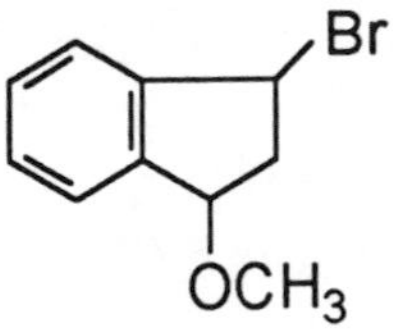 Br, OCH_3

b) 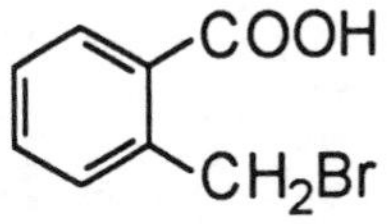 COOH, CH_2Br

c)

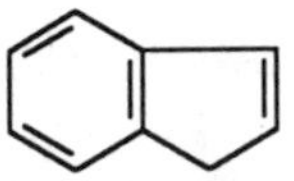

d) 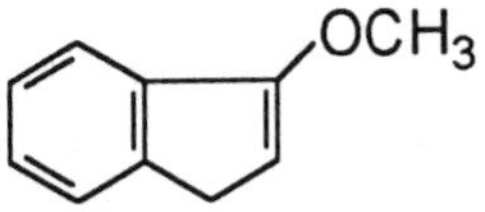 OCH_3

576. What is the most likely product from the following series of reactions?

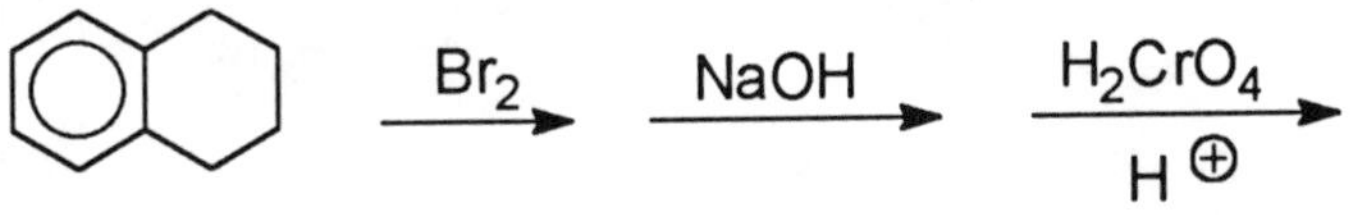

Br_2 → NaOH → H_2CrO_4 / $H^{\oplus}$ →

a)

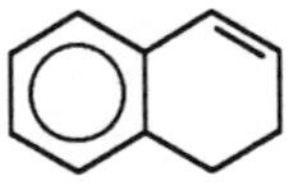

b) 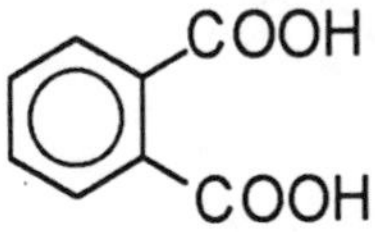 COOH, COOH

c) 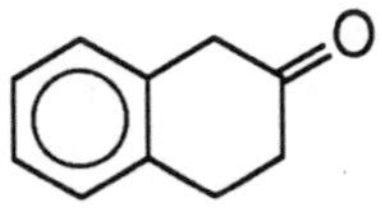 O

d) 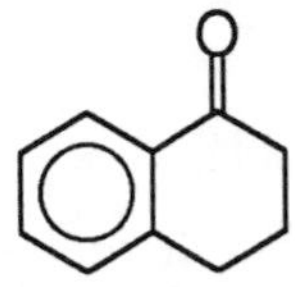 O

577. What is the most likely product from the following reaction?

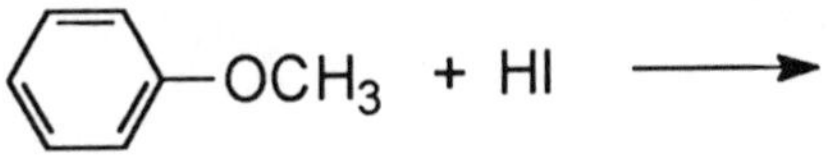

$-OCH_3$ + HI →

a) 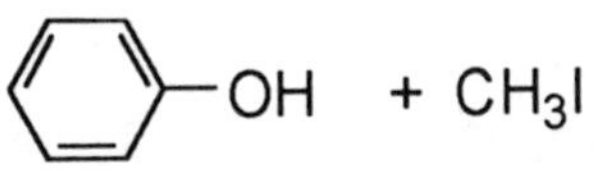 OH + CH_3I

b) 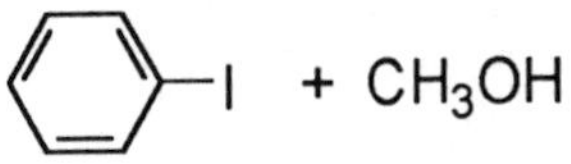 I + CH_3OH

c) 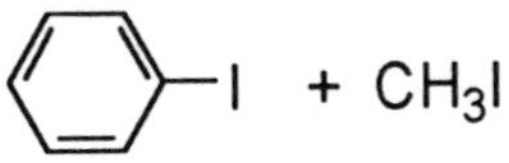 I + CH_3I

d) 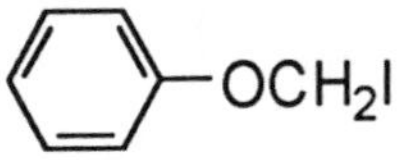 OCH_2I

578. Which of the following reactions is the best procedure for preparing ethyl phenyl ether?

a) sodium ethoxide plus bromobenzene
b) ethanol plus phenol plus acid
c) phenol plus acid followed by ethyl bromide plus acid
d) phenol plus base followed by ethyl bromide

TIP Mechanism is the key to questions 577 and 578. Both involve displacements, which could be either S_N2 or S_N1. Critically, an aryl carbon with a substituent cannot be involved in either possibility. Whatever reactions that do occur for the aryl portion, do not involve the aryl carbon. In question 577, this rules out answers b and c. In question 578, this rules out answers a, b and c.

579. What is the product from the following reaction?

$KMnO_4$; heat, $^{\ominus}OH$

a) 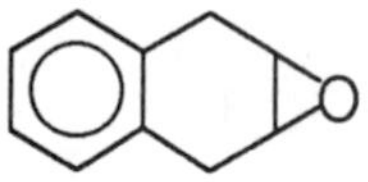 b) 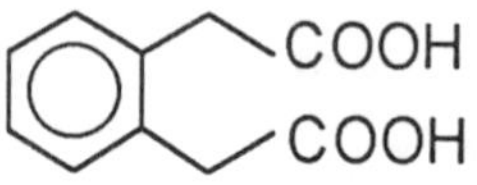 COOH COOH

c) 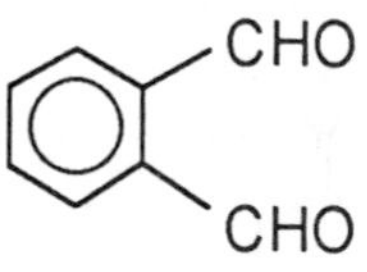 CHO CHO d) 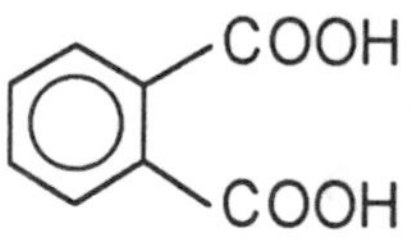 COOH COOH

CHAPTER 16 AROMATICS II

A. ELECTROPHILIC AROMATIC SUBSTITUTION

580. Which of the following substituents on benzene are activating for electrophilic aromatic substitution?

$-Br$	$-NO_2$	$-OCH_3$	$-COCH_3$
I	II	III	IV
$-N(CH_3)_2$	$-C(CH_3)_3$	$-\overset{\oplus}{N}(CH_3)_3$	$-SO_3H$
V	VI	VII	VIII

a) I, II, VII b) II, IV, V c) II, VII, VIII d) III, V, VI

581. Which of the following groups are not meta directing?

$-Br$	$-NH_2$	$-OH$	$-CHO$	$-\overset{\oplus}{N}(CH_3)_3$	$-SO_3H$
I	II	III	IV	V	VI

a) I, III b) II, IV c) III, IV, VI d) I, III, V

582. Which of the following are reactive species in the halogenation, nitration and sulfonation reactions with benzene?

$Br^{\oplus}$	HNO_3	Br_2FeBr_3	$\overset{\oplus}{N}O_2$
I	II	III	IV
$\overset{\oplus}{K}\ \overset{\ominus}{BF_4}$	$\overset{\oplus}{HSO_3}$	$\overset{\ominus\ominus}{SO_4}$	
V	VI	VII	

a) I, IV, VI b) I, IV, V c) III, V, VII d) II, VI, VII

583. Which of the following resonance forms is the sigma complex that is formed in the nitration of nitrobenzene?

a) 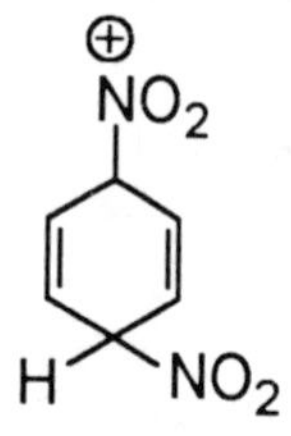 b) 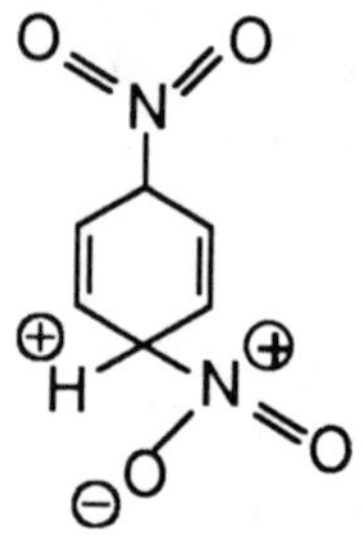 c) 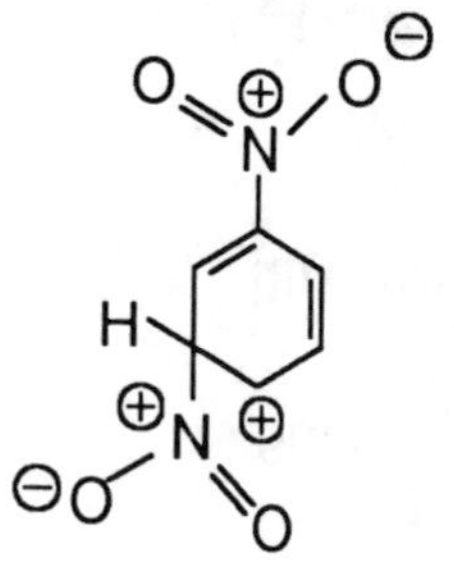 d)

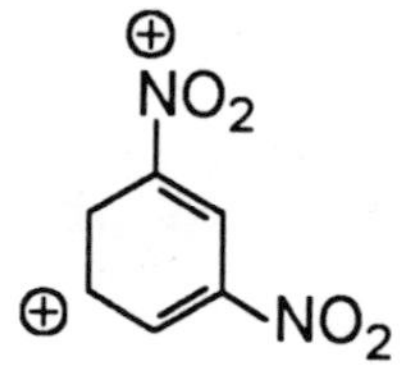

584. Which of the following is the most important resonance form for the complex formed from the electrophilic reaction with toluene?

a) 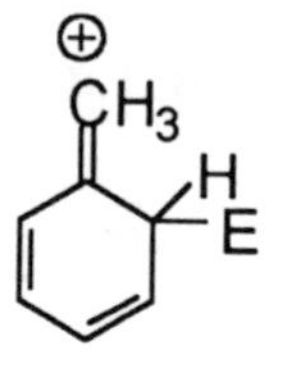 b) 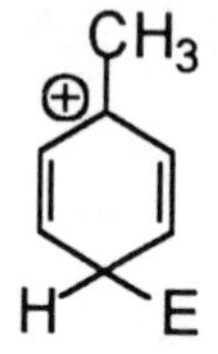 c) 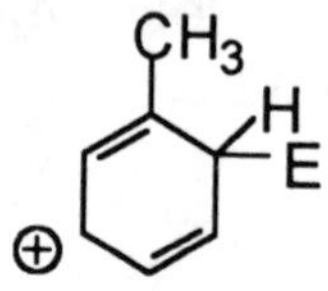 d)

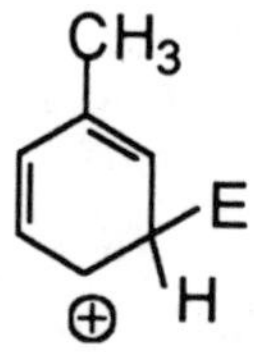

TIP Since a methyl group is ortho-para directing answer d) can be rejected immediately. Distinction among the remaining answers is determined by carbocation stability, which follows the order tertiary>secondary>primary. Only the contributing structure b) is a tertiary carbocation.

585. A benzylic carbocation is stabilized most by which of the following groups?

a) para-nitro b) para-methyl c) meta-methoxy d) meta-chloro

586. What is the major product from the catalytic bromination of naphthalene?

a) 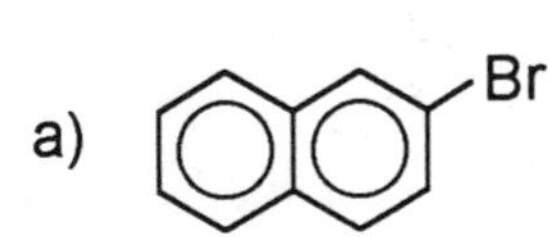

b) Br

c) Br

d) Br

587. What is the order of increasing reactivity for electrophilic substitution of the following compounds (least first)?

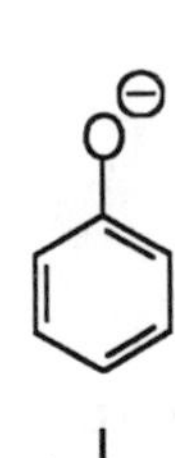

I

$OCCH_3$ (O)

II

OH

III

$COCH_3$ (O)

IV

a) IV, III, II, I b) I, II, III, IV c) IV, II, III, I d) III, II, IV, I

588. Which resonance structure is most important for the complex formed from an electrophilic reaction with phenoxide ion?

a) 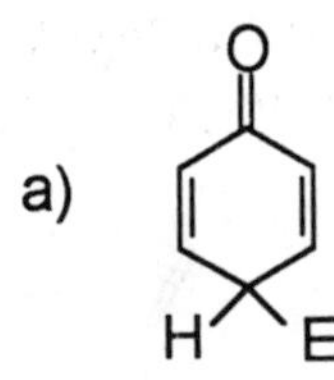

b) 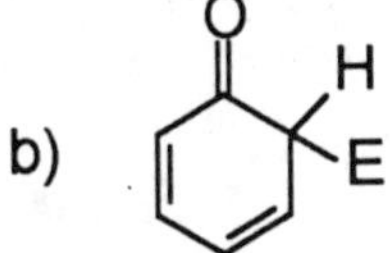

c) 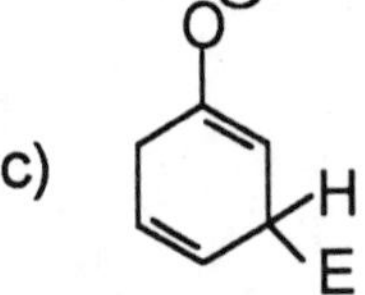

d) 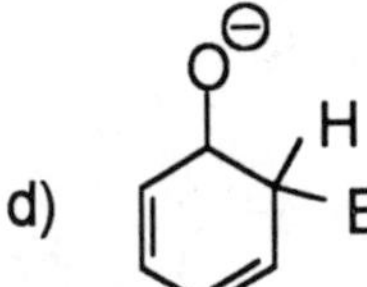

589. What is the order of decreasing reactivity for the following compounds in the nitration reaction (greatest first)?

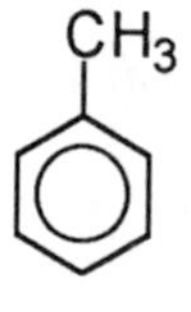

CH_3 (I) Cl (II) OCH_3 (III) NO_2 (IV)

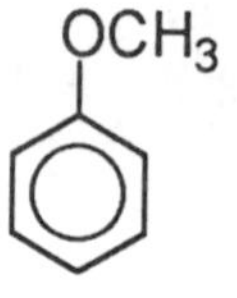

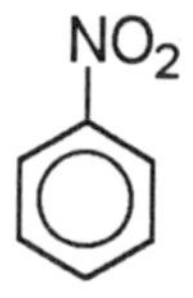

a) III, I, II, IV b) II, III, I, IV c) I, II, IV, III d) III, IV, I, II

TIP For this problem we have to separate activating groups from deactivating groups. Note that there are two examples of each of the two groups. In a second step we have to estimate the relative effects of each substituent within the group. Both methoxy and methyl are activating, but the methoxy with a large resonance contribution is far more reactive. Both chloro and nitro are deactivating, but the nitro group has a major inductive effect (three electronegative atoms) plus a resonance effect and is therefore far less reactive than the chloro group.

590. What is the most likely product from the reaction of phenol with bromine?

a) 4-bromophenol b) 2,4-dibromophenol
c) 2,6-dibromophenol d) 2,4,6-tribromophenol

591. What is the order of decreasing reactivity for electrophilic substitution of the following compounds (greatest first)?

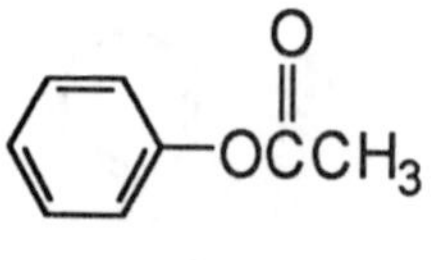

$-OCCH_3$ (O above C) (I) CN (II) $-CH_2-$ (III)  (IV)

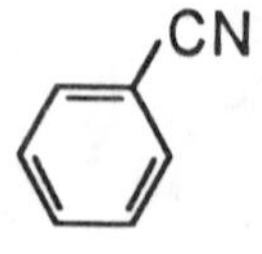

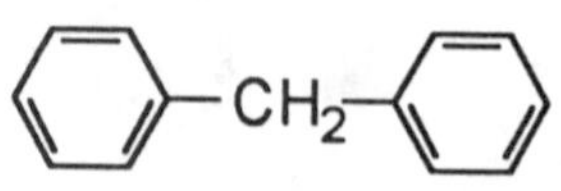

a) II, I, IV, III b) I, II, III, IV c) IV, III, I, II d) I, III, IV, II

592. Which of the following compounds undergoes nitration least readily?

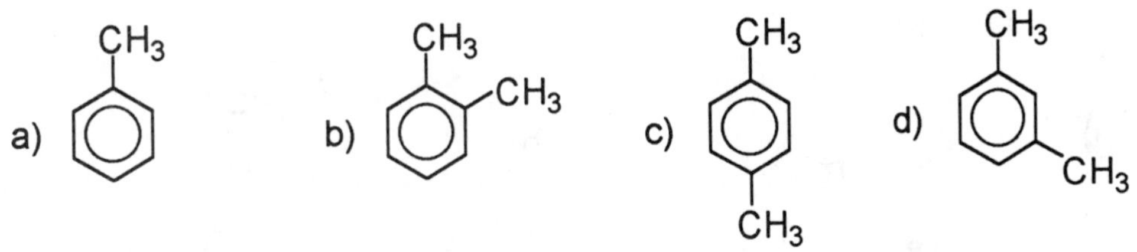

a) [toluene] b) [o-xylene] c) [p-xylene] d) [m-xylene]

CH_3 CH_3 CH_3 CH_3 CH_3 CH_3 CH_3

593. What is a major product from the following reaction?

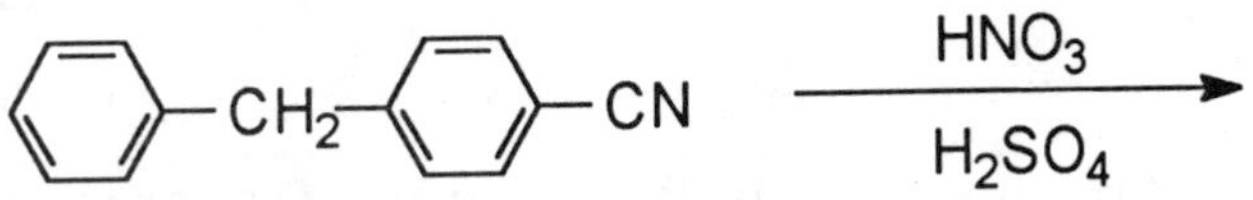

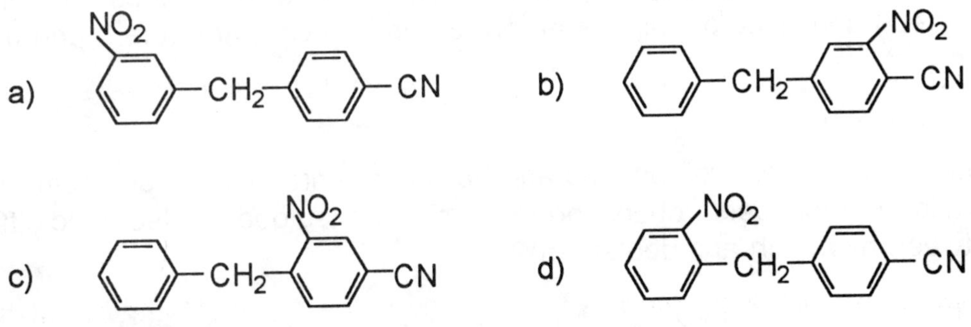

C_6H_5–CH_2–C_6H_4–CN $\xrightarrow[H_2SO_4]{HNO_3}$

a) NO_2 –CH_2– –CN b) –CH_2– NO_2 –CN

c) –CH_2– NO_2 –CN d) NO_2 –CH_2– –CN

594. Which of the indicated positions in the following compound is brominated when the compound is treated with bromine and ferric bromide?

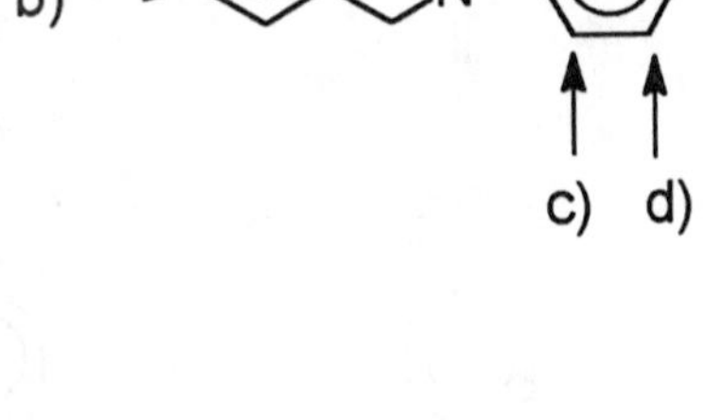

a) → b) → N– ↑ ↑ c) d)

595. What is the major product of the reaction of nitrobenzene with ferric bromide and two moles of bromine?

a) NO_2, Br, Br

b) NO_2, Br, Br

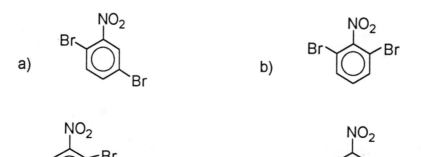

c) NO_2, Br

d) NO_2, Br, Br

TIP The key to this problem is to note that after the first bromine is on the ring the subsequent bromine placement will be directed by the bromine not the nitro group. Thus the first step gives meta bromo nitrobenzene (thus answer c is incorrect) and since bromine is an ortho para director, answers b and d are incorrect.

596. How many different compounds are there in the product mixture from the nitration of monodeuterobenzene (assuming no isotope or electronic differences between hydrogen and deuterium)?

a) 1 b) 2 c) 3 d) 4

597. Which of the following benzene derivatives react readily with methyl chloride and a Lewis acid in a Friedel-Crafts alkylation?

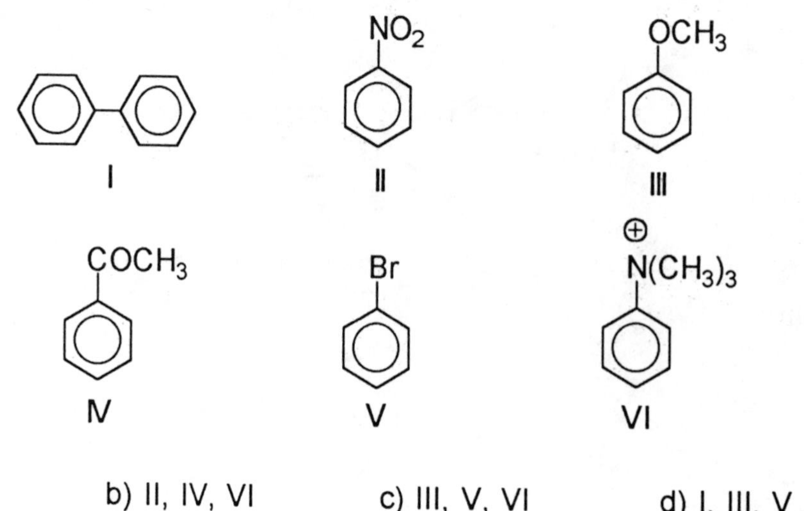

a) I, III, IV b) II, IV, VI c) III, V, VI d) I, III, V

598. What is the major product from the following reaction?

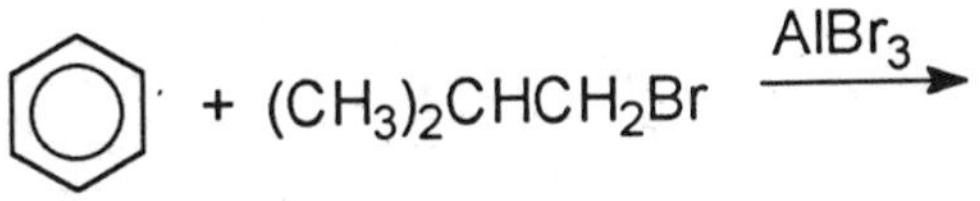

$$C_6H_6 + (CH_3)_2CHCH_2Br \xrightarrow{AlBr_3}$$

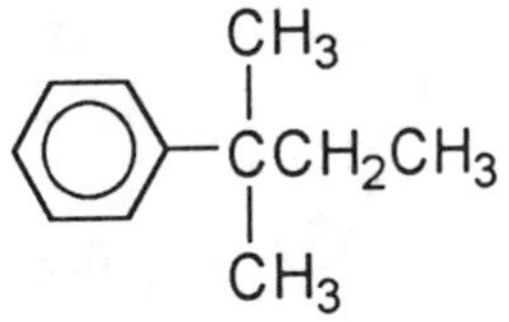

a) 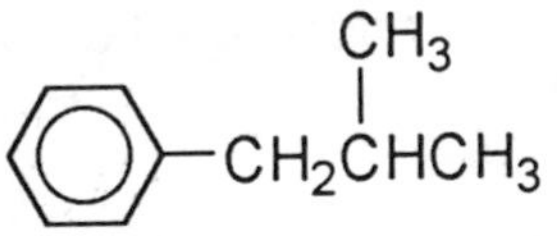 $C_6H_5-CH_2CH(CH_3)CH_3$

b) $C_6H_5-C(CH_3)_2CH_2CH_3$

c) $C_6H_5-C(CH_3)_2CH_3$

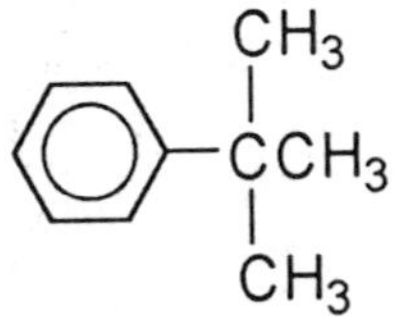

d) 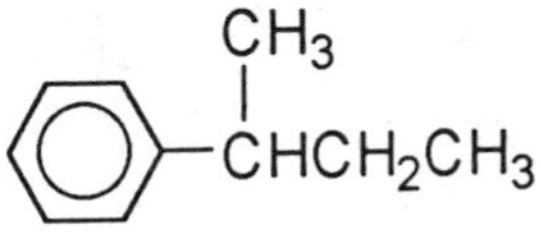 $C_6H_5-CH(CH_3)CH_2CH_3$

599. Which of the following schemes give isopropylbenzene in a Friedel-Crafts alkylation reaction with benzene?

I: $CH_3CH_2CH_2Cl + AlCl_3 \longrightarrow$

II: $(CH_3)_3CBr + FeBr_3 \longrightarrow$

III: $CH_3CH{=}CHCH_3 + H^{\oplus} \longrightarrow$

IV: $CH_2{=}CHCH_3 + HF \longrightarrow$

V: $CH_3CH_2CH_2OH + BF_3 \longrightarrow$

VI: $CH_3CHOHCH_2OH + H_3PO_4 \longrightarrow$

a) I, IV, V b) II, V, VI c) III, IV, VI d) I, III, VI

TIP Isoproprопyl benzene will come from electrophilic attack of the 2-propyl carbocation on benzene. This carbocation will come from reactions I and V, after hydride shift to give the more stable carbocation, and from IV, after Markovnikov protonation. Other carbocations derived from reactions II, III and VI lead to other products. Do II and III lead to the same product? For the answer, determine the structures of the carbocations.

600. Which of the following reacts slowest in a bromination reaction with ferric bromide?

a) anisole b) ethylbenzene c) nitrobenzene d) benzenesulfonic acid

601. Which of the following reacts fastest in a bromination reaction with ferric bromide?

a) anisole b) ethylbenzene c) nitrobenzene d) benzenesulfonic acid

602. What is the most likely product from the following reaction?

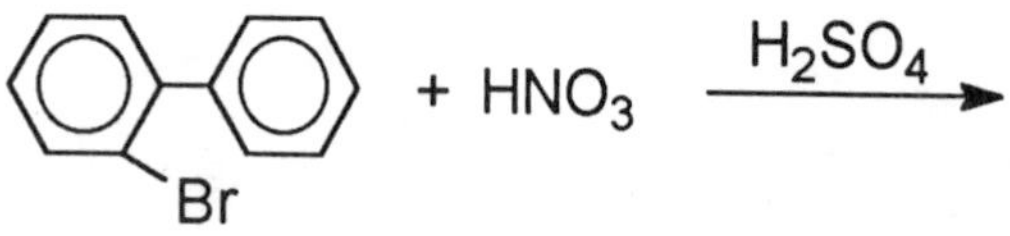

Br + HNO_3 $\xrightarrow{H_2SO_4}$

a) NO_2 Br

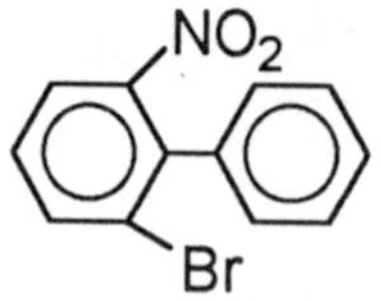

b) O_2N Br

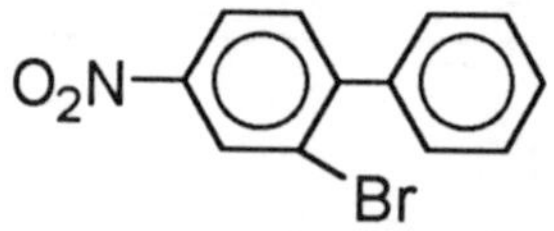

c) Br NO_2

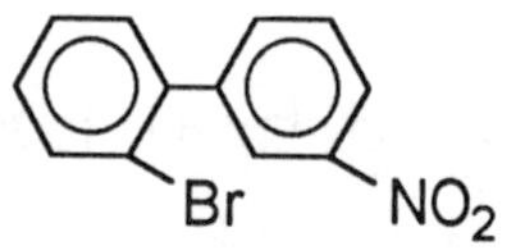

d) NO_2 Br

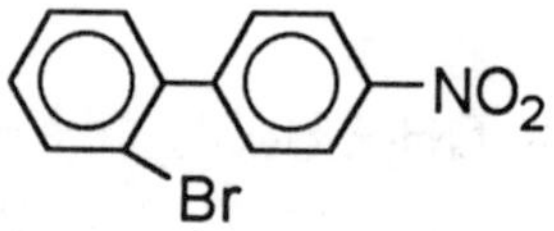

603. What is the order of increasing rates of nitration for the two compounds, toluene and nitrobenzene, at the indicated positions (slowest first)?

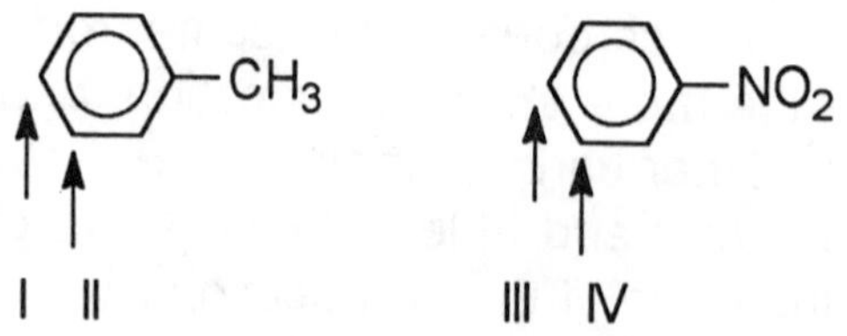

CH_3 NO_2

I II III IV

a) I, II, III, IV b) III, IV, II, I c) III, I, IV, II d) IV, II, III, I

604. Which of the following compounds reacts most readily with ferric chloride and chlorine?

a) toluene b) ortho-xylene c) meta-xylene d) para-xylene

TIP Since a methyl group is activating, the xylenes with two methyl groups are clearly more reactive than toluene with one methyl group. To distinguish among the xylenes, we have to ask whether the methyl groups are in positions on the ring that will be additive or reinforcing. Since a methyl group is an ortho/para director, it is only when the two methyl groups are meta to each other that they will be reinforcing.

605. Which of the following compounds <u>cannot</u> be made from toluene in one step?

a) $C_6H_5-CH_2Cl$ b) $H_3C-C_6H_4-Cl$

c) $H_3C-C_6H_4-SO_3H$ d) C_6H_5-COCl

606. What is the order of increasing rate of addition of HCl for the following compounds (slowest first)?

I: $C_6H_5-CH{=}CH_2$

II: $O_2N-C_6H_4-CH{=}CH_2$

III: $H_3C-C_6H_4-CH{=}CH_2$

IV: $CH_3O-C_6H_4-CH{=}CH_2$

a) I, IV, III, II b) II, I, III, IV c) I, II, III, IV d) III, IV, I, II

B. NUCLEOPHILIC AROMATIC SUBSTITUTION

607. What is the most important resonance form for the sigma complex in a nucleophilic aromatic substitution reaction?

a) 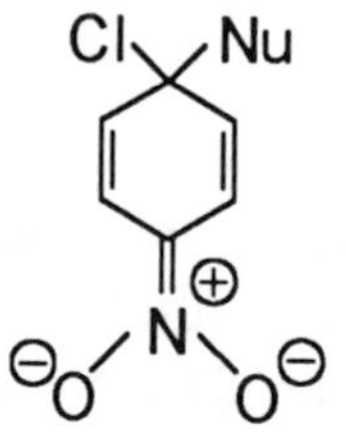 b)

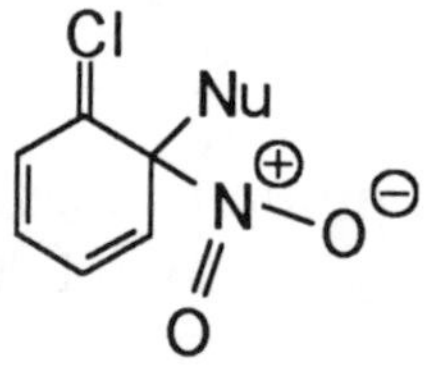

c) 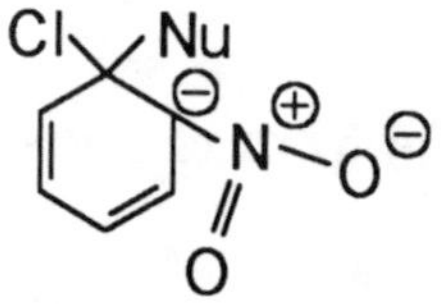 d)

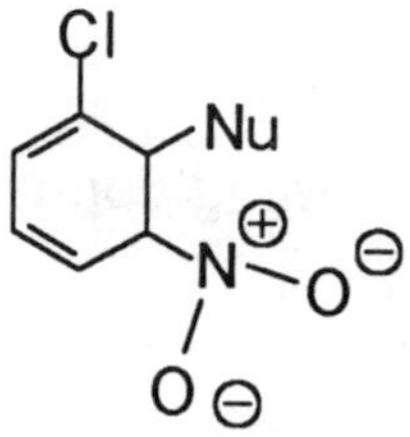

608. Which of the following compounds reacts fastest in a hydrolysis reaction with aqueous base?

a) 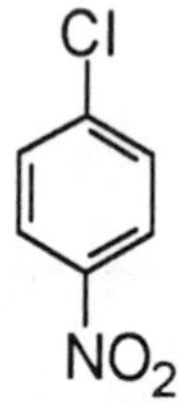 b) 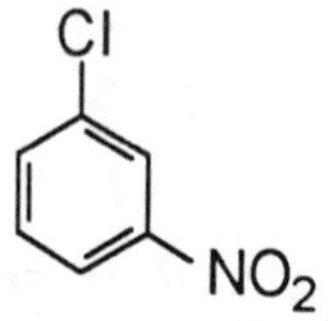 c) 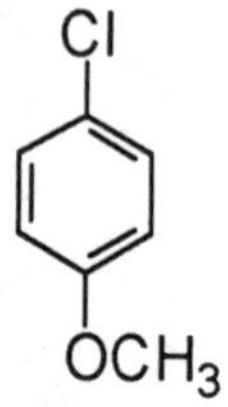 d)

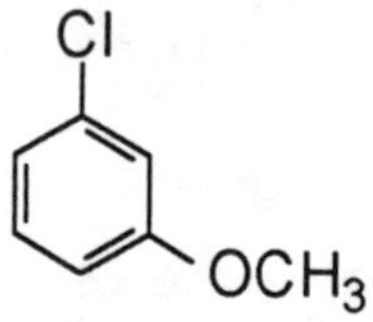

609. Which of the following compounds is most reactive in hot aqueous sodium hydroxide?

a) 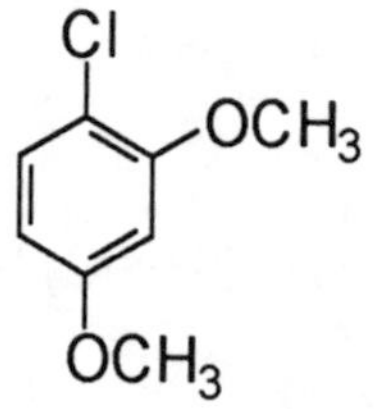 b)

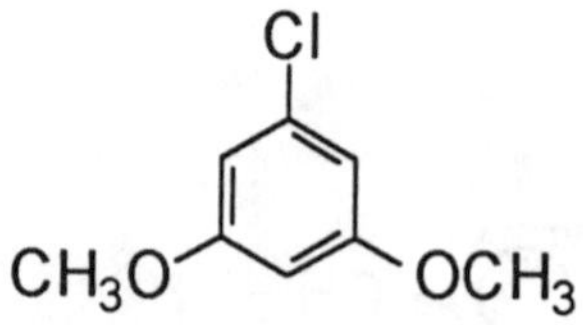

c) 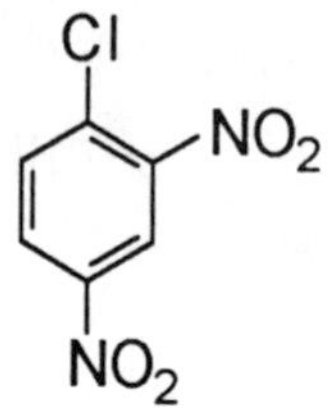 d)

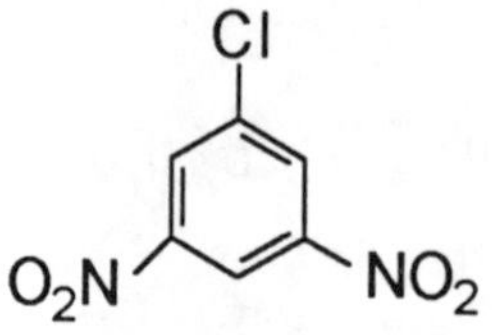

610. Nucleophilic aromatic substitution by hydroxide ion will occur most rapidly with which of the following compounds?

a) 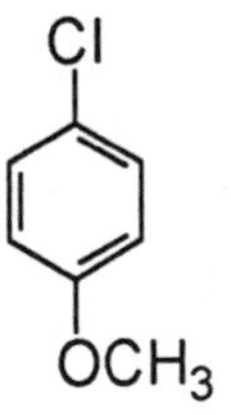 b) 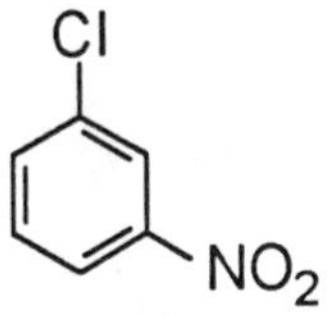 c) 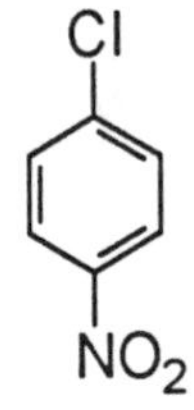 d)

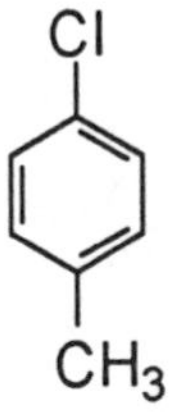

Cl Cl Cl Cl

NO_2

OCH_3 NO_2 CH_3

611. What is the principal product from the following reaction?

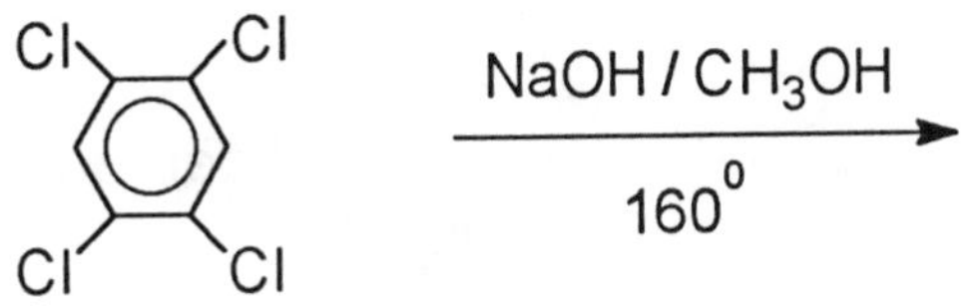

Cl Cl

$NaOH / CH_3OH$

160^0

Cl Cl

a) 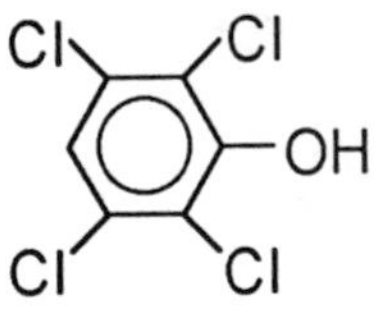 b) 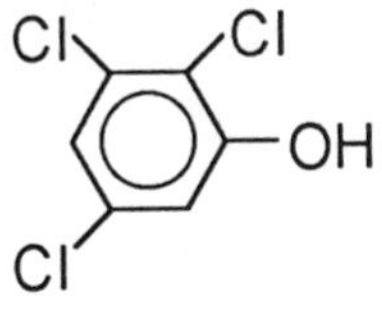 c) 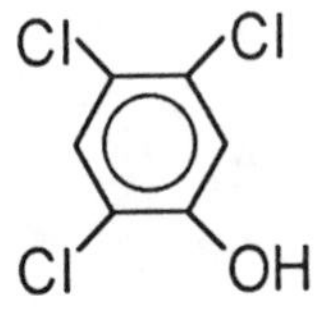 d)

Cl Cl Cl Cl Cl Cl Cl Cl

OH OH OH

Cl Cl Cl Cl OH Cl

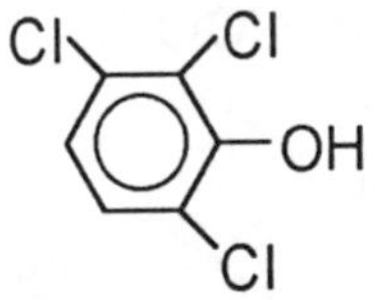

TIP For the nucleophilic aromatic substitution reaction, electron withdrawing groups ortho and para to the leaving groups stabilize the negatively charged sigma complex intermediate. For this compound the chlorine that will be replaced by hydroxide will therefore have one chlorine in an ortho position and one chlorine in a para position. Working backwards this is only fullfilled by product c).

612. What are the products from the reaction of labeled chlorobenzene (^{14}C in the para position) with sodium amide in liquid ammonia?

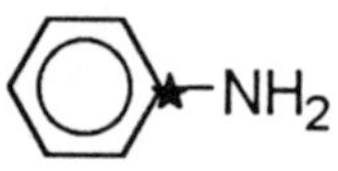

I: ★ at C1 (bearing NH_2)
II: ★ at ortho position
III: ★ at meta position
IV: ★ at para position

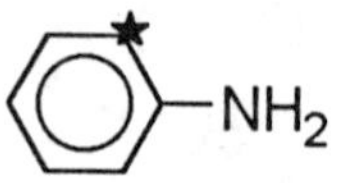

a) I, II b) II, IV c) III, IV d) I, IV

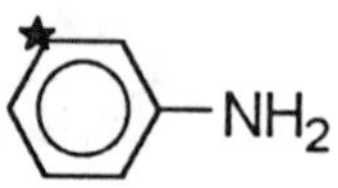

C. REACTIONS

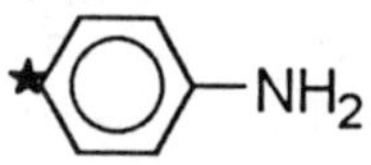

613. Which of the following compounds cannot be prepared from benzene using two successive electrophilic substitution reactions?

a) 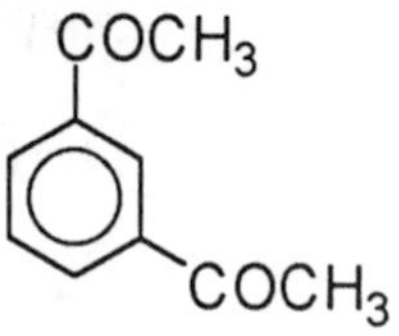 1,3-di($COCH_3$)benzene b) 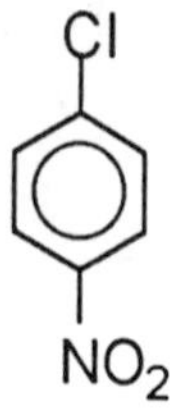 1-Cl-4-NO_2-benzene c) 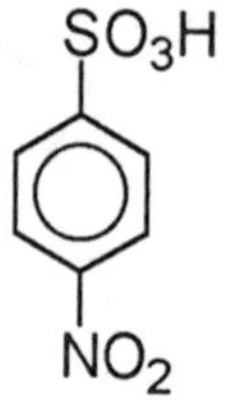 1-SO_3H-4-NO_2-benzene d) 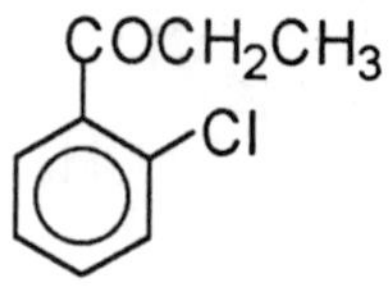 1-$COCH_2CH_3$-2-Cl-benzene

614. What is the principal product from the following reaction?

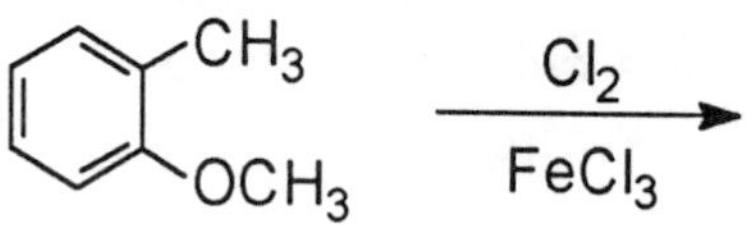

2-methylanisole (CH_3, OCH_3) $\xrightarrow[FeCl_3]{Cl_2}$

a) 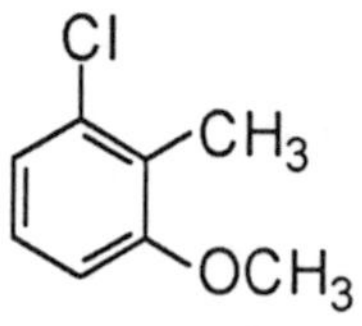 Cl ortho to CH_3 (CH_3, OCH_3, Cl)
b) 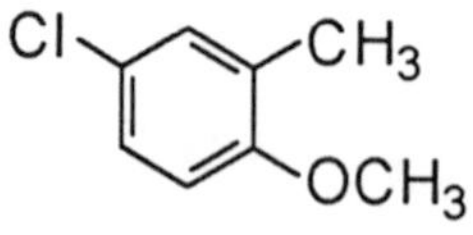 Cl para to CH_3 (CH_3, OCH_3, Cl)
c) 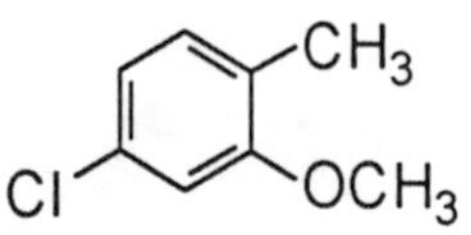 Cl para to OCH_3 (CH_3, OCH_3, Cl)
d) 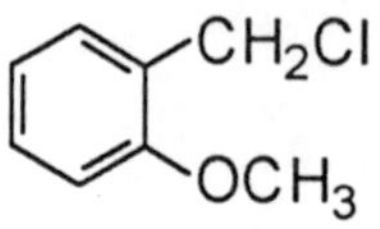 CH_2Cl, OCH_3

615. What are the best conditions for preparing meta-bromoethylbenzene from benzene?

a) $\xrightarrow[AlCl_3]{CH_3COCl}$ $\xrightarrow[FeBr_3]{Br_2}$ $\xrightarrow[HCl]{Zn(Hg)}$

b) $\xrightarrow[AlCl_3]{CH_3CH_2Cl}$ $\xrightarrow[FeBr_3]{Br_2}$

c) $\xrightarrow[AlBr_3]{CH_3COBr}$ $\xrightarrow[HBr]{Zn(Hg)}$ $\xrightarrow[FeBr_3]{Br_2}$

d) $\xrightarrow[FeBr_3]{Br_2}$ $\xrightarrow[AlBr_3]{CH_3CH_2Br}$

616. What is the best procedure for preparing para-bromobenzoic acid from toluene?

a) $\xrightarrow[FeBr_3]{Br_2}$ $\xrightarrow[OH^{\ominus}]{KMnO_4}$

b) $\xrightarrow[OH^{\ominus}]{KMnO_4}$ $\xrightarrow[FeBr_3]{Br_2}$

c) $\xrightarrow[light]{Br_2}$ $\xrightarrow[OH^{\ominus}]{KMnO_4}$

d) $\xrightarrow[OH^{\ominus}]{KMnO_4}$ $\xrightarrow[light]{Br_2}$

617. What is the best procedure for the following synthesis?

$C_6H_5-CH_3 \rightarrow$ Cl–C_6H_3(Br)–COOH

a) $\xrightarrow[FeCl_3]{Cl_2}$ $\xrightarrow[FeBr_3]{Br_2}$ $\xrightarrow[^{\ominus}OH]{KMnO_4}$ $\xrightarrow{H_3O^{\oplus}}$

b) $\xrightarrow[FeCl_3]{Cl_2}$ $\xrightarrow[^{\ominus}OH]{KMnO_4}$ $\xrightarrow{H_3O^{\oplus}}$ $\xrightarrow[FeBr_3]{Br_2}$

c) $\xrightarrow[^{\ominus}OH]{KMnO_4}$ $\xrightarrow{H_3O^{\oplus}}$ $\xrightarrow[FeBr_3]{Br_2}$ $\xrightarrow[FeCl_3]{Cl_2}$

d) $\xrightarrow[^{\ominus}OH]{KMnO_4}$ $\xrightarrow{H_3O^{\oplus}}$ $\xrightarrow[FeCl_3]{Cl_2}$ $\xrightarrow[FeBr_3]{Br_2}$

TIP For problems 615-617, it is the order in which the different steps are done in the reaction sequence that is critical. Furthermore, the directing effect is solely determined by the substituent on the ring. That is, the entering electophile has no effect on position. In these problems we will be converting meta directors to ortho/para directors or vice versa to take advantage of selective reactions.

In question 615, we first put the meta directing acetyl group on the ring as in answer a) and c), but then we brominate to direct the bromine to the meta position before reducing the acetyl group to an ethyl group which is an ortho-para directing group. So b) d) and c) all give an ortho or para product instead of the desired meta product.

In question 616, we brominate first to take advantage of the ortho/para directing methyl group before oxidation as in answer a). Answer b) would be the method of choice for meta bromobenzoic acid.

In question 617, we combine the reinforcing effects of a para chloro and a meta carboxyl by following the sequence in answer b).

618. What is the major product from the following series of reactions?

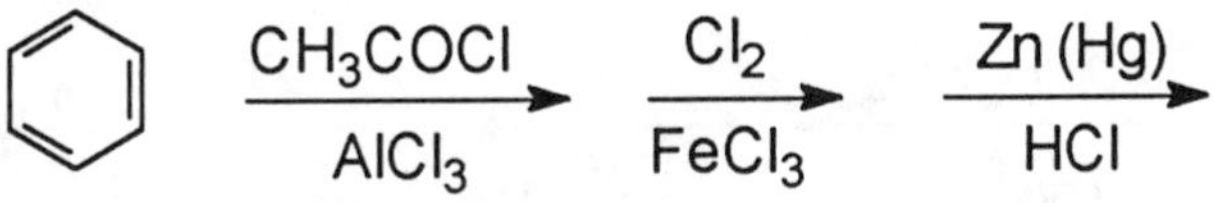

Benzene $\xrightarrow[AlCl_3]{CH_3COCl}$ $\xrightarrow[FeCl_3]{Cl_2}$ $\xrightarrow[HCl]{Zn(Hg)}$

a) 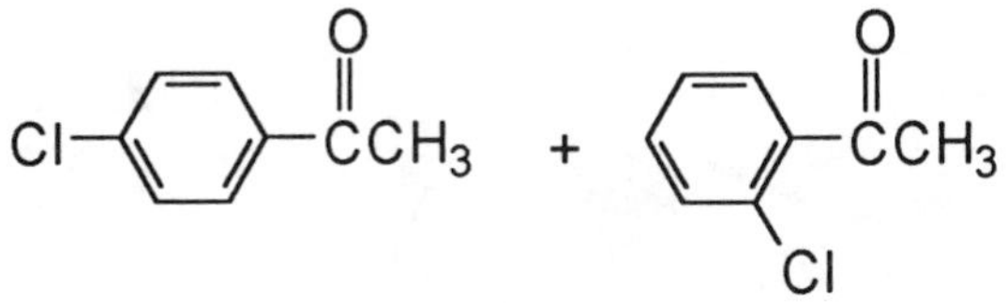 $Cl-C_6H_4-\overset{O}{\overset{\|}{C}}CH_3$ (para) + ortho-chloro isomer, $C_6H_4(Cl)-\overset{O}{\overset{\|}{C}}CH_3$

b) meta-chloro $C_6H_4(Cl)-\overset{O}{\overset{\|}{C}}CH_3$

c) ortho-chloro $C_6H_4(Cl)-CH_2CH_3$ + $Cl-C_6H_4-CH_2CH_3$ (para)

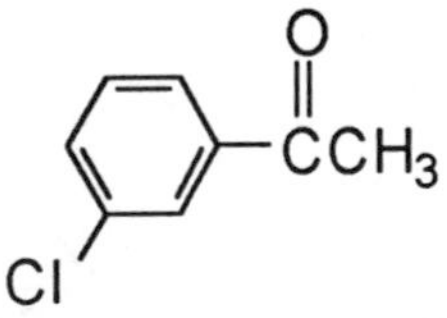

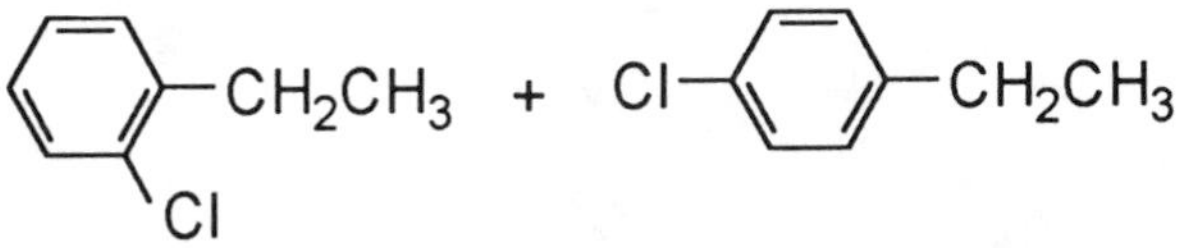

d) 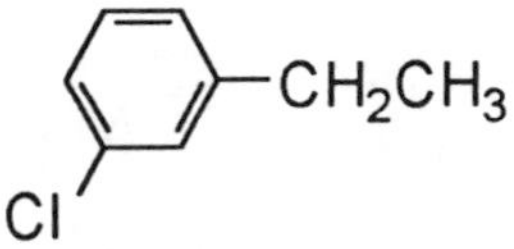 meta-chloro $C_6H_4(Cl)-CH_2CH_3$

619. What is the major product from the following reactions?

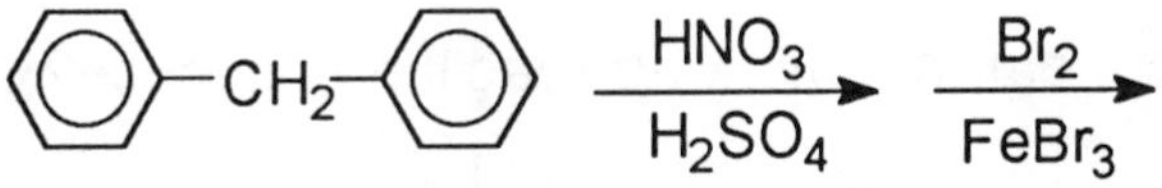

$C_6H_5-CH_2-C_6H_5$ $\xrightarrow[H_2SO_4]{HNO_3}$ $\xrightarrow[FeBr_3]{Br_2}$

a) 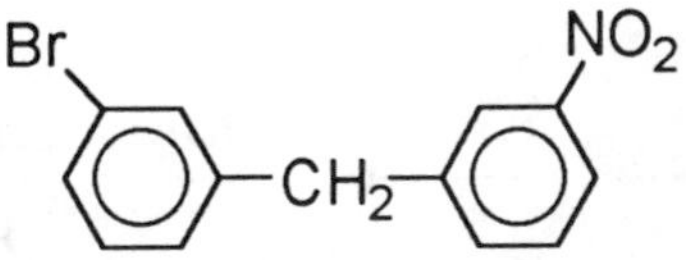 Br (meta) on one ring $-CH_2-$ NO_2 (meta) on the other ring

b) $C_6H_5-CH_2-$ ring bearing Br (ortho to CH_2) and $-NO_2$ (para to CH_2)

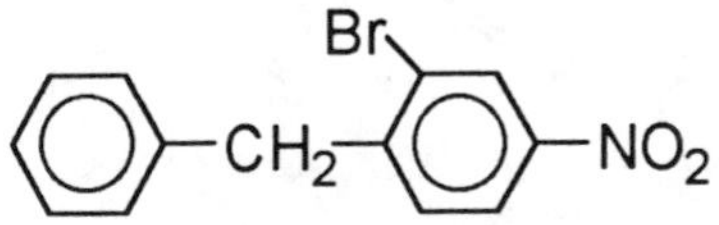

c) $Br-C_6H_4-CH_2-C_6H_4-NO_2$ (both para)

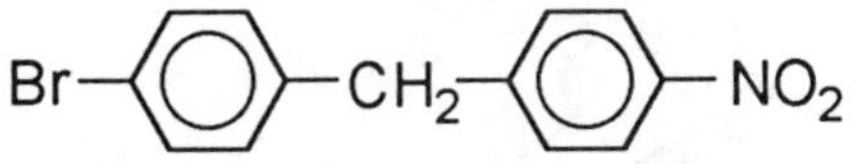

d) 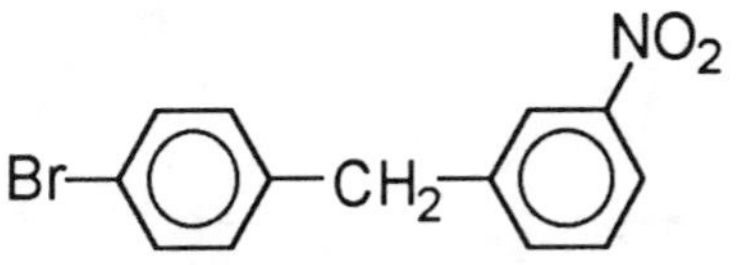 $Br-C_6H_4-CH_2-$ ring with NO_2 (meta)

620. What is the major product from the following reactions?

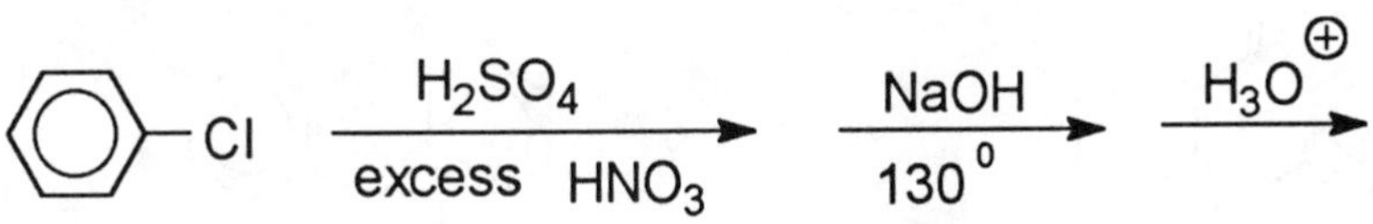

C_6H_5-Cl $\xrightarrow[\text{excess } HNO_3]{H_2SO_4}$ $\xrightarrow[130^0]{NaOH}$ $\xrightarrow{H_3O^{\oplus}}$

a)  OH, Cl (para)

b) 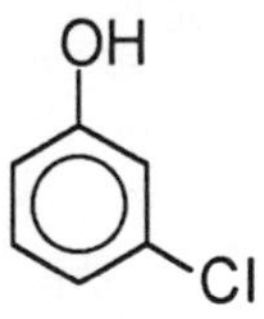 OH, Cl (meta)

c) OH, NO_2 (meta)

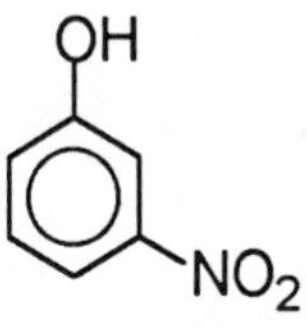

d) 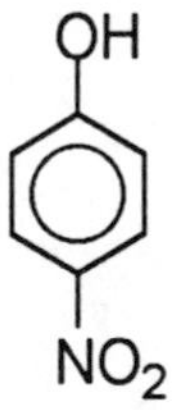 OH, NO_2 (para)

621. What is the major product from the following series of reactions?

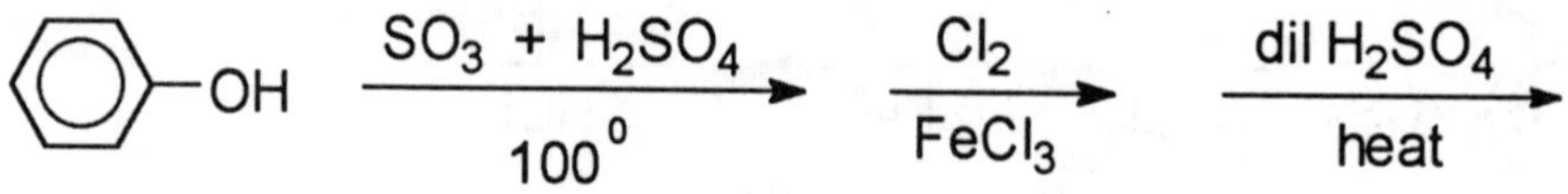

$C_6H_5OH \xrightarrow[100^0]{SO_3 + H_2SO_4} \xrightarrow[FeCl_3]{Cl_2} \xrightarrow[heat]{dil\ H_2SO_4}$

a) HO_3S, OH, Cl (substituted benzene)

b) Cl–C_6H_4–OH

c) Cl, OH, Cl (substituted benzene)

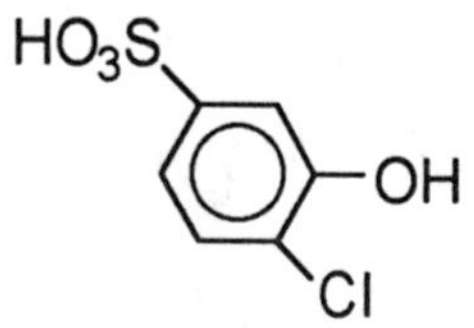

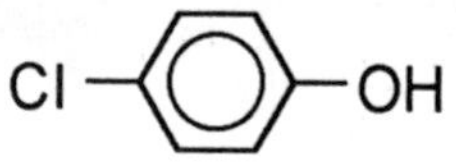

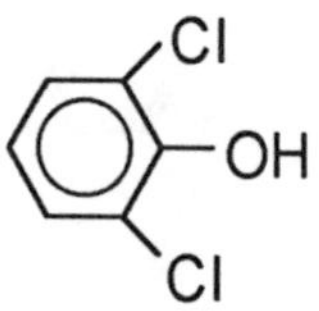

d) 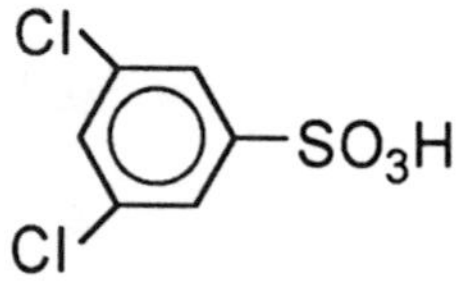 Cl, SO_3H, Cl (substituted benzene)

622. What is the final product from the following reaction sequence?

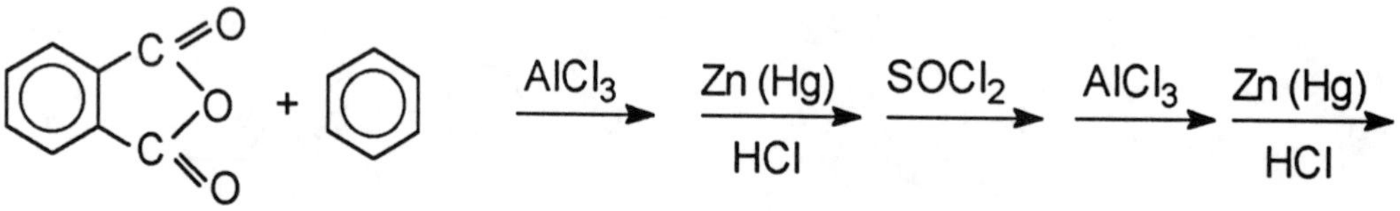

Phthalic anhydride + benzene $\xrightarrow{AlCl_3} \xrightarrow[HCl]{Zn(Hg)} \xrightarrow{SOCl_2} \xrightarrow{AlCl_3} \xrightarrow[HCl]{Zn(Hg)}$

a)

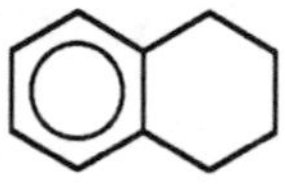

b)

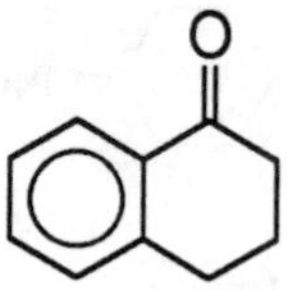

c)

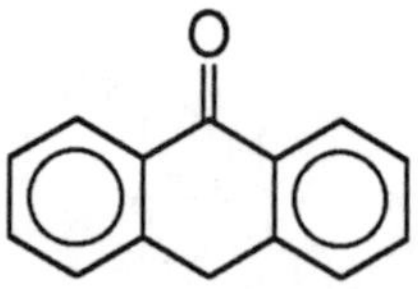

d)

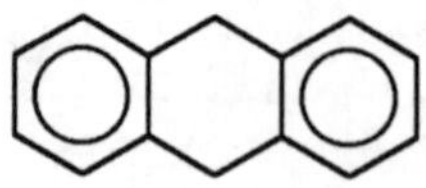

623. What is the major product from the following reaction?

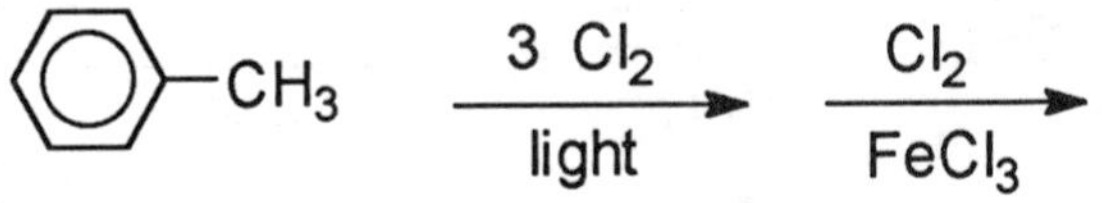

$C_6H_5-CH_3$ $\xrightarrow[\text{light}]{3\ Cl_2}$ $\xrightarrow[FeCl_3]{Cl_2}$

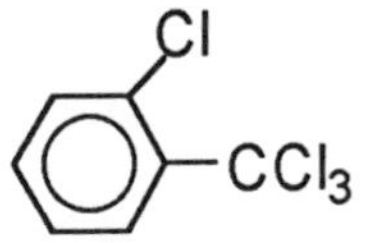

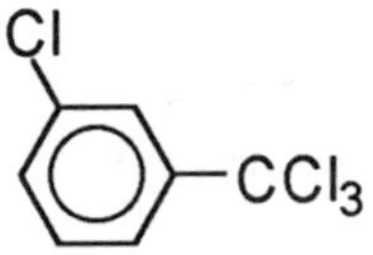

a) 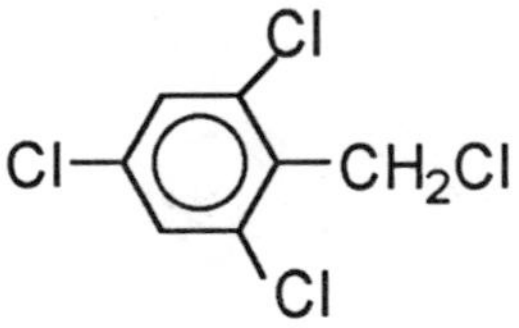 Cl, Cl, Cl, CH_2Cl b) Cl, CCl_3

c) Cl, CCl_3 d) 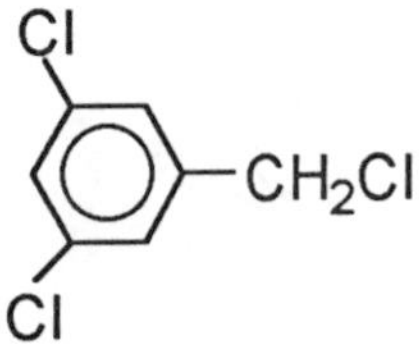 Cl, Cl, CH_2Cl

TIP Unlike a methyl group which is activating and an ortho/para director, the trichloromethyl group with three highly electronegative chlorines is both deactivating and a meta director. Therefore three free radical substitutions of chlorine for hydrogen convert the methyl group to the trichloromethyl group and answer c) is correct. Would answer b) be correct if the order was reversed? Yes.

624. What is the most likely product from the following reaction?

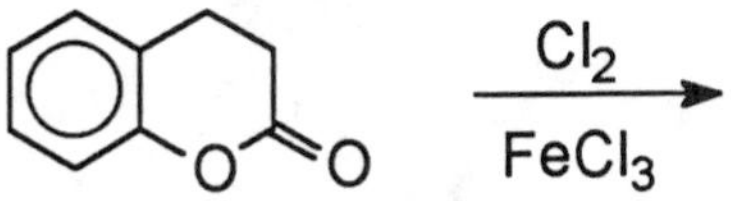

O, O $\xrightarrow[FeCl_3]{Cl_2}$

a) 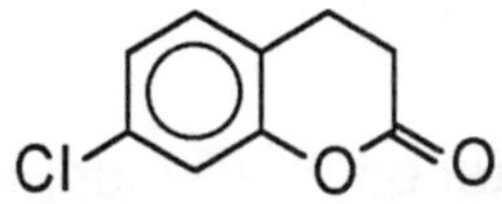 Cl, O, O b) 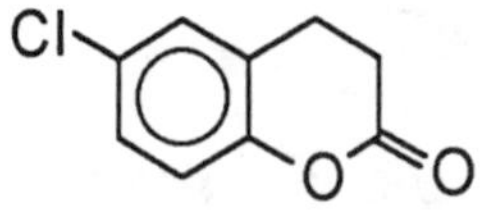 Cl, O, O

c) 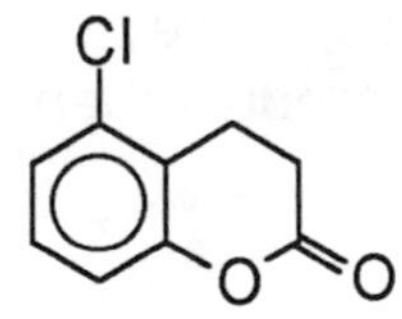 Cl, O, O d) 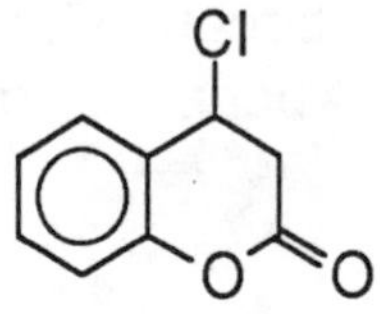 Cl, O, O

625. What is the major product from the following reaction sequence?

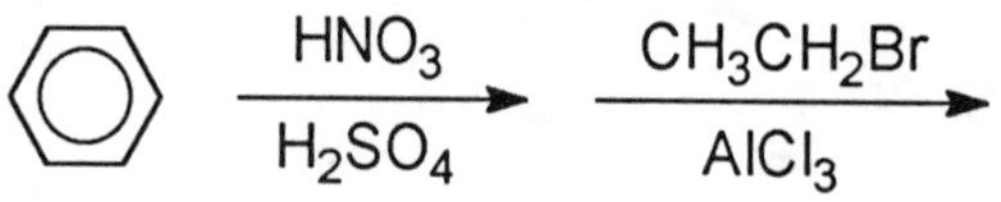

Benzene $\xrightarrow[H_2SO_4]{HNO_3}$ $\xrightarrow[AlCl_3]{CH_3CH_2Br}$

a) 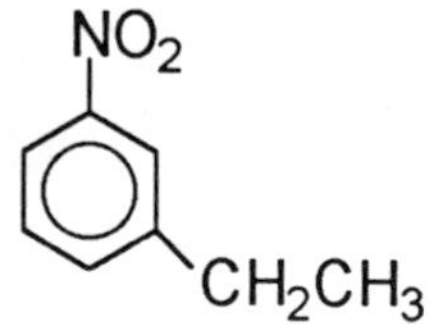 NO_2 / CH_2CH_3 (meta-substituted benzene)

b) 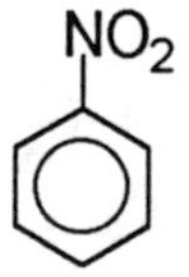 NO_2 (nitrobenzene)

c) 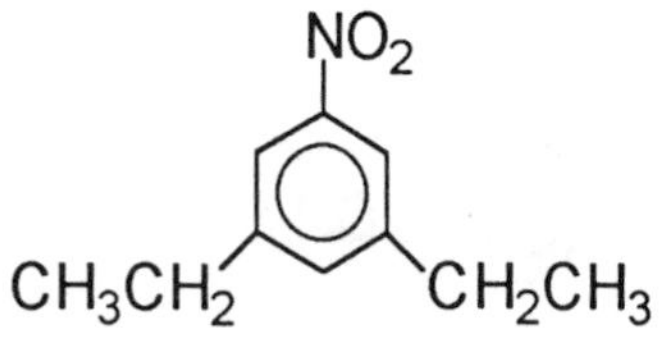 NO_2 / CH_3CH_2 / CH_2CH_3 (1-nitro-3,5-diethylbenzene)

d) 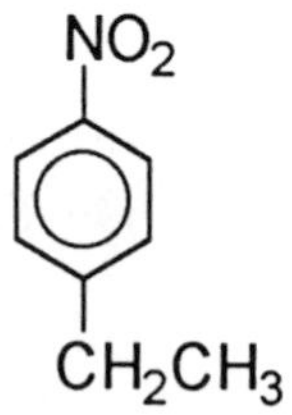 NO_2 / CH_2CH_3 (para-substituted benzene)

626. What is the best procedure for preparing diphenylmethane?

a) 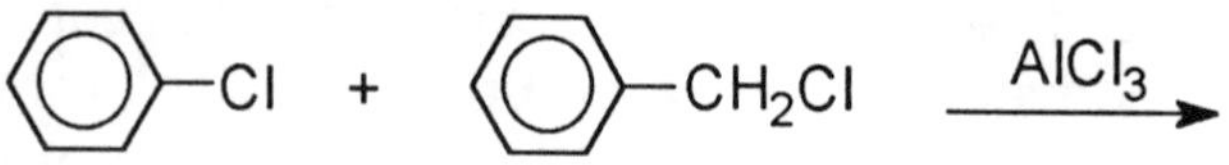 Ph–Cl + Ph–CH_2Cl $\xrightarrow{AlCl_3}$

b) 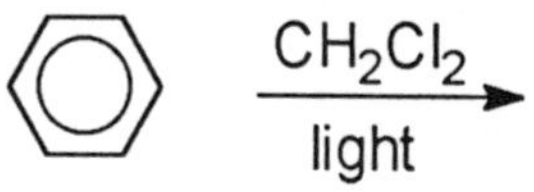 Benzene $\xrightarrow[light]{CH_2Cl_2}$

c) 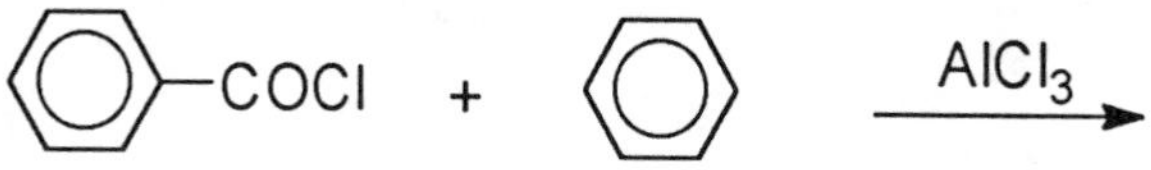 Ph–COCl + Benzene $\xrightarrow{AlCl_3}$

d) 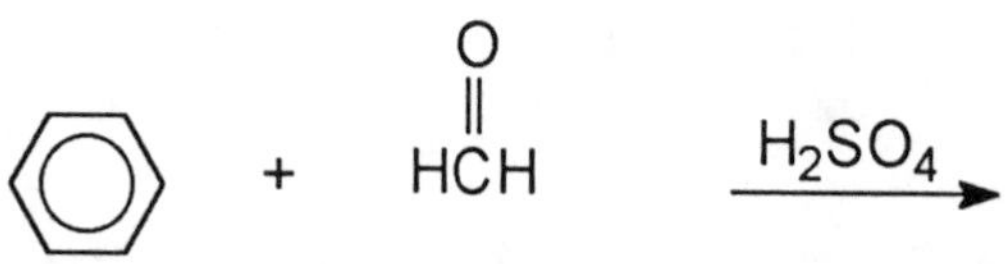 Benzene + HC(=O)H $\xrightarrow{H_2SO_4}$

TIP Answers a, b and c are the most common mistakes that students make in recognizing Friedel-Craft chemistry correctly. In this case the correct answer is harder to see because it comes about in two steps, however, once you recognize that benzyl alcohol is the intermediary product then answer d is logical.

A. NOMENCLATURE and PROPERTIES

627. Which of the following matches of names and structures are correct?

I. acetone	A. $H_2C{=}O$
II. formaldehyde	B. $CH_3CH{=}O$
III. methyl isobutyl ketone	C. $CH_3CH_2\overset{O}{\overset{\|\|}{C}}CH_2CH_3$
IV. acetaldehyde	D. $C_6H_5\overset{O}{\overset{\|\|}{C}}CH_3$
V. acetophenone	E. $CH_3\overset{O}{\overset{\|\|}{C}}CH_2CH(CH_3)_2$

a) I and A, III and E

b) II and A, V and C

c) III and E, V and D

d) I and B, IV and A

628. Which of the following structures have the correct names?

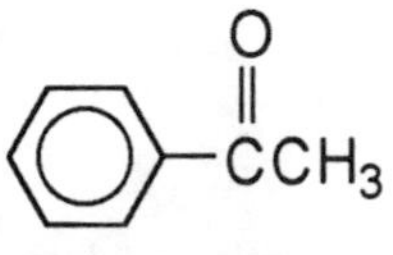

$C_6H_5\overset{O}{\overset{\|\|}{C}}CH_3$	C_6H_5–CHO	$CH_3CHClCH_2\overset{O}{\overset{\|\|}{C}}H$
benzophenone	benzaldehyde	3-chlorobutanal
I	II	III
$CH_3\overset{O}{\overset{\|\|}{C}}CH_3$	CH_3CHO	C_6H_5–CH=CHCHO
acetophenone	acetaldehyde	cinnamaldehyde
IV	V	VI

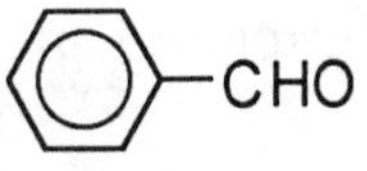

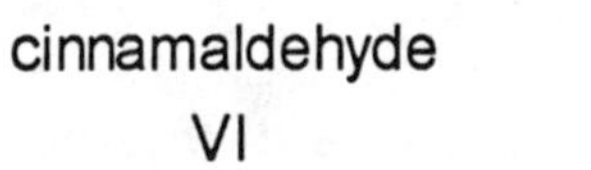

a) I, II, IV, VI b) I, II, V, VI c) II, III, V, VI d) III, IV, V, VI

629. Which of the following IUPAC names are correct for these structures?

I. $CH_3CHClCH_2CHO$ gamma-chloro-butyraldehyde

II. $H_2C{=}O$ methanal

III. $C_6H_5CH_2CHO$ phenylacetaldehyde

IV. $CH_3CH(CH_2CH_3)CH_2C(=O)CH(CH_3)_2$ 2,5-dimethyl-3-heptanone

V. CH_2ClCH_2CHO 2-chloropropanal

a) I, III, IV b) I, II, III c) I, II, IV d) II, III, IV

630. What is the correct IUPAC name for the following structure?

[Structure: cyclohexanone ring (O=) bearing Br, Cl and CHO substituents]

a) 2-bromo-5-chloro-4-formylcyclohexanone
b) 5-bromo-2-chloro-4-oxo-cyclohexanal
c) 5-bromo-2-chloro-3-formylcyclohexanone
d) 1-bromo-3-formyl-4-chlorocyclohexanone

631. Which of the following mixtures can be separated with a bisulfite addition compound?

a) benzophenone and benzyl alcohol b) 2-pentanone and 2-butanone
c) benzaldehyde and acetaldehyde d) pentanal and diethyl ether

TIP Bisulfite is a large nucleophile that reacts with aldehydes and ketones with low steric requirements. Thus both compounds in b) and c) will react, but in a) neither the hindered benzophenone nor the alcohol will react. Diethyl ether will not react but pentanal will, so answer d is correct.

632. Which of the following reagents can be used to distinguish between pentanal and 3-pentanone?

DNPH	NaOH and I_2	$AgNO_3$ in NH_3	$NaHSO_3$
I	II	III	IV

a) II or IV b) II or III c) III or IV d) I or III

633. The Tollens' test will distinguish between which of the following?

I. benzaldehyde and acetophenone
II. formaldehyde and glutaraldehyde
III. propenal and cyclopentenecarbaldehyde
IV. acetone and cinnamaldehyde

a) I and II b) II and III c) I and IV d) III and IV

634. In the following molecule, which of the indicated carbon atoms absorbs furthest upfield in the ^{13}C NMR?

$$CH_3-CH_2-\overset{\overset{\displaystyle O}{\|}}{C}-CH_2-CHO$$

a) CH_3 b) CH_2 (C-2) c) $C=O$ d) CHO

635. In the following molecule, which of the indicated carbon atoms absorbs furthest downfield in the ^{13}C NMR?

$$CH_3-CH_2-\overset{\overset{\displaystyle O}{\|}}{C}-CH_2-CHO$$

a) CH_3 b) CH_2 (C-2) c) $C=O$ d) CHO

636. Which of the following are pairs of resonance structures?

I. $CH_3CH_2CCH_3$ (C=O) and $CH_3\overset{\ominus}{C}HCCH_3$ (C–OH)

II. $CH_2{=}CH\overset{\ominus}{C}HCCH_3$ (C=O) and $CH_2{=}CHCH{=}CCH_3$ (C–$O^{\ominus}$)

III. $CH_2{=}CHCH{=}CCH_3$ (C–$O^{\ominus}$) and $\overset{\ominus}{C}H_2CH{=}CHCCH_3$ (C=O)

IV. $CH_2{=}CHCCH_3$ (C–$O^{\ominus}$) and $\overset{\ominus}{C}H_2CHCCH_3$ (C=O)

a) I, II b) II, III c) III, IV d) II, IV

637. Which of the following are pairs of resonance structures?

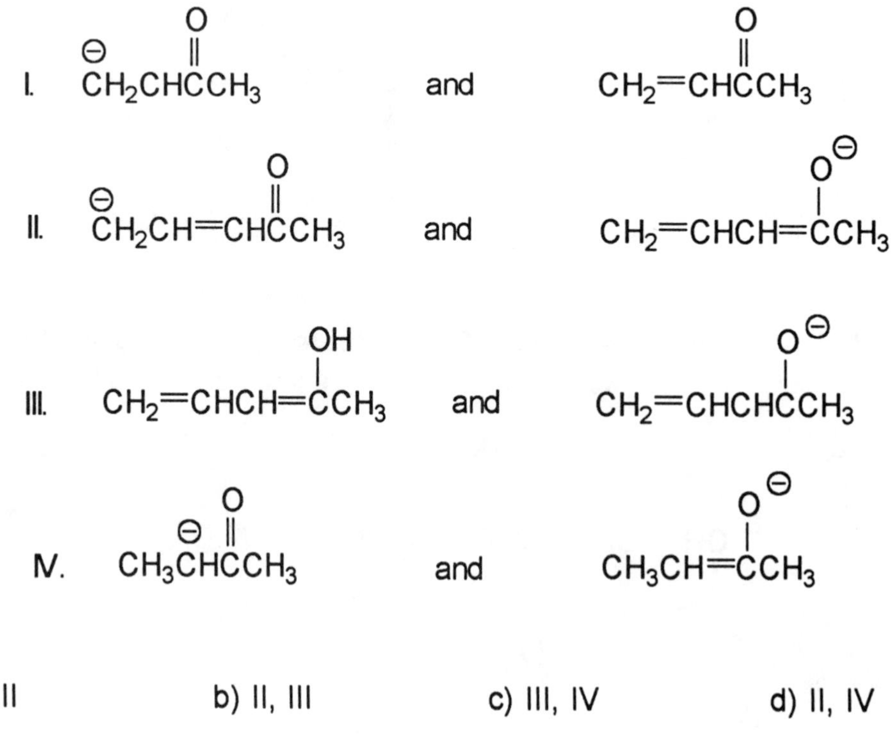

a) I, II b) II, III c) III, IV d) II, IV

B. ENOLS and ENOLATE IONS

638. Arrange the following in the order of increasing acidity (least first).

CH_3CH_2OH	H_2O	$^{\ominus}OH$	$CH_3C(=O)CH_3$
I	II	III	IV

a) III, II, IV, I b) III, IV, I, II c) IV, III, II, I d) IV, I, III, II

639. Arrange the following in the order of increasing acidity (least first).

CH_3CH_2OH	H_2O	$CH_3C(=O)CH_3$	$CH_2{=}C(OH)CH_3$
I	II	III	IV

a) III, I, II, IV b) III, II, IV, I c) II, III, I, IV d) IV, I, III, II

640. Which of the following compounds has the greatest rate of deuterium exchange when treated with D_2O and ^{-}OD?

a) 2,2-dimethylcyclohexane-1,3-dione (H_3C, CH_3) b) cyclohexane-1,3-dione c) cyclohexane-1,4-dione d) cyclohexane-1,2-dione

TIP For questions 638-640. Proton removal and anion stability are key to acidity. Clearly removal of a proton from the hydroxide ion is far more difficult than from neutral compounds. Furthermore, oxygen acids are normally more acidic than carbon acids (oxygen is more electronegative than carbon). For oxygen acids, the electron donating effect of ethyl makes ethanol a weaker acid than water. Similarly the inductive and resonance effects of the vinyl group make IV in question 639 a stronger acid than water. The carbon acids in these problems are considerably more acidic than typical carbon acids because of the resonance and inductive effects of the adjacent carbonyl. In question 640, only the anion derived of proton loss in b) is resonance stabilized by two carbonyls and is therefore the most stable.

641. Which of the following compounds has the greatest amount in the enol form?

a) cyclobutenone b) cyclopentenone c) cyclohexa-2,4-dienone d) cyclohexanone

642. Which of the following are present in the equilibrium mixture of acetone and water at pH 3?

I: CH_3COCH_3 II: $^{\ominus}CH_2COCH_3$ (enolate) III: $CH_3C(OCH_3)_2CH_3$ IV: $CH_3C(OCH_3)(OH)CH_3$

V: $CH_3C(=\overset{\oplus}{O}H)CH_3 \longleftrightarrow CH_3\overset{\oplus}{C}(OH)CH_3$

a) I, IV b) II, III c) I, V d) IV, V

643. Which of the following are present in the equilibrium mixture of acetone and water at pH 9?

I: CH_3COCH_3 II: $^{\ominus}CH_2COCH_3$ (enolate) III: $CH_3C(OCH_3)_2CH_3$ IV: $CH_3C(OCH_3)(OH)CH_3$

V: $CH_3C(=\overset{\oplus}{O}H)CH_3 \longleftrightarrow CH_3\overset{\oplus}{C}(OH)CH_3$

a) I, II b) II, III c) I, V d) IV, V

644. Which of the following characterizes the reaction that forms the new C-C bond in the base-catalyzed aldol condensation of acetaldehyde?

a) $\overset{\ominus}{C}H_2\overset{O}{\overset{\|}{C}}H$ + $CH_3\overset{O}{\overset{\|}{C}}H$ b) $\overset{\ominus}{C}H_2\overset{O}{\overset{\|}{C}}H$ + $CH_3\overset{\oplus OH}{\overset{\|}{C}}H$

c) $CH_2{=}\overset{OH}{\overset{|}{C}}H$ + $CH_3\overset{\oplus OH}{\overset{\|}{C}}H$ d) $CH_3{=}\overset{OH}{\overset{|}{C}}H$ + $CH_3\overset{O}{\overset{\|}{C}}H$

645. Which of the following characterizes the reaction that forms the new C-C bond in the acid-catalyzed aldol condensation of acetaldehyde?

a) $\overset{\ominus}{C}H_2\overset{O}{\overset{\|}{C}}H$ + $CH_3\overset{O}{\overset{\|}{C}}H$ b) $\overset{\ominus}{C}H_2\overset{O}{\overset{\|}{C}}H$ + $CH_3\overset{\oplus OH}{\overset{\|}{C}}H$

c) $CH_2{=}\overset{OH}{\overset{|}{C}}H$ + $CH_3\overset{\oplus OH}{\overset{\|}{C}}H$ d) $CH_3{=}\overset{OH}{\overset{|}{C}}H$ + $CH_3\overset{O}{\overset{\|}{C}}H$

TIP For questions 644 and 645. The aldol reaction is both acid and base catalyzed and always involves a nucleophilic attack on a carbonyl carbon. In base, the nucleophile is the enolate anion which is powerful enough to add to the neutral carbonyl, which is answer a). Is the protonated carbonyl in b) likely to be present in base?

In acid, the nucleophile is the enol which is neutral and weaker. Reaction requires proton activation of the carbonyl as in answer c). Is the enolate anion likely to be present in an acidic solution?

C. ACETALS and KETALS

646. Which of the following are present in the equilibrium mixture of acetone and methanol at pH 3?

I: $CH_3C(OCH_3)(OH)CH_3$ II: $CH_3C(OCH_3)_2CH_3$ III: $^{\ominus}CH_2C(=O)CH_3 \leftrightarrow CH_2=C(O^{\ominus})CH_3$ IV: $CH_3C(=\overset{\oplus}{O}H)CH_3 \longleftrightarrow CH_3\overset{\oplus}{C}(OH)CH_3$

a) I, II, III b) I, II, IV c) I, II d) II, III, IV

647. Which of the following are present in the equilibrium mixture of acetone and methanol at pH 9?

I: $CH_3C(OCH_3)(OH)CH_3$ II: $CH_3C(OCH_3)_2CH_3$ III: $^{\ominus}CH_2C(=O)CH_3 \leftrightarrow CH_2=C(O^{\ominus})CH_3$ IV: $CH_3C(=\overset{\oplus}{O}H)CH_3 \longleftrightarrow CH_3\overset{\oplus}{C}(OH)CH_3$

a) I, II b) I, III c) II, IV d) III, IV

648. Which of the following compounds will not react with hydroxide ion?

I: $CH_3C(OCH_3)(OH)CH_3$ II: $CH_3C(OCH_3)_2CH_3$ III: $CH_3C(CH_3)_2OCH_3$ IV: $CH_3CH{=}CHOCH_3$

V: 1,3,5-trioxane (six-membered ring, O–CH₂–O–CH₂–O–CH₂) VI: $CH_3C(CH_3){=}NCH_3$ VII: cyclohexane ring bearing CN and OH on the same carbon

a) I, III, IV, V b) II, III, IV, V c) III, IV, V, VI d) II, IV, VI, VII

649. Which of the following react with aqueous acid to give an aldehyde?

a) 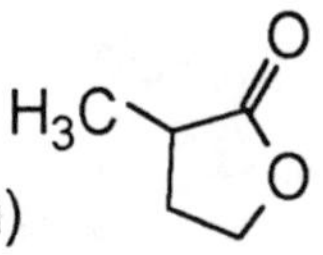 H_3C, O, O b) 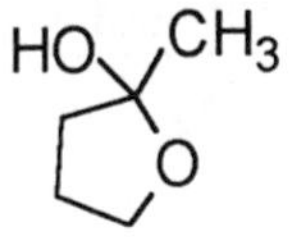 HO, CH_3, O c) 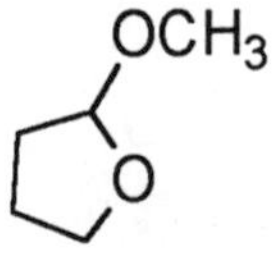 OCH_3, O d) 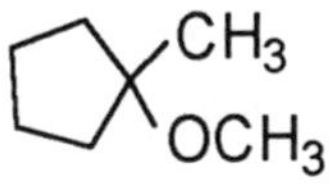 CH_3, OCH_3

650. Which of the following react with base to give a ketone?

a) 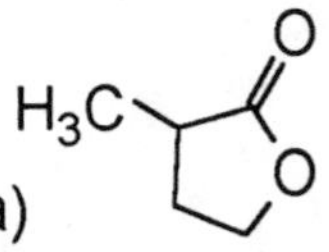 H_3C, O, O b) 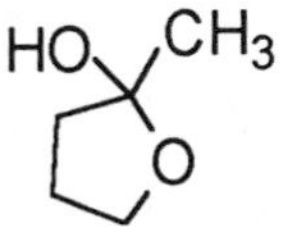 HO, CH_3, O c) 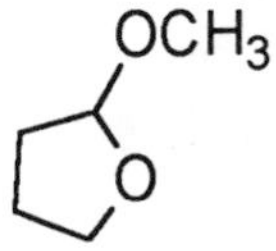 OCH_3, O d) 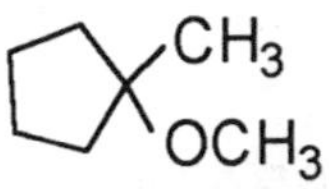 CH_3, OCH_3

651. What is the product of the reaction of acetophenone with methanol in an acidic solution?

a) 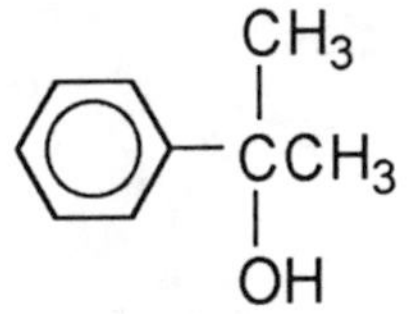 CH_3, CCH_3, OH b) 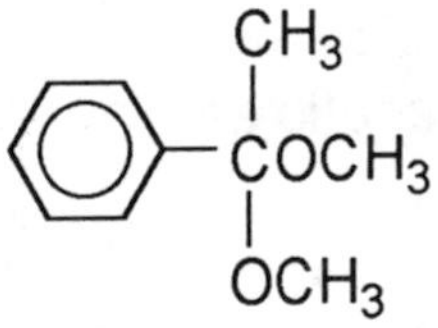 CH_3, $COCH_3$, OCH_3

c) 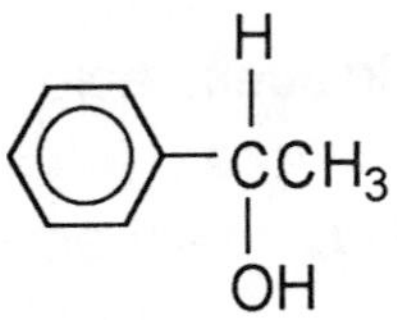 H, CCH_3, OH d) 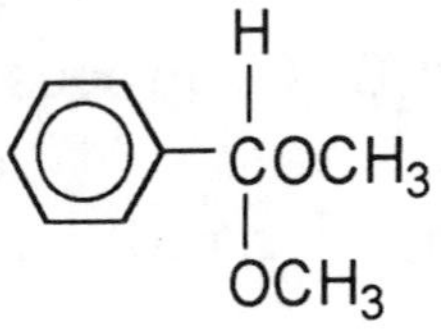 H, $COCH_3$, OCH_3

TIP For questions 646-651. Compounds that contain carbon singly bonded to two oxygens are both important and often confusing. If the carbon contains a hydrogen then they are acetal derivatives, and if the carbon has no hydrogen they are ketal derivatives. Hydrolysis in acid gives the carbonyl compound and the relevant alcohols. In 649 and 650, compounds a) and d) can be eliminated since they do not fulfill the first requirement (carbon singly bonded to two oxygen atoms). Inspection of the structures b) and c) in both problems then focuses directly on whether the carbon has a hydrogen or not. In c) the hydrogen is hard to "see" since it often left out in structures. For 651 we will be looking for a ketal since acetophenone is a ketone.

652. What is the product from the following reduction?

CH_3 / OCH_3 (cyclic structure) $\xrightarrow[H_2O]{\text{dil HCl}}$

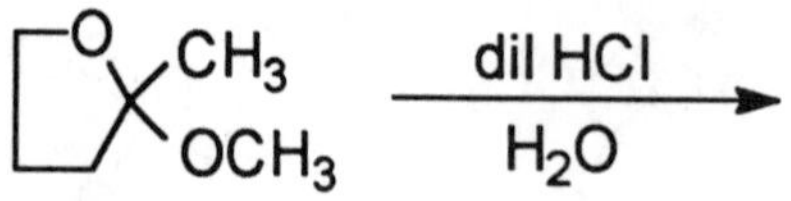

a) $CH_3OCH_2CH_2CH_2CHO$

b) $HOCH_2CH_2CH_2COCH_3$

c) $HOCH_2CH_2CH_2CHO$

d) $CH_3OCH_2CH_2CH_2CH_2OH$

D. PREPARATIONS

653. Friedel-Crafts acylation is a convenient method for preparing which of the following compounds?

a) aromatic ketones
b) acid chlorides
c) acetoacetic esters
d) acyloins

654. Which of the following reaction conditions is best for preparing hexanal from hexanol?

a) $Na_2Cr_2O_7$ in aqueous H_2SO_4

b) Zn in acetic acid

c) CrO_3 and pyridine in chloroform

d) $HgSO_4$ in aqueous H_2SO_4

655. What are the best conditions for preparation of hexanal by hydration of 1-hexyne?

a) $Na_2Cr_2O_7$ in aqueous H_2SO_4

b) $(sia)_2BH$ then H_2O_2 and NaOH

c) CrO_3 and pyridine in chloroform

d) $HgSO_4$ in aqueous H_2SO_4

656. What are the best reducing conditions for preparing hexanal from hexanoyl chloride?

a) NaH
b) $LiAlH_4$
c) $LiAlH(OMe_3)_3$
d) $NaBH_4$

657. Which of the following reagents yield carbonyl compounds when reacted with CH_3COCl?

C_6H_5–Li	$LiAlH_4$	$(CH_3)_2Cd$	$LiAlH(OCMe_3)_3$
I	II	III	IV

a) I, III b) II, IV c) III, IV d) I, II

658. Which reagents can be used for the following conversion?

C_6H_{11}–C(=O)Cl ⟶ C_6H_{11}–C(=O)CH_3

Zn / acetic acid	CH_3Cl / $AlCl_3$	CH_3Li	CH_3MgBr
I	II	III	IV

$LiAlH(OCMe_3)_3$	$(CH_3)_3CuLi$	$(CH_3)_2Cd$
V	VI	VII

a) I, II, III b) VI, VII c) V, VII d) IV, VI, VII

TIP For questions 656-658. Selectivity (reactions that proceed to only a certain product or step) is the key to these three problems. Since aldehyde and ketone carbonyls are reactive with nucleophiles, reactions that generate aldehydes or ketones from nucleophilic reagents demand "selectively tuned" nucleophiles. In 656 the tuning comes from the sterically hindered nucleophile in c). In 657, the same principal applies in IV, and for III, the selectivity comes from the less nucleophilic cadmium reagent. Finally, in 658 the cadmium and copper organometallics both provide the required tuning by being less nucleophilic.

659. Which of the following combination of compounds will give cinnamaldehyde ($C_6H_5CH = CHCHO$) by a condensation reaction?

a) benzaldehyde and acetaldehyde b) formaldehyde and phenyl acetaldehyde
c) formaldehyde and acetophenone d) benzaldehyde and acetone

660. Which of the following combinations of starting materials and organometallic reagents are best for synthesizing 2-butanone?

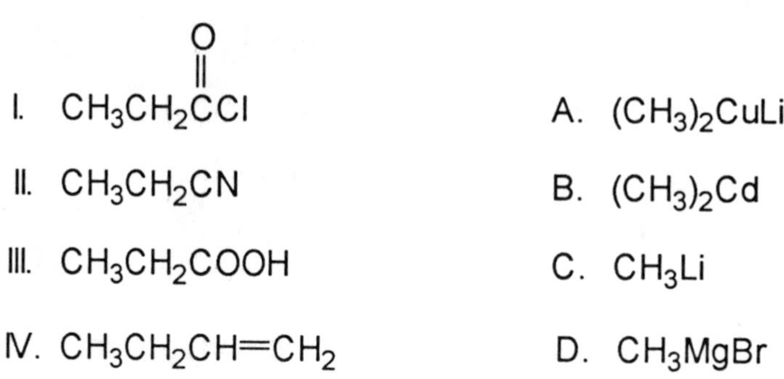

a) I and C, II and A
b) III and D, IV and A
c) III and C, II and D
d) I and B, II and D

661. Which of the following reactions is not a good method for preparing acetophenone?

a) C_6H_5–COOH $\xrightarrow{SOCl_2}$ $\xrightarrow{(CH_3)_2CuLi}$

b) C_6H_5–COCl $\xrightarrow[AlCl_3]{CH_3Cl}$

c) C_6H_5–MgBr $\xrightarrow{CdCl_2}$ $\xrightarrow{CH_3COCl}$

d) C_6H_5–CH=CH$_2$ $\xrightarrow[\text{dil } H_2SO_4]{H_2O}$ $\xrightarrow{H_2CrO_4}$

662. Which of the following reactions can be used to prepare acetophenone?

I. benzene and aluminum chloride and propionyl chloride
II. benzoyl chloride and diethyl copper lithium
III. benzene and aluminum chloride and acetic anhydride
IV. diphenyl cadium and acetyl chloride

a) I, II b) III, IV c) I, III d) II, IV

663. Hexanal can be prepared by ozonolysis and reductive hydrolysis by starting with which of the following compounds?

a) 1-hexene
b) 2-methyl-1-hexene
c) 2-methyl-1-heptene
d) 1-heptene

664. Which of the following reactions is the best method for preparation of benzaldehyde?

a) $C_6H_5-CH_3 \xrightarrow{KMnO_4} \xrightarrow{SOCl_2} \xrightarrow{(CH_3)_2Cd}$

b) $C_6H_5-CH_2OH \xrightarrow{SOCl_2} \xrightarrow{LiAlH[OC(CH_3)_3]_3}$

c) $C_6H_5-Br \xrightarrow[ether]{Mg} \xrightarrow{CO_2} \xrightarrow{H_3O^{\oplus}}$

d) $C_6H_5-CH_2Cl \xrightarrow{NaOH} \xrightarrow[pyridine]{CrO_3}$

665. What two different combinations of starting materials and reagents can be used to prepare benzaldehyde?

I. benzyl alcohol	A. $LiAlH[OC(CH_3)_3]$
II. benzoyl chloride	B. CrO_3 and pyridine
III. styrene	C. B_2H_6 then H_2O_2 and NaOH
IV. phenyl acetylene	D. Sia_2BH then H_2O_2 and NaOH

a) II and A, IV and D
b) III and D, IV and B
c) I and B, II and A
d) I and B, III and C

TIP As in questions 656-658, selectivity is the name of the game with aldehydes. Chromic acid with pyridine oxidizes alcohols to aldehydes, thus answers with I and B (c and d) are correct, however, if you count carbon atoms in styrene you will discover that d) cannot be correct!

666. What is the best method for preparing the following compound?

$C_6H_{11}-C(=O)C(CH_3)_3$

a) sodium dichromate oxidation of $C_6H_{11}-CH(OH)C(CH_3)_3$

b) ozonolysis and hydrolysis of $C_6H_{11}-CH{=}C(CH_3)_2$

c) hydration of $C_6H_{11}-C{\equiv}CCH_3$

d) sulfuric acid oxidation of $C_6H_{11}-C(OH)(CH_3)-C(OH)(CH_3)CH_3$

E. REACTIONS: GRIGNARD

667. Which of the following halides can be used to make Grignard reagents?

$CH_3CH_2CH_2I$	C_6H_5-Br	$CH_2{=}CHCl$	$(CH_3)_2CHBr$	$(CH_3)_3CBr$
I	II	III	IV	V

a) I, IV, V b) I, II, IV, V c) II, IV, V d) all

668. What is the major product from the following reaction?

$$C_6H_5C(=O)CH_3 + CH_3MgBr \xrightarrow{\text{ether}} \xrightarrow{H_3O^{\oplus}}$$

a) $C_6H_5C(OH)(CH_3)CH_3$

b) $C_6H_5CH_2CH(OH)CH_3$

c) $C_6H_5CH(CH_3)CH_2OH$

d) $C_6H_5CH(OH)CH_2CH_3$

669. What is the best procedure for preparing pure o-deuteriotoluene from toluene?

a) $\xrightarrow[FeBr_3]{Br_2} \xrightarrow[\text{ether}]{Mg} \xrightarrow{D_2O}$

b) $\xrightarrow[H_2SO_4]{SO_3} \xrightarrow[FeBr_3]{Br_2} \xrightarrow[\text{ether}]{Mg} \xrightarrow{D_2O} \xrightarrow[\text{heat}]{\text{dil } H_2SO_4}$

c) $\xrightarrow[H_2SO_4]{SO_3} \xrightarrow[FeBr_3]{Br_2} \xrightarrow[\text{heat}]{\text{dil } H_2SO_4} \xrightarrow[\text{ether}]{Mg} \xrightarrow{D_2O}$

d) $\xrightarrow[FeBr_3]{Br_2} \xrightarrow[\text{heat}]{\text{dil } H_2SO_4}$

TIP This is an example of the use of the Grignard reaction to prepare isotopically tagged molecules and the use of a blocking group (SO_3H) to direct a synthesis. In this case, the direction is to the ortho position. When SO_3H is placed in the para position, the next reaction, the bromination, is directed to the ortho position. Importantly the blocking group can be removed since the sulfonation is reversible with dilute sulfuric acid. Then the Grignard reagent can be made and reacted with D_20. Answer a) gives a mixture of ortho and para. Answer b) violates the Grignard sensitivity to acidic hydrogens. Answer d) gives another product.

670. A Grignard reagent is prepared by reacting cyclopentanol with first thionyl chloride and then magnesium in ether. The Grignard reagent is then reacted with acetaldehyde and the reaction mixture is acidified. What is the most likely final product?

a) cyclopentyl–CH(OH)CH$_3$ b) 1-hydroxy-1-(CH_2CH_3)cyclopentane (OH, CH_2CH_3)

c) cyclopentyl–CH_2CH_2OH d) cyclopentyl–CH_2OCH_3

671. What is the product from the following series of reactions?

4-Br-C_6H_4-CHO $\xrightarrow[CH_3CH_2OCH_2CH_3]{CH_3CH_2MgBr}$ $\xrightarrow{H_3O^{\oplus}}$ $\xrightarrow[H_3O^{\oplus}]{K_2Cr_2O_7}$ $\xrightarrow[\text{conc HCl}]{Zn(Hg)}$

a) Br–C_6H_4–$CH_2CHOHCH_3$ b) Br–C_6H_4–CH_2CH_3

c) Br–C_6H_4–$CHOHCH_2CH_3$ d) Br–C_6H_4–$CH_2CH_2CH_3$

672. What is the product from the reaction of ethyl Grignard and propiophenone?

a) $C_6H_5-C(CH_2CH_3)(OH)CH_2CH_3$ (phenyl–C bearing CH_2CH_3 above, CCH_2CH_3, OH below)

b) $C_6H_5-C(CH_2CH_3)(OH)CH_3$ (phenyl–C bearing CH_2CH_3 above, CCH_3, OH below)

c) $C_6H_5-C(CH_3)(OH)CH_2CH_3$ (phenyl–C bearing CH_3 above, CCH_2CH_3, OH below)

d) $C_6H_5-C(CH_3)(OH)CH_3$ (phenyl–C bearing CH_3 above, CCH_3, OH below)

673. Compound A has the molecular formula, C_4H_{10}. It reacts with chlorine and light to give Compound B, C_4H_9Cl. This compound forms a solution with magnesium in dry ether and is then reacted with D_2O to yield a gaseous hydrocarbon having a proton NMR spectrum consisting of a single peak. What is the most likely structure for Compound A?

a) butane b) cyclobutane c) isobutane d) isobutylene

TIP Only answers a) and c) are consistent with the molecular formula. If butane is correct, then Compound B is likely 2-chlorobutane (or less likely 1-chlorobutane). A Grignard from either of those would be quenched with deuterium oxide to give 2-deuteriobutane or 1-deuteriobutane. Importantly each of these would have multiple signals in the proton nmr. With a single peak for the final product what must Compound B be? How many peaks must be present in the proton NMR of Compound B?

F. REACTIONS: ALDOL

674. Which of the following compounds can be used in an aldol-type condensation to make a product with 9-carbon atoms?

I. acetaldehyde II. acetone III. benzaldehyde IV acetophenone

a) I, II b) III, IV c) I, III d) II, IV

675. Which of the following compounds can be used in an aldol-type condensation to make a product with 5-carbon atoms?

I. acetaldehyde II. acetone III. benzaldehyde IV acetophenone

a) I, II b) III, IV c) I, III d) II, IV

676. Which of the following compounds can be used in an aldol-type condensation to make a product with the highest possible molecular weight?

I. acetaldehyde II. acetone III. benzaldehyde IV acetophenone

a) I, II b) III, IV c) I, III d) II, IV

677. What is the most likely product from the reaction of 1,6-cyclodecanedione with hot aqueous NaOH?

a) 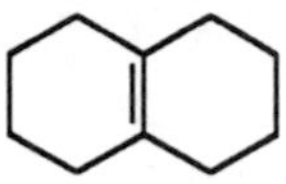 b) 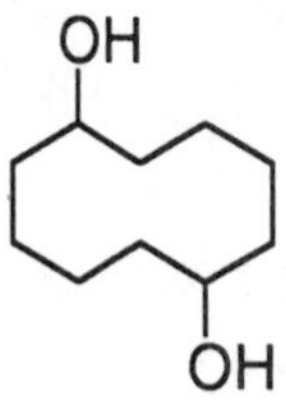 OH OH c) 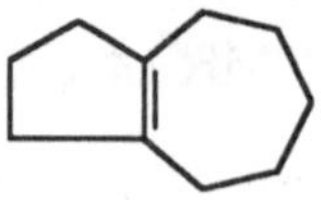 d) 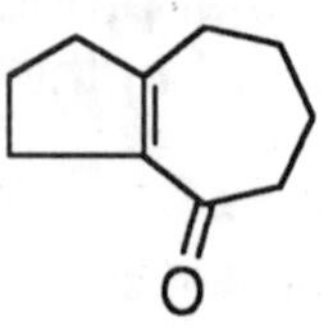 O

678. What is the product from an intramolecular condensation of 2,8-nonadione?

a) 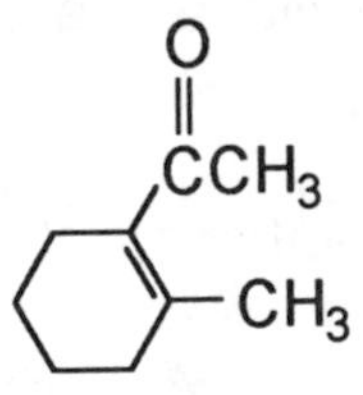 [structure] CCH_3 (C=O), CH_3 on cyclohexene ring

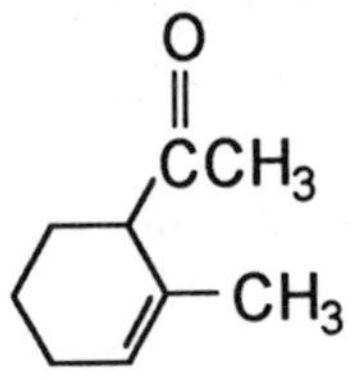

b) [structure] CCH_3 (C=O), CH_3 on cyclohexene ring

c) [structure] CH_3, =O on cyclooctene ring

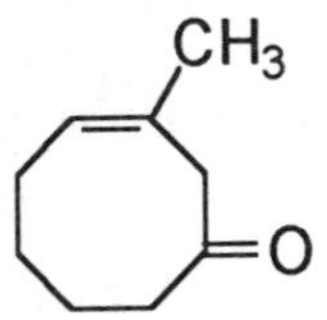

d) 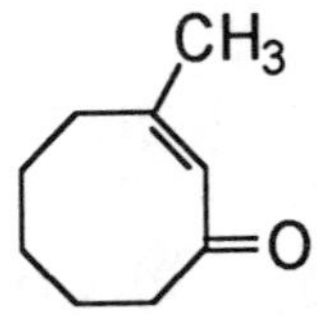 [structure] CH_3, =O on cyclooctene ring

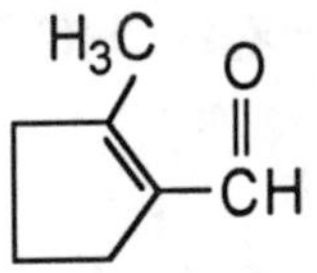

679. What is the major product from the following reaction?

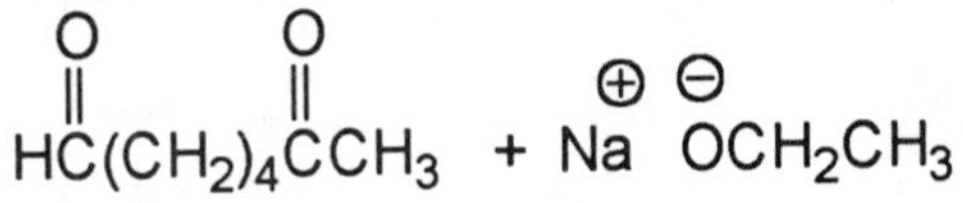

$$HC(=O)(CH_2)_4C(=O)CH_3 + Na^{\oplus}\ {}^{\ominus}OCH_2CH_3$$

a) 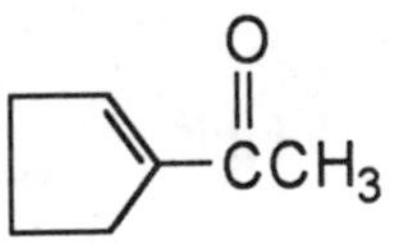 [structure] cyclopentene–$C(=O)CH_3$

b) [structure] cyclopentene with H_3C and $C(=O)H$

c) [structure] cycloheptenone (=O)

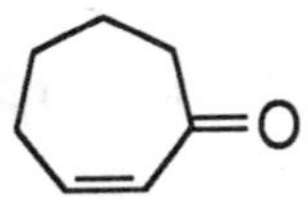

d) 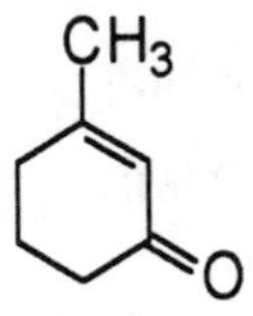 [structure] cyclohexenone with CH_3

TIP This is an intramolecular aldol reaction and two considerations are important. First, aldehyde carbonyls are more reactive to nucleophilic attack than ketone carbonyls. Second, the most favored ring sizes are five and six. Answer c) is ruled out because forming the seven membered ring is not favorable when other paths are available. Answer d) is not a possible aldol product. Answer b) violates the first consideration, and utilizes the carbonyl of the ketone not the aldehyde.

G. REACTIONS: NAME

680. What is the product from a Wolff-Kishner reaction with acetaldehyde?

a) CH_3CH_3 b) $CH_3CH_2CH_3$ c) $CH_2{=}CH_2$ d) CH_3CH_2OH

681. What is the product from a Baeyer-Villiger oxidation of acetone?

a) $CH_3\overset{O}{\overset{\|}{C}}OCH_3$ b) $CH_3\overset{O}{\overset{\|}{C}}OOH$ c) $CH_3\overset{O}{\overset{\|}{C}}OH$ d) $CH_3\overset{O}{\overset{\|}{C}}OCH_2CH_3$

682. Which of the following is a possible product from a Wittig synthesis?

a) $CH_2{=}CH\overset{NH_2}{\overset{|}{C}}HCOOH$

b) 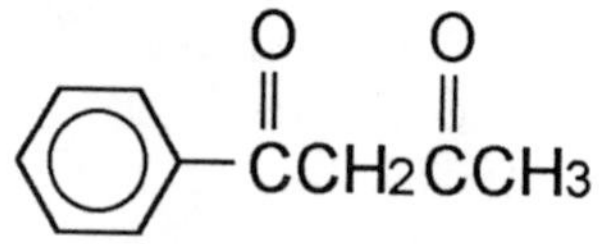 $C_6H_5-\overset{O}{\overset{\|}{C}}CH_2\overset{O}{\overset{\|}{C}}CH_3$

c) $CH_3\overset{O}{\overset{\|}{C}}CH_2CN$

d) $CH_3\overset{NH_2}{\overset{|}{C}}HCH_3$

683. Which of the following is a major product from the Clemmensen reaction of benzaldehyde?

a) 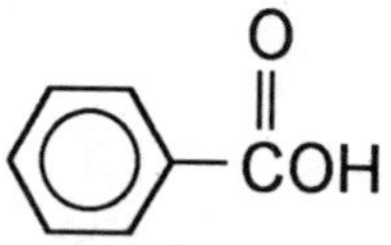 $C_6H_5-\overset{O}{\overset{\|}{C}}OH$

b) $C_6H_5-CH_2OH$

c) $C_6H_5-CH_3$

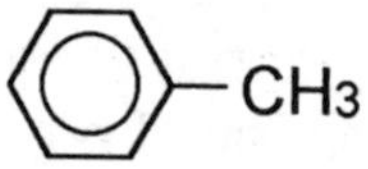

d) 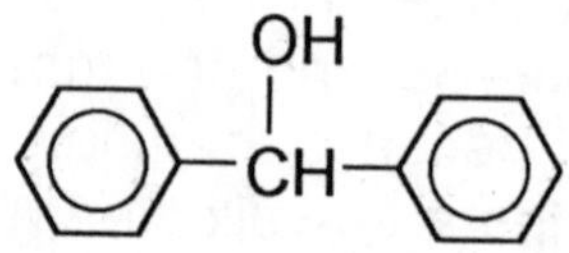 $C_6H_5-\overset{OH}{\overset{|}{C}}H-C_6H_5$

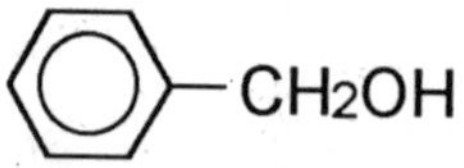

684. In which of the following reactions are enols or enolate ions not involved as reaction intermediates?

a) the Aldol condensation
b) the iodoform reaction
c) the Claisen reaction
d) the Clemmensen reaction

685. Which of the following reactions does not result in the formation of a new C-H bond?

a) Raney nickel desulfurization
b) the Wolff-Kishner reaction
c) the Williamson synthesis
d) the Clemmensen reduction

686. Which of the following reactions does not result in the formation of a new C-C bond?

a) the Kolbe reaction
b) the Grignard reaction
c) the Reformatsky reaction
d) the Clemmensen reaction

687. Which of the following reactions yield a product that contains a new C-H bond?

I. Clemmensen
II. Reformatsky
III. Wolff-Kishner
IV. Baeyer-Villiger
V. Claisen
VI. Wittig

a) I, III
b) IV, V
c) III, V
d) II, VI

688. Which of the following reactions yield a product that contains a new C-C bond?

I. Clemmensen
II. Reformatsky
III. Wolff-Kishner
IV. Bayer-Villiger
V. Claisen
VI. Wittig

a) I, II, III
b) IV, V, VI
c) I, IV, VI
d) II, V, VI

TIP For questions 684-688. Name reactions are alternative ways of mentally storing important steps. In a sense this is akin to mnemonics. Grouping reactions by the fundamental conversion processes is a different way to categorize. We have rather few methods of creating certain kinds of bonds (C-H and C-C) and a clear understanding of these methods gives new power to solving synthetic problems.

689. Which of the name reactions is best for the synthesis of the following compound?

$=CH_2$

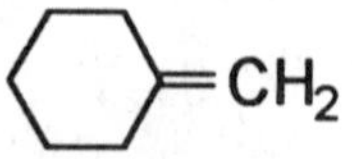

a) Wittig b) Wolff-Kishner c) Clemmensen d) Grignard

690. Which of the name reactions is best for the synthesis of the following compound?

$$CH_3CH(OH)C(CH_3)_2COOCH_2CH_3$$

a) Reformatsky
b) Baeyer-Villiger
c) aldol condensation
d) Wittig

691. Which of the name reactions is best for the synthesis of the following compound?

$$HOH_2C-C(CH_2OH)_2-CHO$$

a) aldol Condensation
b) Grignard / Clemmensen
c) Wittig
d) Baeyer-Villiger

692. Which of the name reactions is best for the synthesis of the following compound?

$=O$

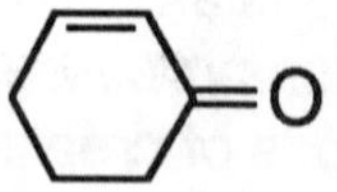

a) Reformatsky b) aldol condensation c) Grignard d) Friedel-Crafts

TIP For questions 689-692. Again, we want to use information that has been organized by the name reactions in order to create new compounds. In question 689, the critical step is the formation of a carbon-carbon double bond. Reaching back into our memories, we note that only the Wittig reaction does this. That same reasoning rules out the Wittig reaction in questions 690 and 691. The presence of the ester in question 690 implicates the carbon-carbon bond forming Reformatsky reaction. In question 691, the presence of a beta hydroxy aldehyde is the signal for an aldol reaction. Finally in question 692, the alpha-beta unsaturated ketone is a clear signal for an aldol condensation.

693. What is the product from the Reformatsky reaction of α-bromoethyl acetate and cyclohexyl methyl ketone?

a) H_3C O
(cyclohexyl)–$CHCHCOCH_2CH_3$
OH

b) H_3C O
(cyclohexyl)–$CCH_2COCH_2CH_3$
OH

c) H_3C CH_3
(cyclohexyl)–$CCHCOOCH_2CH_3$
HO

d) H_3C OH
(cyclohexyl)–$CCHCH_2COOCH_2CH_3$
HO

694. What is the product from the reaction of methyl vinyl ketone with the Wittig reagent ($CH_2 = PPh_3$)?

a) b) c) d)

695. What is the product of the Clemmensen reaction with C_6H_5CDO?

a) $C_6H_5CH_3$ b) $C_6H_5CD_2H$ c) $C_6H_5CDH_2$ d) $C_6H_5CD_3$

696. What is the product from the Baeyer-Villiger oxidation of cyclohexanone?

a) $HOOC-(CH_2)_4-COOH$ b) 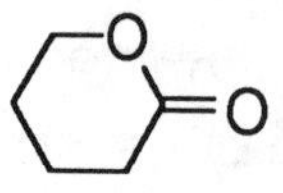 [six-membered lactone ring with C=O] c) [seven-membered lactone ring with C=O] d) $HOOC-(CH_2)_3-COOH$

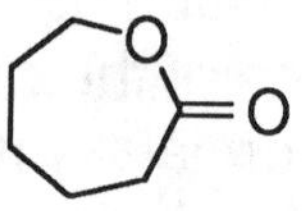

H. REACTIONS: COMBINATIONS

697. What is the product from the following reaction?

$$CH_3CHO \xrightarrow{HCN} \xrightarrow[heat]{H_2SO_4}$$

a) $CH_2{=}CHCOOH$ b) $CH_3CH(OH)COOH$

c) $CH_2(OH)CH_2COOH$ d) $CH_3C(CH_3)COOH$

698. What is the most likely product from the reaction of acetone heated with sulfuric acid?

a) acetophenone b) mesitylene
c) pentaerythritol d) 3-hydroxy-2-methylpentanal

699. What are the best conditions for the following conversion?

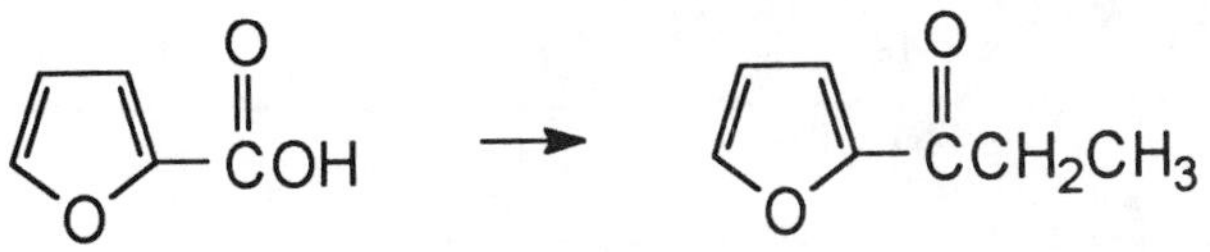

(2-furyl)$-\overset{O}{\overset{\|}{C}}OH$ → (2-furyl)$-\overset{O}{\overset{\|}{C}}CH_2CH_3$

a) $\xrightarrow{CH_3CH_2MgBr}$ $\xrightarrow[H_2O]{H_3O^{\oplus}}$

b) $\xrightarrow{SOCl_2}$ $\xrightarrow{CH_3CH_2MgBr}$ $\xrightarrow[H_2O]{H_3O^{\oplus}}$

c) $\xrightarrow{SOCl_2}$ $\xrightarrow{(CH_3CH_2)_2Cd}$

d) $\xrightarrow{SOCl_2}$ $\xrightarrow{LiAlH[OC(CH_3)_3]_3}$ $\xrightarrow{CH_3CH_2MgBr}$ $\xrightarrow[H_2O]{H_3O^{\oplus}}$

TIP This type of conversion is often an area of confusion in synthesis even though it looks simple. Answer a), among other problems, violates the Grignard sensitivity to acidic hydrogens. Answers b) and d) lead to alcohols not ketones.

700. What is the product from the lithium aluminum hydride reduction of acetophenone?

a) 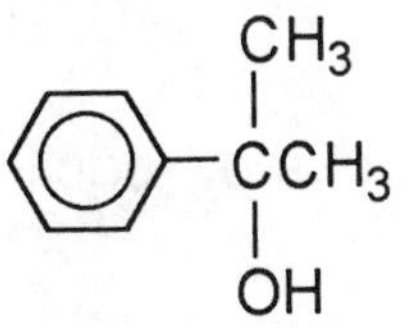 $C_6H_5-C(CH_3)(CH_3)OH$

b) $C_6H_5-C(CH_3)(OCH_3)OCH_3$

c) $C_6H_5-C(H)(OCH_3)OCH_3$

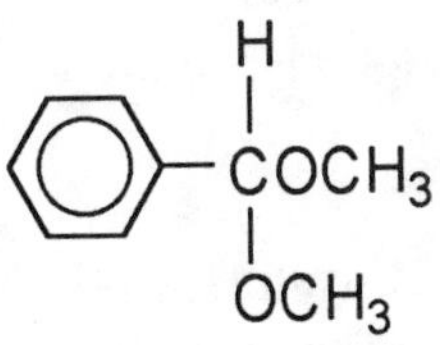

d) 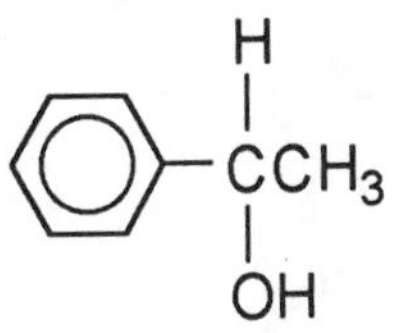 $C_6H_5-C(H)(CH_3)OH$

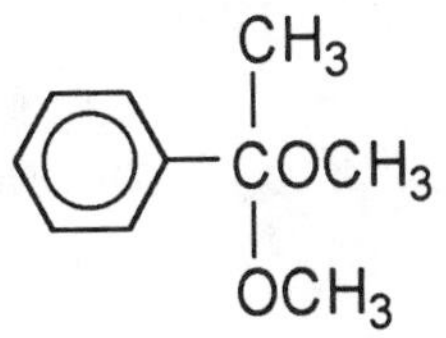

701. What is the most likely product from the following series of reactions?

$$CH_3CH_2CHO \xrightarrow{OH^{\ominus}} \xrightarrow[\text{heat}]{H_3O^{\oplus}} \xrightarrow[HCl]{Zn(Hg)} \xrightarrow[H_2]{Pt}$$

a) $CH_3CH(CH_3)CH(CH_3)_2$ b) $CH_3CH_2CH_2CH_2CH_2CH_3$

c) $CH_3CH_2CH_3$ d) $CH_3CH_2CH_2CH(CH_3)_2$

702. What is the most likely product from the following series of reactions?

$$C_6H_5\text{-}CHO \xrightarrow[NH_3]{Ag(NH_3)_2} \xrightarrow{H_3O^{\oplus}} \xrightarrow{SOCl_2} \xrightarrow[AlCl_3]{\text{benzene}}$$

a) acetophenone b) benzoyl chloride
c) benzophenone d) phenylbenzaldehyde

703. What is the most likely product from the following series of reactions?

$$CH_3I + (C_6H_5)_3P \longrightarrow \xrightarrow{C_4H_9Li} \xrightarrow{C_6H_5CHO}$$

a) $C_6H_5\text{-}CH{=}CHCH_3$ b) $C_6H_5\text{-}C(CH_3){=}CH_2$

c) $C_6H_5\text{-}CH(CH_3){=}CHCH_3$ d) $C_6H_5\text{-}CH{=}CH_2$

704. What is the most likely product from the following series of reactions?

$$C_6H_5\text{-}CH_2Cl + LiCN \longrightarrow \xrightarrow{CH_3CH_2MgBr} \xrightarrow{H_3O^{\oplus}}$$

a) acetone b) 2-phenylbutanal
c) 1-phenyl-2-butanone d) phenylacetaldehyde

705. What are the most likely products from the following reaction?

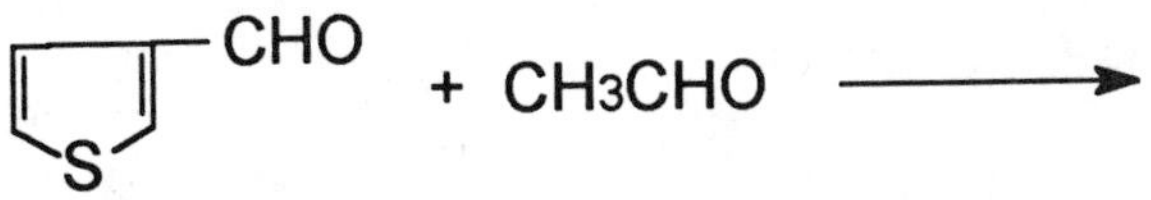

(thiophene-3-CHO) + CH_3CHO ⟶

a) 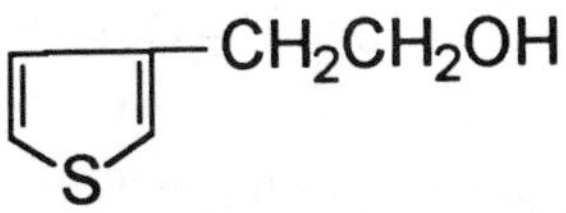 (thiophene-3-)CH_2CH_2OH

b) 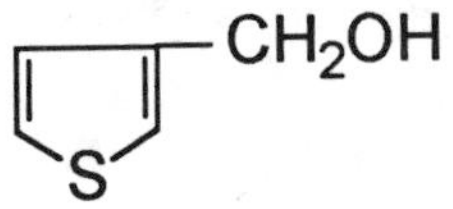 (thiophene-3-)CH_2OH

c) 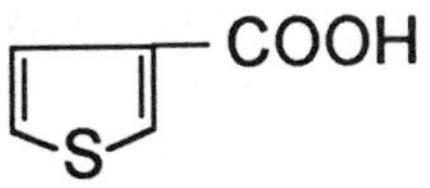 (thiophene-3-)COOH

d) 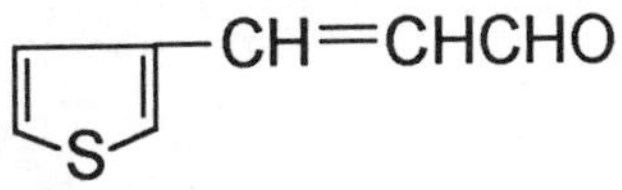 (thiophene-3-)CH=CHCHO

706. Which reagent is best for the following conversion?

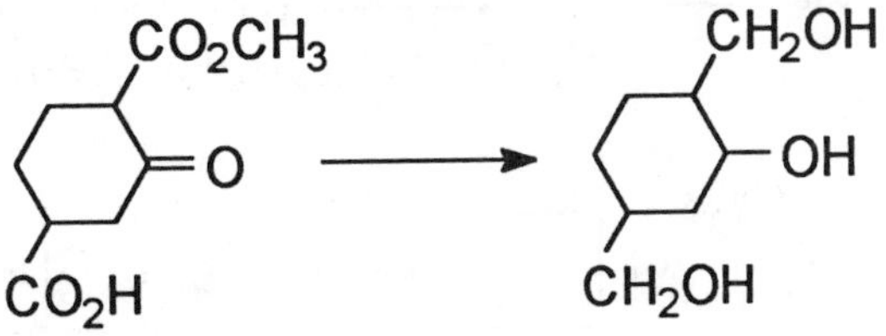

(cyclohexanone bearing CO_2CH_3 and CO_2H) ⟶ (cyclohexanol bearing two CH_2OH groups and OH)

a) $NaBH_4$ b) $LiAlH_4$ c) Na / liquid NH_3 d) $K_2Cr_2O_7$ / H_2SO_4

707. The following keto ester cyclizes upon treatment with sodium ethoxide to give which product?

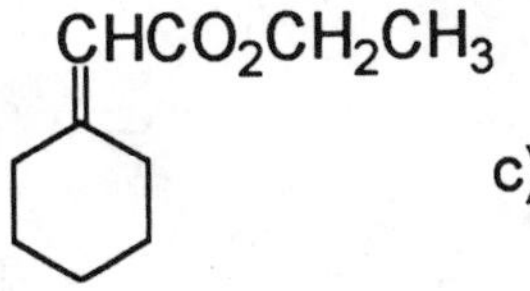

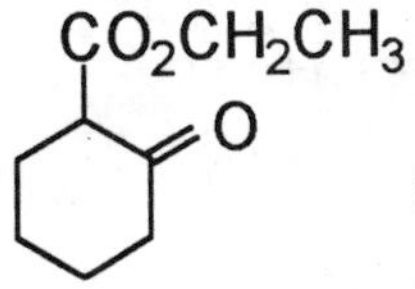

$CH_3\overset{O}{\overset{\|}{C}}(CH_2)_4\overset{O}{\overset{\|}{C}}OCH_2CH_3$

a) 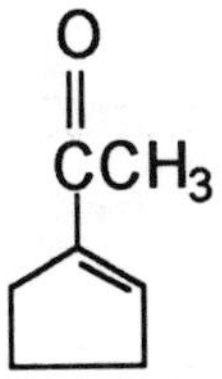 (cyclopentene)-$\overset{O}{\overset{\|}{C}}CH_3$ b) (cyclohexylidene)=$CHCO_2CH_2CH_3$ c) (2-oxocyclohexyl)-$CO_2CH_2CH_3$ d) 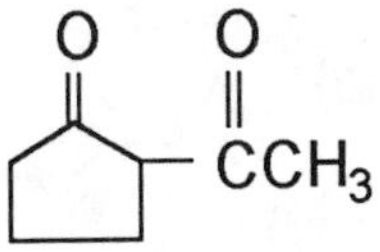 (2-oxocyclopentyl)-$\overset{O}{\overset{\|}{C}}CH_3$

TIP The keto ester cyclization with sodium ethoxide is the Claisen-Schmidt reaction. Two things are central, the ketone will provide the enolate and the ester the carbonyl to give a beta diketone as the product. This effectively rules out answers a), b) and c).

708. What are the best conditions for the following conversion?

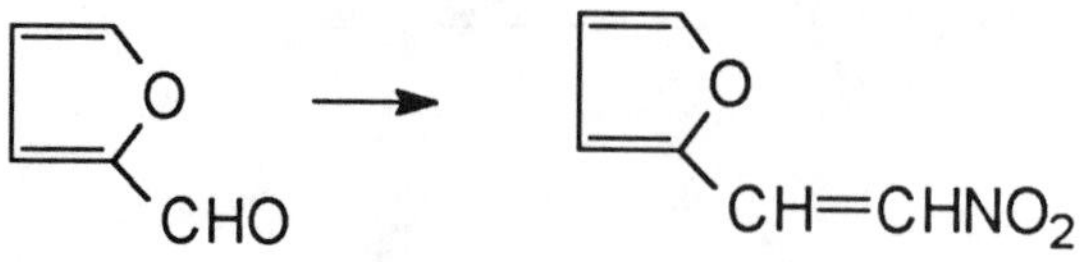

2-furaldehyde (furan–CHO) → furan–$CH=CHNO_2$

a) NH_2OH

b) CH_3NO_2 + NaOH + CH_3OH

c) $CH_3CH_2NO_2$ + $H_3O^{\oplus}$ + H_2O

d) CH_3NHOH

709. What is the major product from the following reaction?

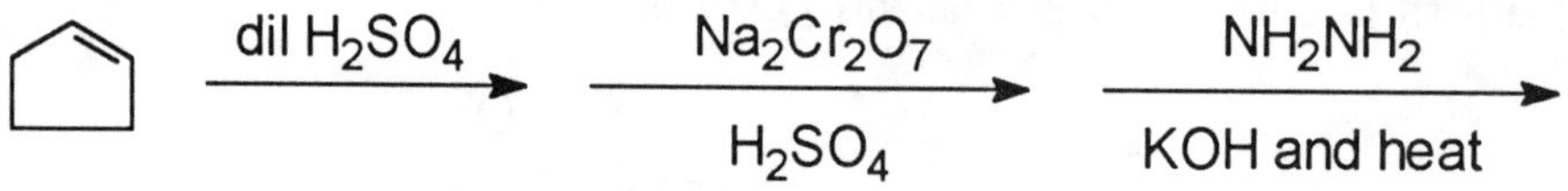

cyclopentene $\xrightarrow{\text{dil } H_2SO_4}$ $\xrightarrow[H_2SO_4]{Na_2Cr_2O_7}$ $\xrightarrow[\text{KOH and heat}]{NH_2NH_2}$

a) 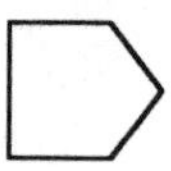 cyclopentane

b) 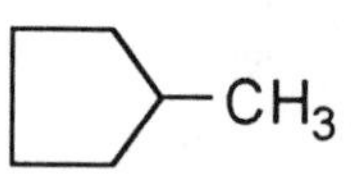 cyclopentane–CH_3

c) 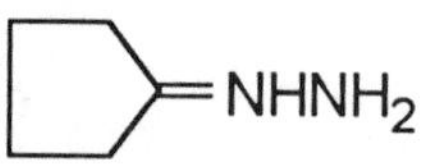 cyclopentane=$NHNH_2$

d) $CH_3(CH_2)_3COO^{\ominus}$

710. What is the most likely product from the following series of reactions?

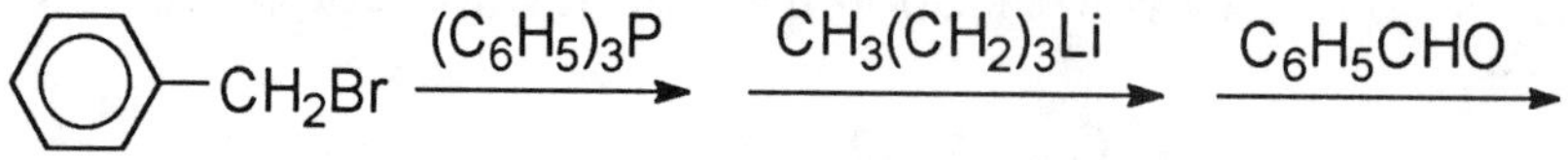

C_6H_5–CH_2Br $\xrightarrow{(C_6H_5)_3P}$ $\xrightarrow{CH_3(CH_2)_3Li}$ $\xrightarrow{C_6H_5CHO}$

a) 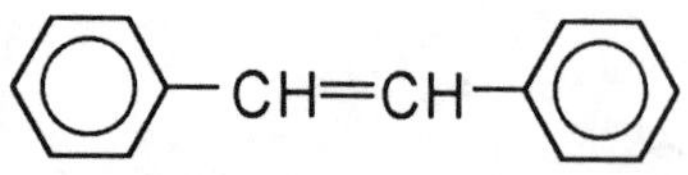 C_6H_5–CH=CH–C_6H_5

b) 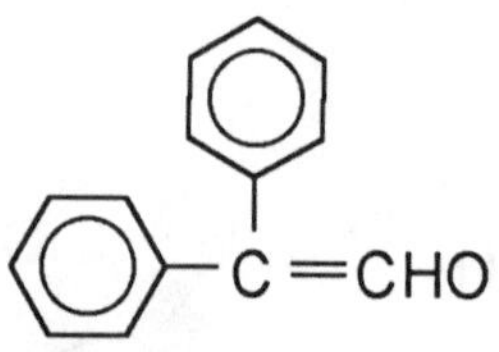 $(C_6H_5)_2$C=CHO

c) $(C_6H_5)_2$C=C$(C_6H_5)_2$

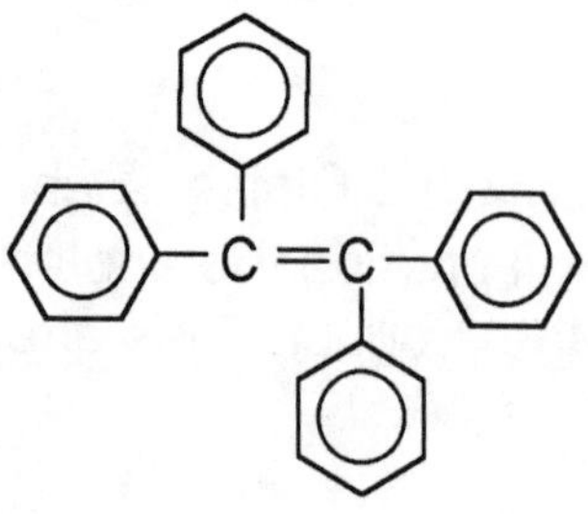

d) 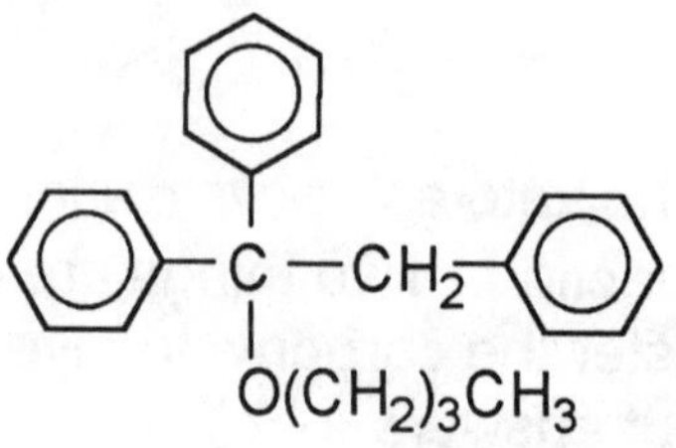 $(C_6H_5)_2$C($O(CH_2)_3CH_3$)–CH_2–C_6H_5

711. What is the product from the following series of reactions?

$CH_3OCH_2Cl + Ph_3P \xrightarrow{BuLi} \xrightarrow{\text{cyclopentanone}} \xrightarrow{H_3O^{\oplus}}$

a) $Ph_3P(OCH_3)(CH_2\text{-}C_5H_8OH)$ (HO-cyclopentyl-CH_2 and CH_3 on PO)

b) cyclopentylidene=CHOCH$_3$

c) cyclopentyl-CHO

d) cyclopentyl-COCH$_3$

TIP This is a Wittig reaction with a very difficult twist. The Wittig itself is straightforward and before hydrolysis the product is the methyl ether derived of methylene cyclopentane (answer b). Hydrolysis will yield a carbocation which is similar to those involved in acetal hydrolysis and thus a carbonyl compound is the answer. When we count carbon atoms we find that answer d) cannot be a correct answer.

712. Which of the following statements is untrue for acetaldehyde?

a) It will undergo an aldol condensation.
b) It will undergo a Clemmensen reaction.
c) It will undergo a haloform reaction.
d) Enolization is catalyzed by acid or base.

713. What is the outcome of the following reaction?

$C_6H_5\text{-}\overset{O}{\overset{\|}{C}}CH_2\text{-}C_6H_5 \xrightarrow[{}^{18}OD^{\ominus}]{D_2{}^{18}O}$

a) exchange of ^{18}O for ^{16}O
b) exchange of D for H
c) exchange of both
d) no exchange

CHAPTER 18 CARBOHYDRATES

A. STRUCTURE and NOMENCLATURE

714. Which of the following structures is an aldose?

a) 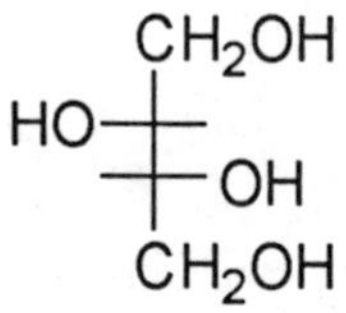 CH_2OH / HO– / –OH / CH_2OH b) 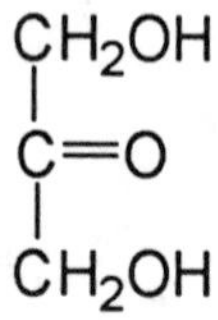 CH_2OH / C=O / CH_2OH c) CHO / HO– / HO– / –OH / CH_2OH d) 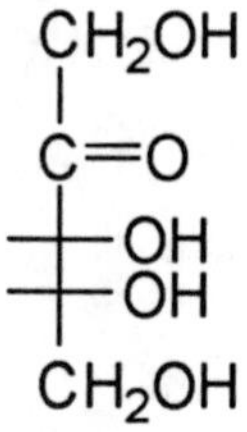 CH_2OH / C=O / –OH / –OH / CH_2OH

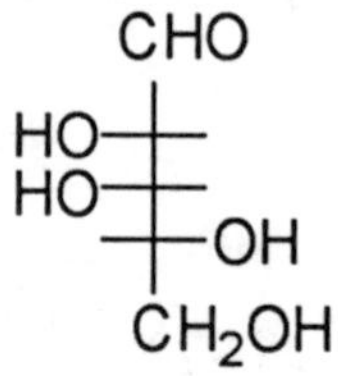

715. Which of the following structures are ketoses?

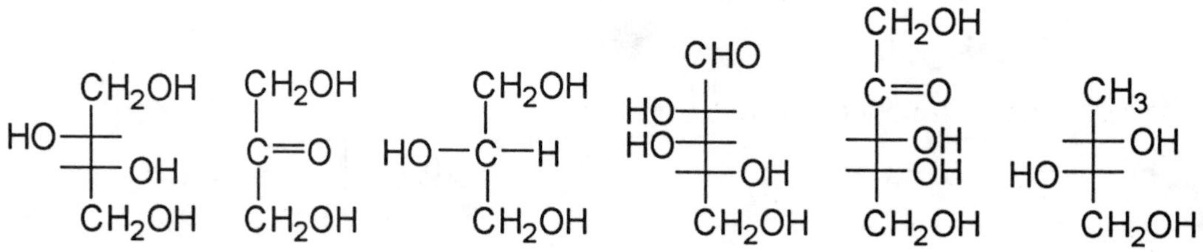

I: CH_2OH / HO– / –OH / CH_2OH
II: CH_2OH / C=O / CH_2OH
III: CH_2OH / HO—C—H / CH_2OH
IV: CHO / HO– / HO– / –OH / CH_2OH
V: CH_2OH / C=O / –OH / –OH / CH_2OH
VI: CH_3 / –OH / HO– / CH_2OH

a) I, IV b) II, V c) III, IV d) I, VI

716. Which of the following structures are D forms?

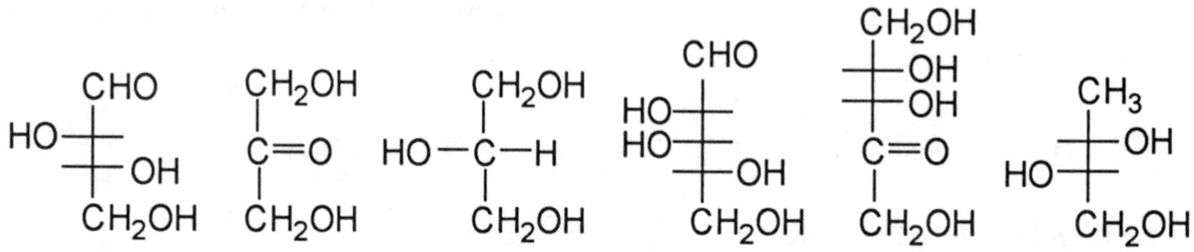

I: CHO / HO– / –OH / CH_2OH
II: CH_2OH / C=O / CH_2OH
III: CH_2OH / HO—C—H / CH_2OH
IV: CHO / HO– / HO– / –OH / CH_2OH
V: CH_2OH / –OH / –OH / C=O / CH_2OH
VI: CH_3 / –OH / HO– / CH_2OH

a) I, IV b) II, V c) III, IV d) I, IV,VI

717. Which of the following structures are L forms?

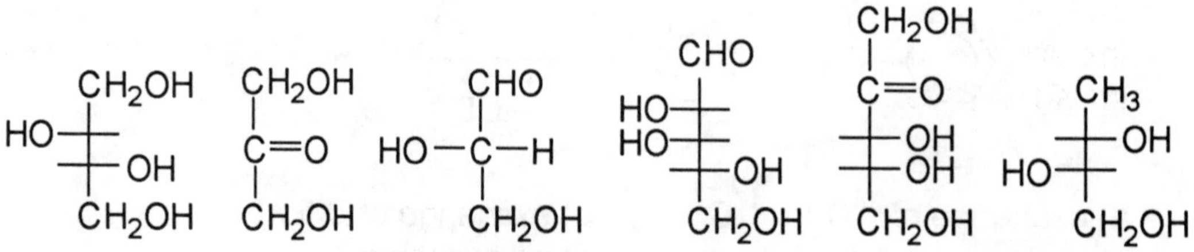

I: CH_2OH / HO– / –OH / CH_2OH
II: CH_2OH / C=O / CH_2OH
III: CHO / HO—C—H / CH_2OH
IV: CHO / HO– / HO– / –OH / CH_2OH
V: CH_2OH / C=O / –OH / –OH / CH_2OH
VI: CH_3 / –OH / HO– / CH_2OH

a) I, IV b) II, V c) III, VI d) I, IV, VI

718. Which of the following structures is a hemiacetal?

a) 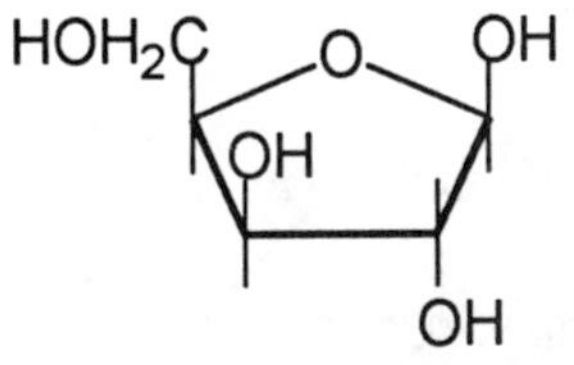 (ring structure: HOH_2C, O, OH, OH, OH)

b) 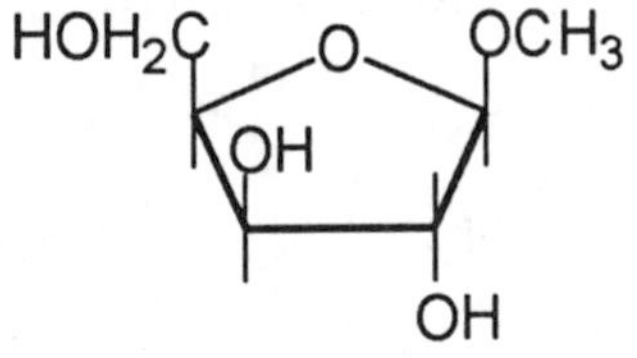 (ring structure: HOH_2C, O, OCH_3, OH, OH)

c) 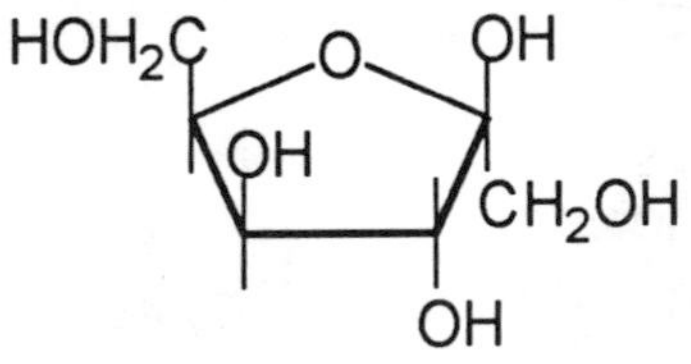 (ring structure: HOH_2C, O, OH, OH, CH_2OH, OH)

d) 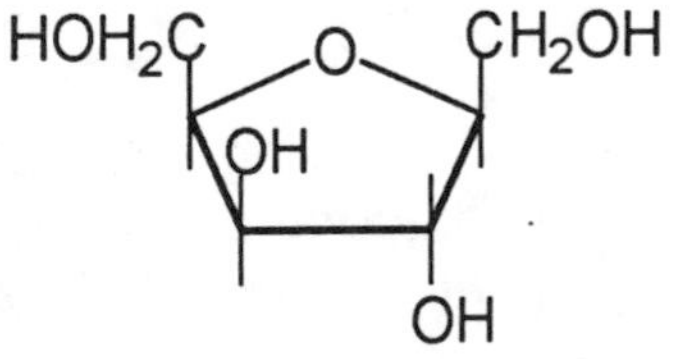 (ring structure: HOH_2C, O, CH_2OH, OH, OH)

719. Which of the following structures is a hemiketal?

a) 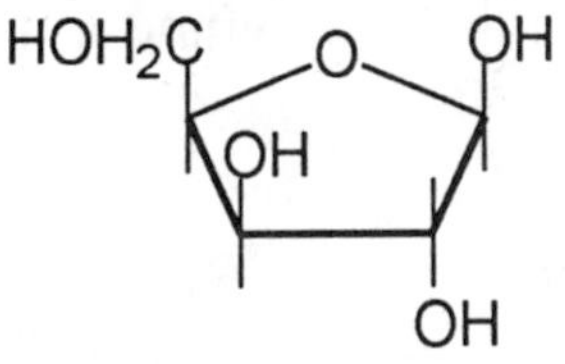 (ring structure: HOH_2C, O, OH, OH, OH)

b) 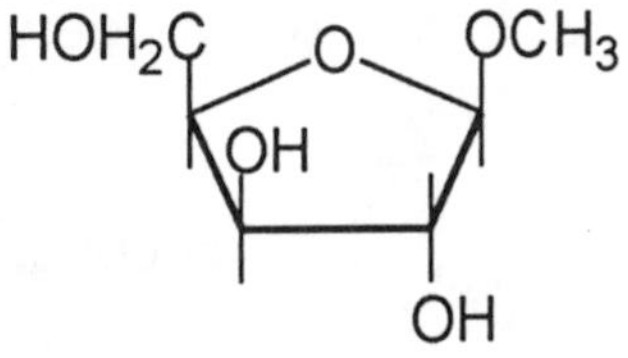 (ring structure: HOH_2C, O, OCH_3, OH, OH)

c) 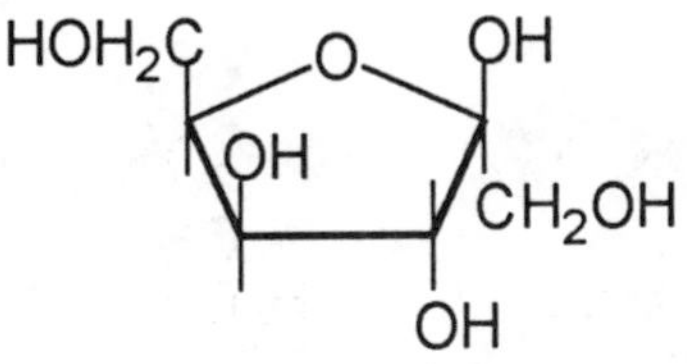 (ring structure: HOH_2C, O, OH, OH, CH_2OH, OH)

d) 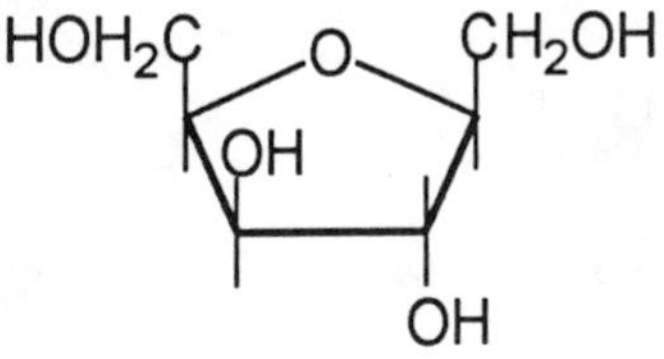 (ring structure: HOH_2C, O, CH_2OH, OH, OH)

TIP For questions 718 and 719. Hemiacetals and hemiketals are characterized by a carbon atom singly bonded to two oxygen atoms and one of the oxygen atoms has a proton. The hemiacetal distinction in addition requires a hydrogen on that carbon atom. In both of these problems, only answers a) and c) fulfill the first rule. Can you find the H on the carbon in a) for the second requirement for a hemiacetal?

720. What is the correct name for the following structure?

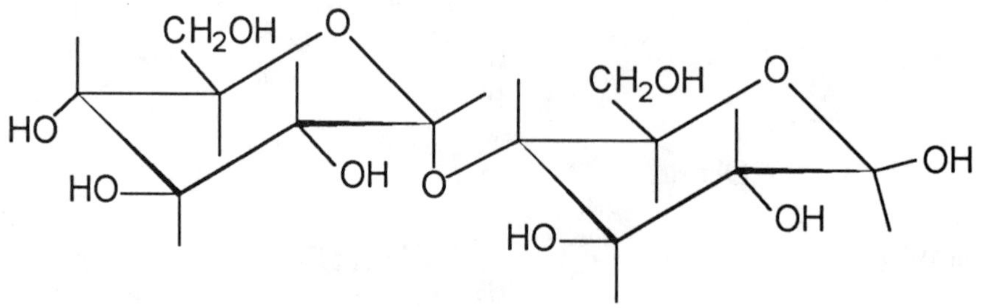

a) 4-0-(α-D-glucopyrosyl)-β-D-glucopyranoside
b) 1-0-(α-D-glucopyrosyl)-β-D-glucopyranoside
c) 4-0-(β-D-glucopyrosyl)-α-D-glucopyranoside
d) 1-0-(β-D-glucopyrosyl)-β-D-glucopyranoside

721. Cellulose, starch and glycogen are polysaccharides containing which of the following sugars?

a) sucrose b) glucose c) fructose d) lactose

B. STEREOISOMERS

722. What is the number of possible D-ketopentoses?

a) 2 b) 4 c) 8 d) 16

723. How many chiral atoms are there in a 2-ketoheptose?

a) 3 b) 4 c) 5 d) 16

724. How many stereoisomers are possible for a D-aldopentose?

a) 2 b) 4 c) 8 d) 16

725. What is the relationship between the following two structures?

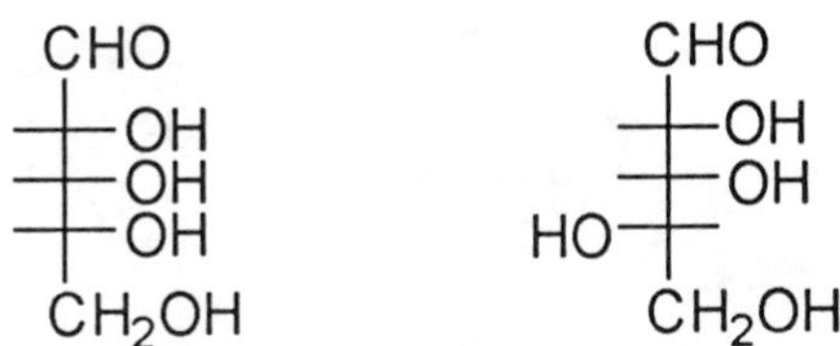

I. enantiomers II. diastereomers III. epimers IV. anomers

a) I, IV b) I, III c) II, III d) II, IV

TIP These structures are clearly not mirror images. Stereoisomers that are not mirror images are diastereomers. In the sugars, diastereomers are often further categorized. Sugars that differ only in the stereochemistry at a single carbon atom are called epimers. Cyclic sugars that differ in the configuarion at the hemiacetal carbon are called anomers. These structures are not cyclic, so only the first two categories pertain.

726. What is the relationship between the following two structures?

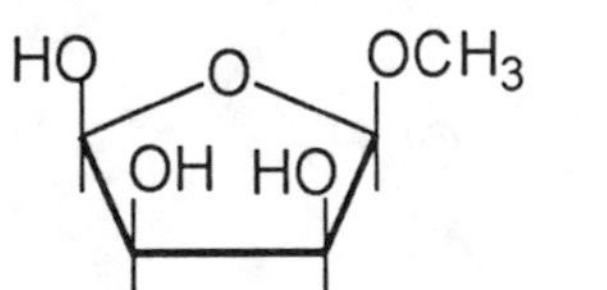

I. enantiomers II. diastereomers III. anomers

a) I, III b) II c) III d) II, III

727. How many stereoisomers are possible for the product from the following reaction?

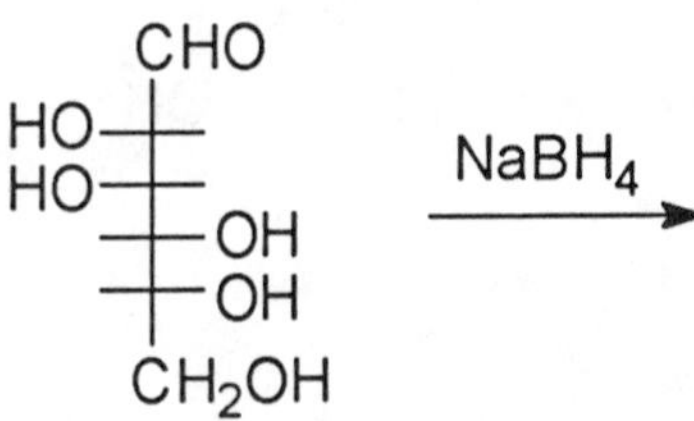

a) 1 b) 2 c) 4 d) 8

728. Which of the following statements is NOT true for the these structures?

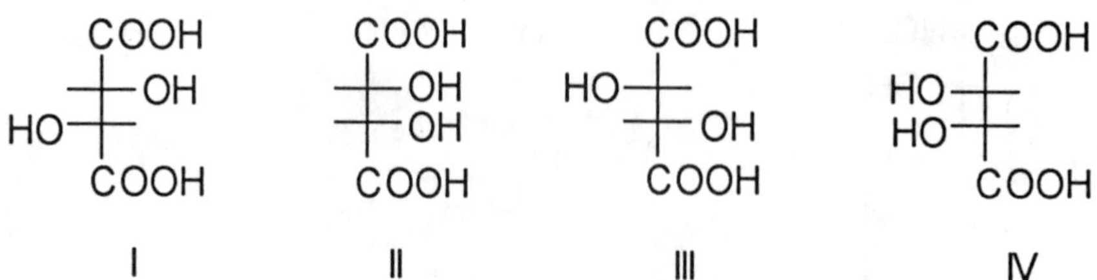

a) I and II are diastereomers
b) I and III are enantiomers
c) II and IV are enantiomers
d) II and III are diastereomers

729. How many stereoisomers are there for the product from the following reaction?

$\xrightarrow{NaBH_4}$

a) 1
b) 2
c) 4
d) 8

730. Which of the following compounds are enantiomers?

I

II

III

IV

a) I, II
b) III, IV
c) II, III
d) II, IV

731. Which of the following is a pair of epimers at carbon atom-2?

a)
CHO / –OH / –OH / –OH / CH_2OH and CHO / HO– / HO– / HO– / CH_2OH

b)
CHO / –OH / –OH / –OH / CH_2OH and CHO / –OH / –OH / HO– / CH_2OH

c)
CHO / –OH / –OH / –OH / CH_2OH and CHO / HO– / –OH / –OH / CH_2OH

d)
CHO / –OH / –OH / –OH / CH_2OH and CHO / –OH / HO– / –OH / CH_2OH

732. Which of the following structures are identical?

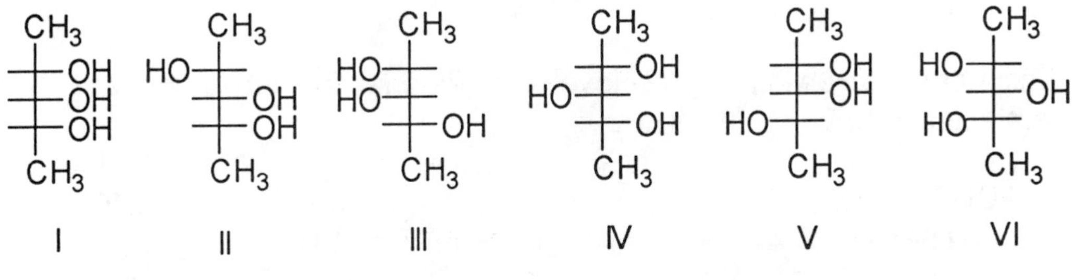

a) I and IV, II and III
b) III and V, IV and VI
c) II and V, I and III
d) II and III, IV and VI

TIP The key to these and related structures is to recall that a 180 degree rotation in the plane of the paper is perfectly valid for Fischer representations. So if two structures are identical after such a rotation, they are identical. For II and V we note that this operation reveals that they are enantiomers, but for II and III that they are identical. For IV and VI the identity is illustrated additionally by noting that they are meso.

733. The following Fischer projection formula for L-idose corresponds to which Haworth projection formula?

CHO
HO–
–OH
HO–
HO–
CH_2OH

a) CH_2OH, OH, OH, HO, HO

b) CH_2OH, HO, HO, OH, OH

c) CH_2OH, OH, OH, HO, OH

d) HO, CH_2OH, OH, HO, OH

734. Which of the following sugars have an equatorial OH substituent at C-3 in the hemiacetal form?

I	II	III	IV	V
CHO	CHO	CHO	CHO	CHO
–OH	–OH	–OH	–OH	HO–
HO–	–OH	–OH	HO–	HO–
–OH	HO–	–OH	HO–	–OH
–OH	HO–	–OH	–OH	HO–
CH_2OH	CH_2OH	CH_2OH	CH_2OH	CH_2OH

a) I, III, V b) I, IV, V c) II, IV, V d) III, IV, V

TIP A trick helps here. In the hemiacetal form of glucose, all substituents are equatorial except the hemiacetal OH at C-1, and this depends upon whether the anomer is alpha or beta. Thus by knowing the glucose configuration in open chain form (and this we have to know or find in the textbook), we can compare other sugars and provide an answer. From that diagnostic, sugars I, IV and V have equatorial OH substituents at C-3.

735. How many equatorial substituents are there in methyl-α-D-mannopyranoside?

a) 2 b) 3 c) 4 d) 5

C. REACTIONS

736. Which reagent is used to convert glucose to glucitol?

a) Tollens b) H_5IO_6 c) $NaBH_4$ d) Br_2 / H_2O

737. Which reagent is used to convert glucose to glucaric acid?

a) Tollens b) H_5IO_6 c) HNO_3 d) Br_2/ H_2O

738. Which reagent is used to convert mannose to formic acid and formaldehyde?

a) Tollens b) H_5IO_6 c) HNO_3 d) Br_2 / H_2O

739. Which pairs of sugars react with sodium borohydride to give identical products?

a) CHO, HO–, HO–, HO–, CH_2OH and CHO, –OH, –OH, –OH, CH_2OH

b) CHO, –OH, –OH, –OH, CH_2OH and CHO, HO–, –OH, –OH, CH_2OH

c) CHO, –OH, –OH, –OH, CH_2OH and CHO, –OH, –OH, HO–, CH_2OH

d) CHO, –OH, –OH, –OH, CH_2OH and CHO, –OH, HO–, –OH, CH_2OH

740. Two sugars give identical products when treated with sodium borohydride. Which of the following are possible structures for the sugars?

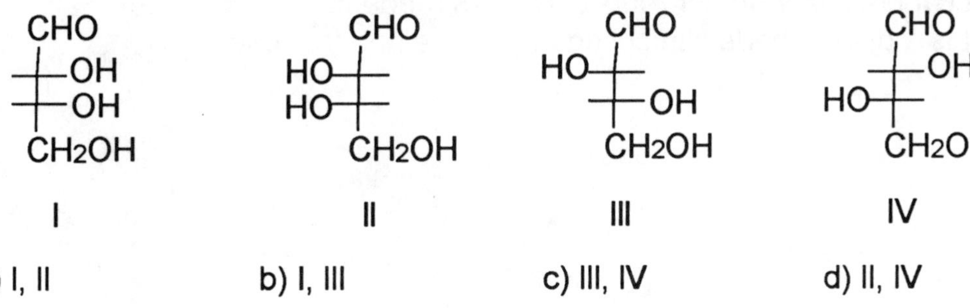

a) I, II b) I, III c) III, IV d) II, IV

741. Which of the following compounds give the same glucaric acid as its enantiomer when oxidized with nitric acid?

a)
CHO
—OH
HO—
HO—
—OH
CH_2OH

b)
CHO
HO—
HO—
—OH
—OH
CH_2OH

c)
CHO
HO—
—OH
HO—
—OH
CH_2OH

d)
CHO
HO—
HO—
HO—
—OH
CH_2OH

TIP Oxidation with nitric acid converts C_1 and C_6 to carboxylic acids. Thus the top and bottom carbon atom configurations are the same. Only for answer a) is the glucaric acid meso. The remaining glucaric acids will be chiral. Does this define answer a) as correct?

742. What is the major product when D-glucose is reacted with aniline?

a)
$CH_2NHC_6H_5$
—OH
HO—
—OH
—OH
CH_2OH

b)
$CH{=}NC_6H_5$
—OH
HO—
—OH
—OH
CH_2OH

c)
$CH{=}NC_6H_5$
—OH
—OH
HO—
—OH
CH_2OH

d)
CHO
—OH
HO—
—OH
—OH
$CH{=}NC_6H$

743. Which of the following statements about sucrose is NOT true?

a) The linkage between anomeric carbon atoms is 1,4'.
b) It is hydrolyzed to monosaccharides by invertases.
c) It cannot muturate nor reduce Tollen's reagent.
d) It is a disaccahride composed of glucose and fructose.

744. Which of the following give a positive Tollens test?

I. D-sorbitol II. D-glucose III. D-fructose IV. sucrose

a) I, II b) III, IV c) II, IV d) II, III

745. Which reagent(s) can determine if a sugar is an aldose or a ketose?

a) Tollens b) HNO_3 c) CH_3I + Ag_2O d) Br_2 + H_2O

746. What is the major product from the following reaction?

CH2OH, O, OH, OH, O, HO $\xrightarrow{HNO_3}$

a)
COOH
—OH
HO—
—OH
—OH
COOH

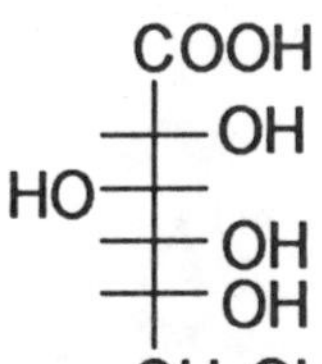

b)

c)
CH_2OH
—OH
HO—
—OH
—OH
CH_2OH

d)
CHO
—OH
HO—
—OH
—OH
CH_2OH

747. If D-glucose is allowed to stand for several days in aqueous calcium hydroxide, which of the following sugars will be in the mixture?

I. mannose II. arabinose III. fructose IV. ribose

a) I, II b) II, III c) II, IV d) I, III

TIP Glucose is an aldehyde and so, in a basic solution, reactions at the acidic beta hydrogens are possible. This leads to two changes. First loss of the acidic proton and reprotonation at C-2 will lead to loss of configuration at C-2 and thus both glucose and mannose are formed. Enolization and subsequent protonation at C-1 now leads to fructose.

748. Which of the following reagents catalyzes the mutarotation of the D-glucopyranoses?

a) acid b) acid or base c) base d) acid and/or base

749. Which of the following is expected to exibit the property of mutarotation and to form an osazone when treated with excess phenylhydrazine?

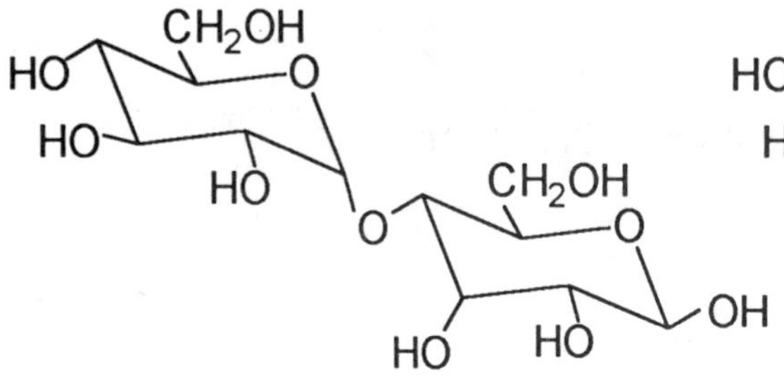

a) 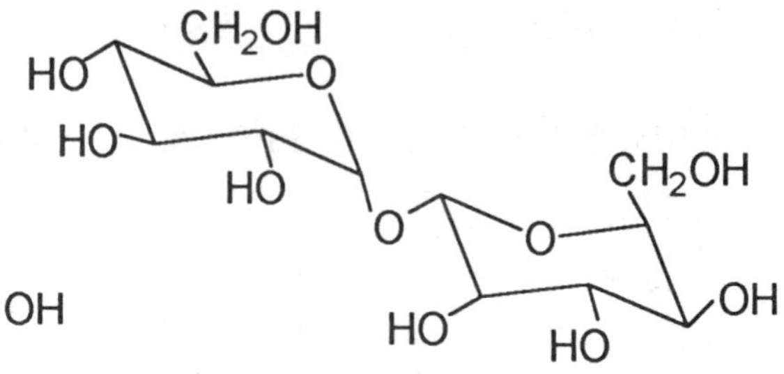 b)

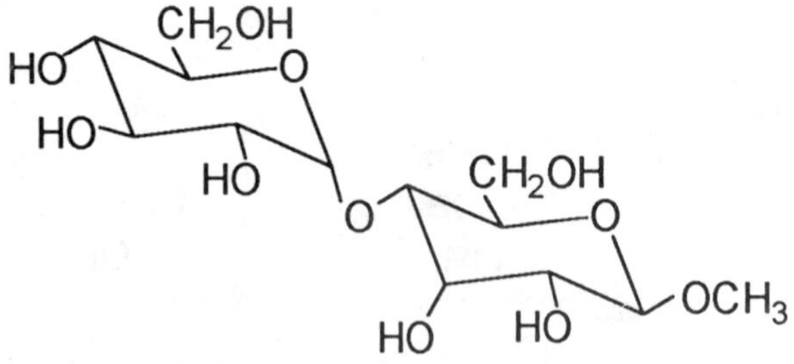

c) 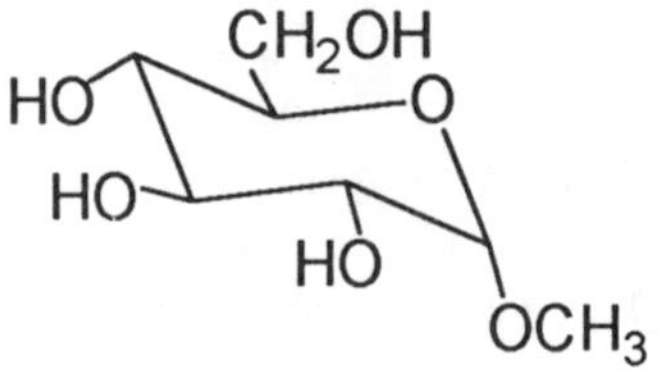 d)

750. Which compound is 2-hydroxypyridine converted to in the mutarotation of glycopyranose?

a) 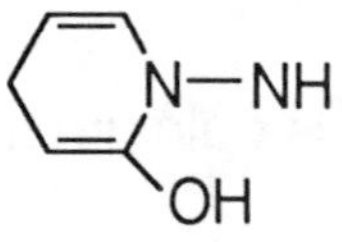 b) 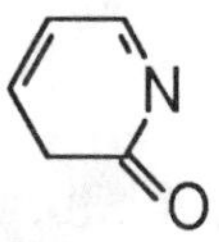 c) 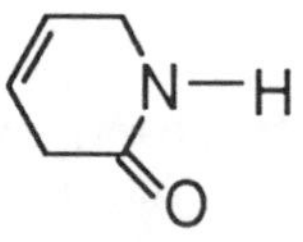 d)

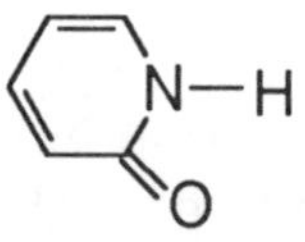

751. Which of the following is the most effective catalyst for mutarotation?

a) phenol b) 2-hydroxypyridine c) pyridine d) pyridine and phenol

TIP Mutarotation (which involves conversion of a hemiacetal to an aldehyde and an alcohol and the reverse reaction) is a process that is both acid and base catalyzed. The conversion is most efficient when both acid and base are present and at the optimum distance from the reaction site (similar to the role that enzymes play). In the 2-hydroxypyridine molecule, the phenolic-like acidic OH group is perfectly placed to protonate the oxygen atom, while the basic N atom in pyridine can accept the proton from the hemiacetal OH group.

D. SYNTHESIS

752. What processes are needed for the conversion of arabinose to glucose?

a) enolization and oxidation
b) carbon atom addition and oxidation
c) carbon atom addition and enolization
d) enolization and reprotonation

753. What processes are needed for the conversion of mannose to glucose?

a) enolization and oxidation
b) carbon atom addition and oxidation
c) carbon atom addition and enolization
d) enolization and reprotonation

754. What is the result of the reaction of D-glyceraldehyde with HCN?

a) a pair of enantiomers
b) a pair of diastereomers
c) a meso compound
d) two pairs of enantiomers

755. Which of the following sugars are key intermediates in the procedure of converting D-glyceraldehyde to D-glucose?

a) D-ribose and D-erythrose
b) D-arabinose and D-mannose
c) D-arabinose and D-erythrose
d) D-arabinose and D-threose

TIP This is an example of building sugars by using defined configurations. D-glyceraldehyde is a three carbon sugar and D-glucose is a six carbon sugar. The four and five carbon sugars related to D-glucose are D-erythrose and D-arabinose.

E. REACTIONS to ESTABLISH STRUCTURE and RELATIVE CONFIGURATION

756. The following compound reacts with X moles of periodic acid to give Y moles of formaldehyde and Z moles of formic acid. What are the correct values for X,Y and Z?

HOH_2C O CH_3
OH HO
CH_2OH

a) 2, 0, 1 b) 2, 1, 0 c) 6, 1, 5 d) 1, 0, 0

757. Which of the following sugars are key intermediates in the procedure of converting D-mannose to D-glyceraldehyde?

a) D-ribose and D-erythrose
b) D-arabinose and D-mannose
c) D-arabinose and D-erythrose
d) D-arabinose and D-threose

758. What is the best procedure for converting glucose to 2,3,4,6-tetra-O-methyl-D-glucose?

a) $\xrightarrow[HCl]{CH_3OH}$ $\xrightarrow[NaOH]{(CH_3)_2SO_4}$

b) $\xrightarrow[HCl]{CH_3OH}$

c) $\xrightarrow[HCl]{CH_3OH}$ $\xrightarrow[NaOH]{(CH_3)_2SO_4}$ $\xrightarrow[HCl]{H_2O}$

d) $\xrightarrow[NaOH]{(CH_3)_2SO_4}$

759. Which of the following compounds give the same products upon oxidation with periodic acid?

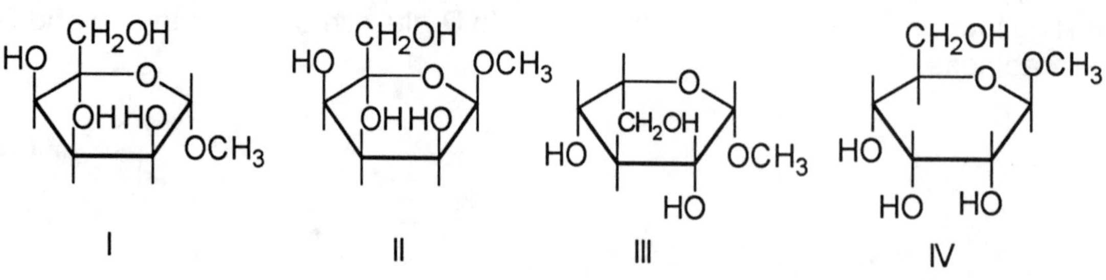

a) I, II b) II, III c) I, III d) II, IV

760. Which of the following compounds give the same products upon acid hydrolysis?

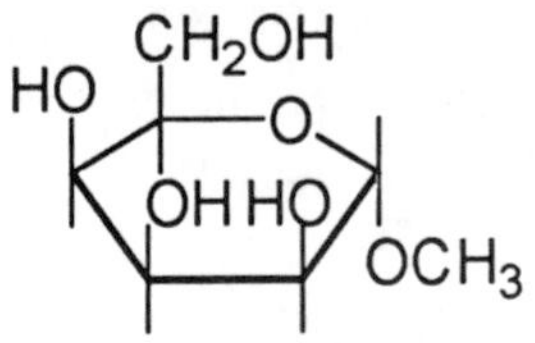

I 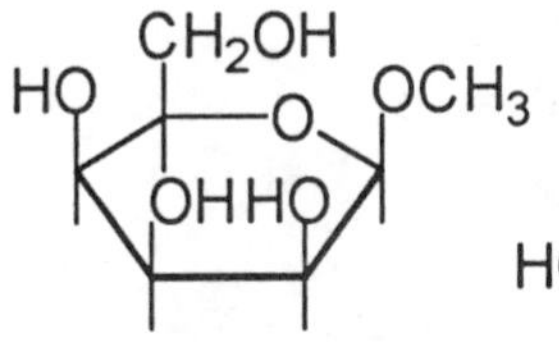 II 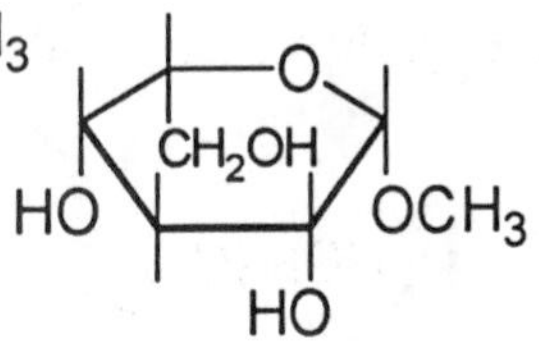 III 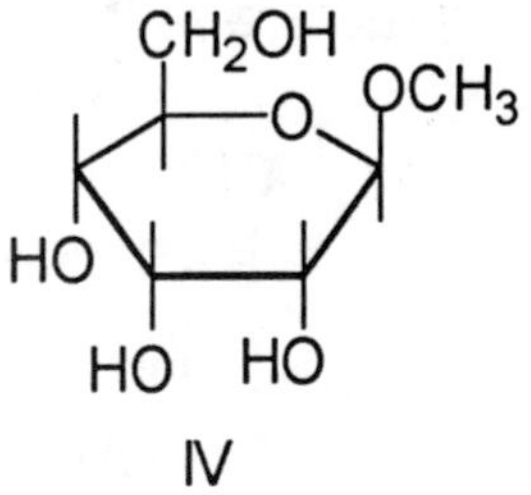 IV

a) I, II b) II, III c) I, III d) II, IV

TIP For questions 759 and 760. Periodic oxidation causes the configuration at carbon atoms 2, 3 and 4 to be lost in structures I, II, and IV. (Compound III is clearly different because of the configuration of the CH_2OH group at carbon atom 5.) Thus configuration at C-1 is critical, and II and IV are identical in that aspect. Acid hydrolysis will lead to loss of configuration at C-1 with no other changes. Note that the structures I and II differ only with respect to C-1.

761. Sucrose and maltose can be distinguished by which reagents?

a) Benedicts solution
b) periodic acid
c) Br_2 / H_2O
d) all of these

762. Periodic acid oxidation of a sugar gives 2 moles of formaldehyde, 1 mole of carbon dioxide and 3 moles of formic acid. What is the sugar?

a) fructose b) arabinose c) mannose d) a ketopentose

763. Periodic acid oxidation of a sugar gives 4 moles of formic acid and 1 mole of formaldehyde. The sugar reacted with nitric acid to give a meso diacid. The final product of successive sugar degradations is D-glyceraldehyde. What is the structure of the sugar?

a) CHO; –OH; –OH; –OH; CH_2OH

b) CHO; HO–; HO–; HO–; CH_2OH

c) CH_2OH; C=O; HO–; HO–; CH_2OH

d) CH_2OH; C=O; –OH; –OH; CH_2OH

764. A sugar was prepared in a 2-step synthesis from L-glyceraldehyde. Treatment of the sugar with nitric acid gave an optically active diacid. What is the structure of the sugar?

a) CHO; –OH; HO–; CH_2OH

b) CHO; HO–; HO–; CH_2OH

c) CHO; HO–; –OH; CH_2OH

d) CHO; C=O; –OH; CH_2OH

765. A D-aldopentose is oxidized to an optically active diacid. The sugar degradation product is an aldotetrose which is oxidized to an optically inactive diacid. What is the structure of the pentose?

a) CHO; –OH; HO–; –OH; CH_2OH

b) CHO; –OH; HO–; HO–; CH_2OH

c) CHO; HO–; –OH; –OH; CH_2OH

d) CHO; –OH; –OH; HO–; CH_2OH

TIP The oxidation step rules out answer a) since the resulting molecule is not chiral. The aldotetrose which gives an optically inactive diacid must have a non meso-like configuration for carbon atoms 3 and 4 in the original pentose. This rules out answers b) and c).

766. A hexose gives on reduction D-glucitol. Upon reaction with 2-methylaniline it yields an imine that is different from the imine derived from D-glucose. What is the hexose?

a)	b)	c)	d)
CH_2OH	CHO	CHO	CHO
C=O	HO–	HO–	–OH
HO–	HO–	HO–	HO–
–OH	–OH	–OH	–OH
–OH	–OH	HO–	–OH
CH_2OH	CH_2OH	CH_2OH	CH_2OH

767. An aldohexose, A, gives an aldopentose, B, upon sugar degradation. When A is subjected to a sugar synthesis, there is produced a mixture of two aldoheptoses, C and D. The glycaric acids from nitric acid oxidation of A, B, C, and D are all chiral. What is the structure for A?

a)	b)	c)	d)
CHO	CHO	CHO	CHO
HO–	–OH	HO–	–OH
–OH	HO–	HO–	HO–
–OH	–OH	–OH	HO–
–OH	–OH	–OH	–OH
CH_2OH	CH_2OH	CH_2OH	CH_2OH

768. Two aldoses, A and B, give the same aldopentose, C, upon sugar degradation. The sugar synthesis starting with B gives a mixture of aldoheptoses, D and E. The glycaric acids from the nitric acid oxidation of A and E are chiral. Those from B, C, and D are achiral. What is the structure of A?

a)	b)	c)	d)
CHO	CHO	CHO	CHO
HO–	HO–	HO–	–OH
–OH	HO–	–OH	–OH
–OH	–OH	HO–	HO–
–OH	–OH	–OH	–OH
CH_2OH	CH_2OH	CH_2OH	CH_2OH

TIP This question is an example of chemical logic in determining the structure of A. The important information we get from the reactions is: 1) A and B can differ only at carbon atom 2 because of the result of the sugar degradation; 2) B must have the meso configuration at carbon atoms 2,3, 4 and 5 because the oxidation product from B is achiral; 3) the sugar synthesis of B to an aldoheptose and subsequent oxidation to one chiral and one achiral product gives the clue to the configuration at C-2 for B. Now that we know the structure of B, we go directly to a) for the structure of A.

CHAPTER 19 CARBOXYLIC ACIDS

A. NOMENCLATURE and STRUCTURE

769. Which of the following compounds are named correctly?

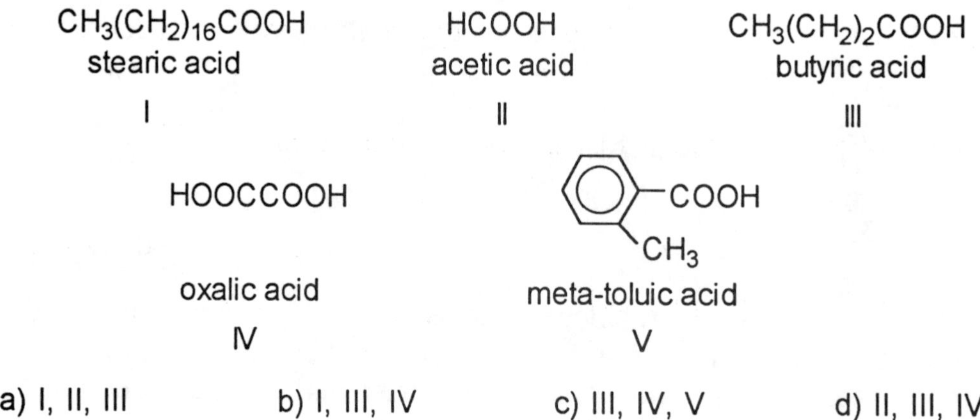

a) I, II, III b) I, III, IV c) III, IV, V d) II, III, IV

770. Which of the following acids have the correct structure?

I. 3,3-dichlorobutanoic acid $CH_3CCl_2CH_2COOH$

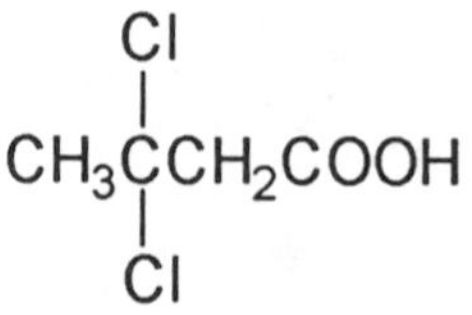

II. 3-methylpentanoic acid $CH_3CH(CH_2CH_3)CH_2CH_2COOH$

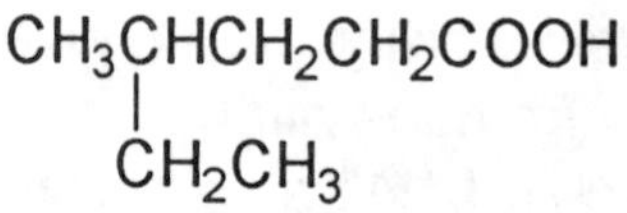

III. 2-hydroxy-3-phenylpropanoic acid $C_6H_5{-}CH_2CH(OH)COOH$

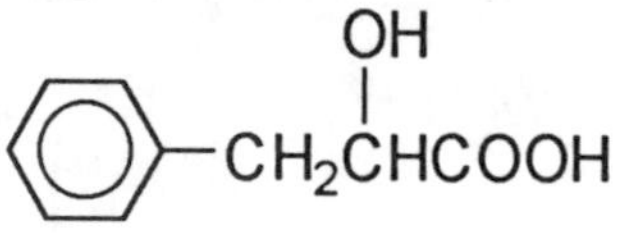

IV. benzoic acid $C_6H_5{-}COOH$

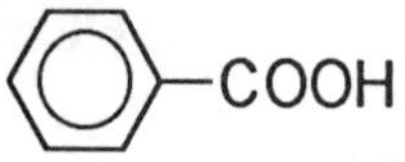

V. phthalic acid $C_6H_4(COOH)(OH)$ (ortho)

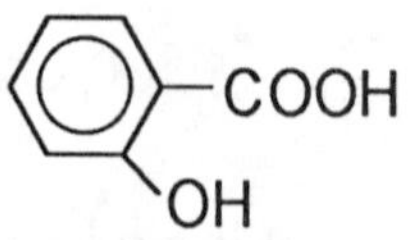

a) I, II, III b) I, III, IV c) II, III, V d) III, IV, V

771. What is the correct name for the following structure?

$HOOCCH_2CH_2COOH$

a) oxalic acid b) succinic acid c) malonic acid d) adipic acid

772. What is the correct structure for Z-3-hexenedioic acid?

a) $CH_3CH_2CH{=}CHCH_2COOH$

b) 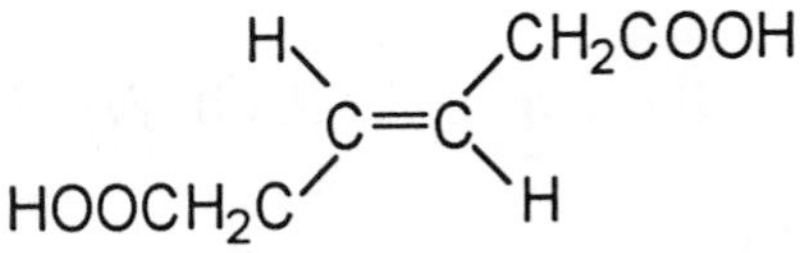 $HOOCH_2C(H)C{=}C(H)CH_2COOH$ (H and CH_2COOH on upper side; $HOOCH_2C$ and H on lower side)

c) $HOOCH_2C(H)C{=}C(H)CH_2COOH$ ($HOOCH_2C$ and CH_2COOH on upper side; H and H on lower side)

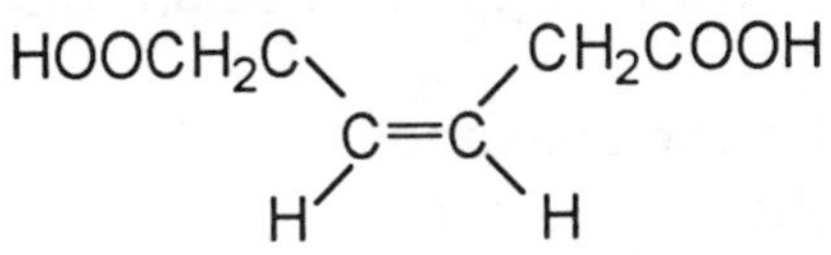

d) 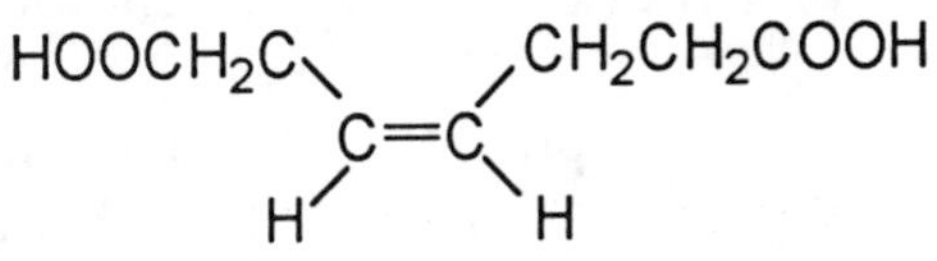 $HOOCH_2C(H)C{=}C(H)CH_2CH_2COOH$ ($HOOCH_2C$ and CH_2CH_2COOH on upper side; H and H on lower side)

773. What is the correct structure for phthalic acid?

a) 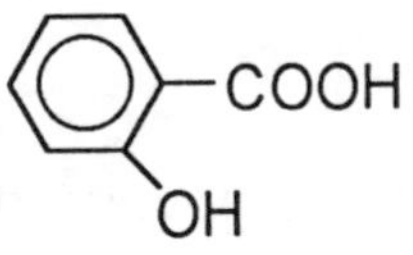 benzene ring with –COOH and ortho –OH

b) 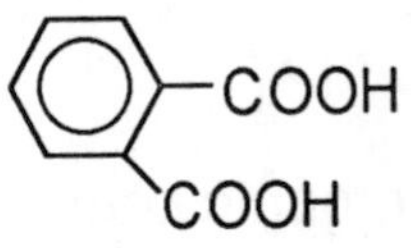 benzene ring with –COOH and ortho –COOH

c)  HOOC–(benzene ring)–COOH (para)

d) 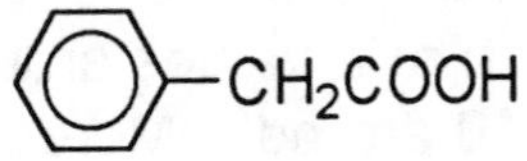 benzene ring with –CH_2COOH

B. PHYSICAL PROPERTIES

774. Arrange the following acids in the order of increasing melting point (lowest first).

I. $CH_3(CH_2)_6COOH$ II. $HOOC(CH_2)_5COOH$

III. $CH_3(CH_2)_2COOH$ IV. $CH_3CH{=}CHCOOH$

a) IV, III, I, II b) III, I, IV, II c) II, III, I, IV d) IV, III, II, I

775. What is the order of increasing polarity for the following acids (lowest first)?

I. CH_3CH_2COOH II. CH_3COOH III. $ClCH_2COOH$ IV. $Cl(CH_2)_2COOH$

a) I, II, III, IV b) I, II, IV, III c) II, I, IV, III d) IV, II, I, III

776. Which of the following compounds are soluble in aqueous sodium bicarbonate?

I. phenol II. phthalic acid III. cresol IV. benzoic acid

a) I, III b) II, IV c) I, II, IV d) II, III

777. Which of the following compounds are soluble in aqueous sodium bicarbonate?

I. phenol II. benzoic acid III. cyclohexanol
IV. cresol V. 2,4-dinitrophenol

a) I, II b) III, IV c) II, V d) IV, V

TIP Solubility of organic compounds in sodium bicarbonate is governed by the acidity of the compound relative to carbonic acid (pKa 6.4). Thus carboxylic acids with pKa values around 5 are soluble, but typical phenols with pKa values around 10 are not. Moreover alcohols with pKa values of 16 are not. In problem 777 a new wrinkle is added. Introduction of two nitro groups on the phenol so stabilizes the anion that the pKa value of 2,4 dinitrophenol is below that of carbonic acid.

778. Which of the following reagents can distinguish between para-toluic acid and para-cresol?

a) $\overset{\oplus}{Ag(NH_3)}\ \overset{\ominus}{OH}$ b) $KMnO_4$ c) NaOH d) $NaHCO_3$

779. Which of the following reactions will proceed as written?

I. $HCl + CH_3COO^{\ominus} Na^{\oplus} \rightarrow CH_3COOH + NaCl$

II. $CH_3COOH + NaOH \rightarrow CH_3COO^{\ominus} Na^{\oplus} + H_2O$

III. $CH_3COOH + NaCl \rightarrow CH_3COO^{\ominus} Na^{\oplus} + HCl$

IV. $CH_3COOH + H_2O \rightarrow CH_3COO^{\ominus} + H_3O^{\oplus}$

V. $HCl + H_2O \rightarrow Cl^{\ominus} + H_3O^{\oplus}$

a) I, II, V b) II, III, IV c) I, IV, V d) II, IV, V

780. Arrange the following compounds in the order of increasing acid strength (lowest first).

I: m-$CF_3C_6H_4COOH$ (COOH on benzene ring, CF_3 meta)
II: p-$CF_3C_6H_4COOH$ (COOH on benzene ring, CF_3 para)
III: p-$CH_3C_6H_4COOH$ (COOH on benzene ring, CH_3 para)
IV: C_6H_5COOH

a) IV, III, I, II b) II, IV, III, I c) III, IV, II, I d) I, II, IV, III

781. Arrange the following compounds in the order of increasing acid strength (lowest first).

I: C_6H_5OH (OH on benzene ring)
II: $C_6H_5CH_2OH$ (CH_2OH on benzene ring)
III: p-$(CH_3)_3N^{\oplus}C_6H_4OH$ (OH on benzene ring, $^{\oplus}N(CH_3)_3$ para)
IV: p-$CH_3C_6H_4COOH$ (COOH on benzene ring, CH_3 para)

a) II. I. III. IV b) II, I, IV, III c) III, II, I, IV d) IV, II, III, I

782. Which of the following is the weakest base?

a) $CH_3CH_2CH_2^{\ominus} Li^{\oplus}$ b) $CH_3CH_2O^{\ominus} Na^{\oplus}$

c) CH_3OCH_3 d) $CH_3OC(=O)CH_3$

783. Which acid is stronger and for what reason?

HOOCCOOH (I) CH_3COOH (II)

a) I due to internal H bonding
b) I due to inductive effect
c) II due to resonance stabilization
d) II due to H bonding

TIP Oxalic acid is a considerably stronger acid than acetic acid because of the inductive effect of the carboxylic acid substituent in stabilizing the anion.

784. What is the order of decreasing base strength for the following compounds (greatest first)?

I: $C_6H_5COO^{\ominus}$ II: $C_6H_5O^{\ominus}$ III: $p\text{-}NO_2C_6H_4COO^{\ominus}$ IV: $C_6H_5CH_2O^{\ominus}$

a) I, III, II, IV
b) III, IV, II, I
c) IV, II, I, III
d) II, IV, I, III

785. What is the order of decreasing acid strength for the following compounds (greatest first)?

I: C_6H_5COOH II: C_6H_5OH III: $p\text{-}NO_2C_6H_4COOH$ IV: $C_6H_5CH_2OH$

a) I, III, II, IV
b) III, I, II, IV
c) IV, II, III, I
d) II, IV, I, III

786. What is the order of decreasing acid strength for the following compounds (greatest first)?

CH_3COOH (I) CH_3CH_2COOH (II) $ClCH_2COOH$ (III)

FCH_2COOH (IV) $C_6H_5CH_2COOH$ (V)

a) IV, III, V, I, II b) V, IV, III, I, II c) IV, V, III, II, I d) I, II, III, IV, V

787. Which of the following compounds is the strongest acid?

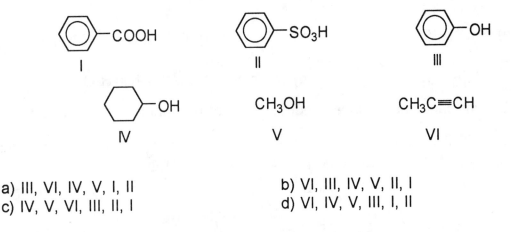

a) benzoic acid (C6H5COOH) b) m-nitrobenzoic acid (COOH, NO_2) c) p-aminobenzoic acid (COOH, NH_2) d) p-nitrobenzoic acid (COOH, NO_2)

788. Arrange the following compounds in the order of increasing acid strength (weakest first).

C_6H_5–COOH (I) C_6H_5–SO_3H (II) C_6H_5–OH (III)

cyclohexyl–OH (IV) CH_3OH (V) $CH_3C{\equiv}CH$ (VI)

a) III, VI, IV, V, I, II b) VI, III, IV, V, II, I
c) IV, V, VI, III, II, I d) VI, IV, V, III, I, II

TIP In problems like this it is often useful to find the strongest and weakest. The strongest of this set is the sulfonic acid, II, and the weakest is the carbon acid, VI. The remaining compounds are all carbon-oxygen acids and resonance factors determine the order.

C. PREPARATIONS

789. Which of the following reactions is best for preparing phenylacetic acid?

a) $C_6H_5CH_2Cl \xrightarrow[\text{ether}]{\text{Mg}} \xrightarrow{CO_2} \xrightarrow{H_3O^{\oplus}}$

b) $C_6H_5CH_2OH \xrightarrow{KMnO_4} \xrightarrow{H_3O^{\oplus}}$

c) $C_6H_5CH_2CH_2Br \xrightarrow[CH_3CH_2OH]{NaCN} \xrightarrow[\text{heat}]{H_3O^{\oplus}}$

d) $C_6H_5\overset{O}{\overset{\|}{C}}CH_3 \xrightarrow[NaOH]{Br_2} \xrightarrow{H_3O^{\oplus}}$

790. What are the best conditions for the following preparation?

1-bromo-1-methylcyclohexane → 1-methylcyclohexanecarboxylic acid (cyclohexane ring bearing CH_3 and Br → cyclohexane ring bearing CH_3 and COOH)

a) NaCN, KOH, $H_3O^{\oplus}$ b) KOH then Ag_2O in NaOH, H_2O then HCl

c) HNO_3, H_2O d) Mg in ether then CO_2 then $H_3O^{\oplus}$

791. Which reaction sequence is not suitable for preparing pentanoic acid?

a) $CH_3(CH_2)_3Cl \xrightarrow[\text{ether}]{\text{Mg}} \xrightarrow{CO_2} \xrightarrow{H_3O^{\oplus}}$

b) $CH_3(CH_2)_4OH \xrightarrow{KMnO_4} \xrightarrow{H_3O^{\oplus}}$

c) $CH_3(CH_2)_4Br \xrightarrow[CH_3CH_2OH]{NaCN} \xrightarrow[\text{heat}]{H_3O^{\oplus}}$

d) $CH_3(CH_2)_3\overset{O}{\overset{\|}{C}}CH_3 \xrightarrow[NaOH]{Br_2} \xrightarrow{H_3O^{\oplus}}$

792. Which of the following reactions will yield the same carboxylic acid?

I. cyclopentyl–CH_2CHO $\xrightarrow{K_2Cr_2O_7}$

II. cyclopentyl–$\overset{O}{\overset{\|}{C}}CH_3$ $\xrightarrow[Cl_2]{NaOH}$

III. cyclopentyl–CH_2Cl $\xrightarrow{KCN}$ $\xrightarrow[heat]{H_2SO_4}$

IV. cyclopentyl–CH_2CH_2Br $\xrightarrow[ether]{Mg}$ $\xrightarrow{CO_2}$ $\xrightarrow{H_3O^{\oplus}}$

a) I, II b) II, III c) III, IV d) I, III

TIP Synthetic methods for carboxylic acids can: 1) increase the number carbon atoms, (reactions III and IV); 2) decrease the number of carbon atoms (reaction II); 3) leave the number of carbon atoms unchanged (reaction I). To arrive at a reasonable answer, we must count carbon atoms in the different products.

793. Which of the following reactions will not prepare naphthanoic acid?

a) 1-naphthyl–OH $\xrightarrow{K_2Cr_2O_7}$

b) 1-naphthyl–CHO $\xrightarrow{Ag(NH_3)_2NO_3}$ $\xrightarrow[H_2O]{HCl}$

c) 1-naphthyl–CN $\xrightarrow[heat]{H_2O,\ HNO_3}$

d) 1-naphthyl–Br $\xrightarrow[ether]{Mg}$ $\xrightarrow{CO_2}$ $\xrightarrow{H_3O^{\oplus}}$

794. What is the best reaction route for the preparation of propionic acid from ethyl bromide?

a) $\xrightarrow[\text{ether}]{\text{Mg}}$ $\xrightarrow{CH_3COBr}$

b) $\xrightarrow[CH_3CH_2OH]{K^{\oplus}\ {}^{\ominus}OCH_2CH_3}$ $\xrightarrow{HOBr}$ $\xrightarrow{H_2SO_4}$

c) $\xrightarrow{H_2CrO_4}$

d) $\xrightarrow[\text{acetone}]{LiCN}$ $\xrightarrow[\text{heat}]{H_3O^{\oplus}}$

D. REACTIONS

795. What are the best conditions for the following transformation?

(cyclohexane)–COOH → (cyclohexane)–COCl

a) Ag_2O ; Cl_2, CCl_4 b) Cl_2 c) $SOCl_2$ d) C_6H_5COCl

796. What is the major product from heating $HOOCCH_2CH_2COOH$?

a) (succinic anhydride: five-membered ring with O and two C=O) b) CH_3CH_2COOH c) $CH_2{=}CHCOOH$ d) (cyclobutane-1,2-dione)

797. Benzene is reacted with acetyl chloride and aluminum chloride. The product is then reacted with bromine and sodium hydroxide, followed by work-up in aqueous acid. What is the final product?

a) C_6H_5–CBr_2CHO (Br₂CCHO on benzene) b) C_6H_5–CH_2COBr (H₂CCOBr on benzene) c) C_6H_5–COOH d) C_6H_4(COCH₃)(CH₃) (COCH₃ and meta CH₃ on benzene)

TIP Friedel craft acetylation followed by a haloform reaction is a useful method of preparing benzoic acids. The initial product is the acetophenone, and then the products from reaction of a methyl ketone with bromine and base are the salt of the benzoic acid and bromoform.

798. What is the product of heating glutaric acid?

a) b)

c) HOOCCHCH2COOH
OOC(CH2)2COOH d)

799. Toluene is reacted with potassium permanganate in basic solution and the product is treated with aqueous acid, followed by reaction with thionyl chloride. What is the final product of these reactions?

a) CH_2OCl b) COCl c) CH_2Cl d) Cl, COOH

800. Phenol is treated with carbon dioxide in basic solution. The product is then reacted with thionyl chloride. What is the principal product from these reactions?

a) COCl, OH b) COOH, Cl c) OCOCl d) COCl

TIP This is the classic Kolbe synthesis used in the commercial preparation of aspirin. In this case, the initial salicylic acid (ortho hydroxy benzoic acid) is converted to the acid chloride by reaction with thionyl chloride. Answers c) and d) are incompatible with the Kolbe reaction, and answer b) is the wrong chemistry for conversion of the phenol OH to a halide.

CHAPTER 20 FUNCTIONAL DERIVATIVES OF CARBOXYLIC ACIDS

A. NOMENCLATURE and STRUCTURE

801. Which of the following is the correct structure for maleic anhydride?

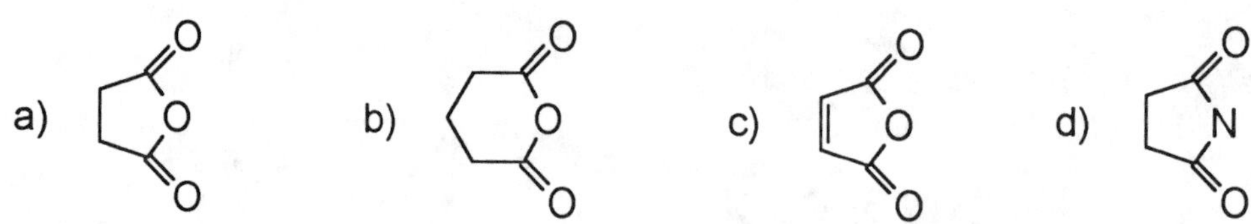

802. Which carboxylic acid derivative is the following compound?

O O
|| ||
$-COCCH_3$

a) ester b) ether c) anhydride d) lactone

803. Which of the following structures have the correct name?

I. $CH_3COOCH_2CH_3$ methyl acetate

II. C_6H_5COCl benzoyl chloride

III. $HOCN(CH_3)_2$ N,N'-dimethylformamide

IV. H_3C— —COONa
methyl-cyclohexyl- acetate

V. O O O
phthalic anhydride

a) I, III, V b) II, IV, V c) I, III, IV d) II, III, V

804. Ethyl carbamate has which carboxylic acid derivative unit(s)?

a) lactam b) ester c) amide d) ester and amide

805. Which functional unit(s) is (are) in the following molecule?

O
NH

a) lactam and imide b) lactam and amide c) amide d) lactam

806. Which functional unit is in the following molecule?

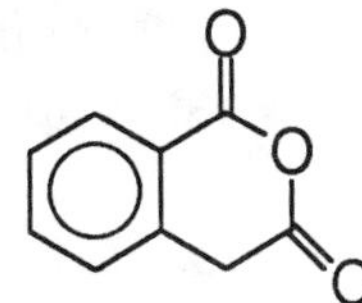

a) ester b) ether c) anhydride d) lactone

807. Which of the following is the structure for aspirin?

a) OH, OH, CH_3 b) $COOCH_3$, OH c) $OCOCH_3$, COOH d) OCH_3, OH

808. Match a numbered item with a lettered item.

I. CH_2O	A. Sibelius and Saarinen
II. parafins	B. used in underground chemistry
III. 2 moles	C. a source of kelp
IV. IO_4 IO_4 IO_4 IO_4	D. the periodic table
V. catalyst	E. transparence
VI. Ma(H)C=C(H)Pa	F. a stockyard inventory

TIP Try II and A; VI and F is definitely not correct.

B. PHYSICAL PROPERTIES

809. Carboxylic acids and amides have higher boiling points and melting points than esters and anhydrides mostly because of what properties?

a) dipolar association
b) resonance stabilization
c) conjugated functional groups
d) hydrogen bonding

810. What is the most likely structure for a compound that has the molecular formula of $C_5H_{10}O_2$ and a proton NMR spectrum with the following resonances:

Doublet	1 ppm	(6 protons)
Septet	2.3 ppm	(1 proton)
Singlet	3.3 ppm	(3 protons)

a) methyl isobutyrate
b) isobutyric acid
c) pentanoic acid
d) methyl butyrate

811. What is the most likely structure for a compound that has an infrared adsorption at about 2200 cm^{-1} and a proton NMR spectrum having no peaks with a chemical shift greater than 3 ppm?

a) 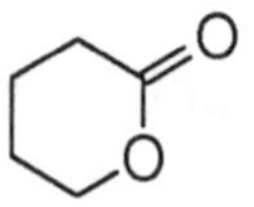 O, O
b) 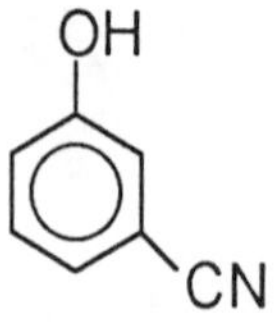 OH, CN
c) 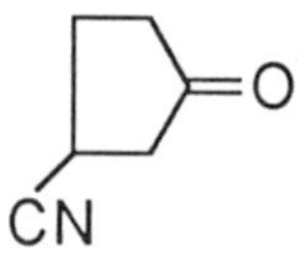 O, CN
d) 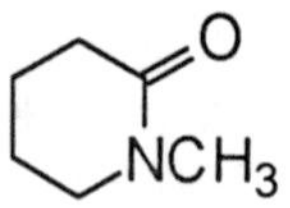 O, NCH3

TIP This infrared absorption is in a highly diagnostic region. Only carbon-carbon and carbon-nitrogen triple bonds have absorption in this region. This rules out answers a) and d). Answer b) has three aromatic protons whose resonances will be in the range of 6-8 ppm and is therefore not the answer.

812. Which property of the functional unit, Z, directs the order of reactivity of nucleophilic substitution for compounds having the following general structure?

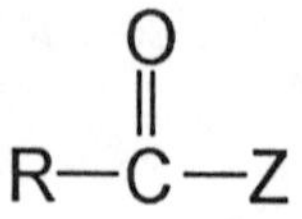

$$R-\overset{\overset{\displaystyle O}{\|}}{C}-Z$$

a) polarity
b) hydrogen bonding
c) ionization potential
d) basicity

813. What is the order of increasing reactivity as a leaving group for the following functional groups (lowest first)?

I. R-COOH II. R-COCl III. R-$CONH_2$
IV. R-COOR V. R-COOCOR

a) III, IV, I, V, II b) V, IV, I, III, II c) I, V, IV, II, III d) III, V, IV, I, II

814. The following reaction is fastest for which functional group, Z?

$$C_6H_5-\overset{O}{\overset{\|}{C}}-Z + \overset{\ominus}{O}H \longrightarrow C_6H_5-\overset{O}{\overset{\|}{C}}-\overset{\ominus}{O} + HZ$$

a) C_6H_5O- b) CH_3- c) CH_3O- d) $CH_3O\overset{O}{\overset{\|}{C}}O-$

TIP For questions 812-814 the leaving group controls reactivity. As in all reactions with leaving groups, reactivity is primarily determined by basicity. Thus the anions of stronger acids are better leaving groups than anions of weaker acids. A simple way to rank groups is to place a proton on each leaving group and then rank the relative acid strengths. In question 813, the protonated leaving groups are I. H_2O, II. HCl. III. NH_3, IV. ROH. V. RCOOH. Now we just put these in increasing order of acidity to get the increasing order of reactivity. In question 814, we do the same. Methane (b) is orders of magnitude a weaker acid than methyl carbonic acid (d), etc.

815. What is the order of increasing reactivity for esterification by acetic acid (least first)?

I. CH_3CH_2OH II. $(CH_3)_2CHCH_2OH$ III. $(CH_3)_3CCH_2OH$

a) I, II, III b) III, II, I c) II, III, I d) II, I, III

C. PREPARATIONS

816. What is the major product from the following reaction sequence?

$$C_6H_6 + CH_3COCl \xrightarrow{AlCl_3} \xrightarrow[Cl_2]{NaOH} \xrightarrow{H_3O^{\oplus}} \xrightarrow[H_3O^{\oplus}]{CH_3OH}$$

a) $C_6H_5-COCH_3$ b) $C_6H_5-COOCH_3$

c) $C_6H_5-OOCCH_3$ d) $C_6H_5-C(OCH_3)_2CH_3$

817. What is the major product when ethylenediamine is heated with dimethyloxalate?

a) $CH_3\overset{O}{\overset{\|}{C}}NHCH_2CH_2CH\overset{O}{\overset{\|}{C}}CH_3$ b) 1,4-dioxane-2,3-dione (six-membered ring: O, O, two C=O)

c) piperazine-2,3-dione (six-membered ring: N, N, two C=O) d) $CH_3\overset{O}{\overset{\|}{C}}NHCH(CH_3)NH\overset{O}{\overset{\|}{C}}CH_3$

818. What is the major product from the following reaction sequence?

$$CH_3(CH_2)_3COCl \xrightarrow{NH_3} \xrightarrow[NaOH]{Br_2}$$

a) $CH_3CH_2CH_2CHBrCONH_2$ b) $CH_3CH_2CH_2C(Br)_2COO^{\ominus}\ NH_4^{\oplus}$

c) succinimide (five-membered ring with NH flanked by two C=O) d) $CH_3CH_2CH_2CH_2NH_2$

819. What is the major product from the following reaction sequence?

Phenol (OH on benzene) $\xrightarrow{NaOH}$ $\xrightarrow{CO_2}$ $\xrightarrow{H_3O^{\oplus}}$ $\xrightarrow{(CH_3CO)_2O}$

a) OH, $COOCH_3$ (ortho-substituted benzene)

b) $OCOCH_3$, $COOH$ (ortho-substituted benzene)

c) benzofuranone ring with =O and OCH_3

d) $OCOCH_3$, $COOH$ (para-substituted benzene)

820. What is the major product from the following reaction?

$$CH_3COOCH_2CH(OOCCH_3)CH_2OOCCH_3 + CH_3OH \xrightarrow{H_3O^{\oplus}}$$

a) $CH_3CH_2OH + CH_3O\overset{O}{\overset{\|}{C}}CH_3$

b) $HOCH_2CH(OH)CH_2OH + CH_3COOH$

c) $HOCH_2CH(OH)CH_2OH + CH_3O\overset{O}{\overset{\|}{C}}CH_3$

d) $H_3CO\overset{O}{\overset{\|}{C}}CH_2CH(COOCH_3)CH_2(COOCH_3) + CH_3CH_2OH$

TIP This is the often confusing acid catalyzed **trans**esterification reaction. The new ester will be composed of the original acid component and the alcohol of the reactants. The remaining product is the alcohol of the original ester. The correct answer requires correct identification of the original ester. Does the hint that the starting ester in this problem is glycerol **triacetate** and that glycerol has three hydroxyl groups guide you to the answer?

821. What is the major product from the following reactions?

$$Cl(CH_2)_2COOH \xrightarrow{LiCN} \xrightarrow{acetone} \xrightarrow[H_2SO_4,\ heat]{CH_3CH_2OH}$$

a) $CH_3COCH_2COOCH_2CH_3$ b) $CH_3CH_2OOC(CH_2)_2COOCH_2CH$

c) $CH_3CH_2OCH_2COOCH_2CH_3$ d) $CH_2(COOCH_2CH_3)_2$

D. REACTIONS

822. What are the best conditions for the following conversion?

$$\triangleright\!-\overset{O}{\overset{\|}{C}}OH \longrightarrow \triangleright\!-\overset{O}{\overset{\|}{C}}CH_2CH_3$$

a) $SOCl_2$; $(CH_3CH_2)_2Cd$ b) $SOCl_2$; CH_3CH_2MgBr

c) $LiAlH_4$; $SOCl_2$; Mg in ether ; CH_3CHO d) CH_3OH, $H_3O^{\oplus}$; $LiAlH_4$

823. Which of the following reagents will convert benzoyl chloride to an alcohol?

I. phenyl magnesium bromide II. lithium aluminum hydride
III. dimethyl cadmium IV. sodium borohydride

a) I, III b) II, IV c) III, IV d) I, II

824. Which conditions are best for the following preparation?

$$\text{cyclopentyl}-COOH \longrightarrow \text{cyclopentyl}-CH_3$$

a) $LiAlH_4$; $H_3O^{\oplus}$; HCl; Mg in ether ; H_2O

b) $SOCl_2$; $LiAlH_4$; $H_3O^{\oplus}$; Zn(Hg), HCl

c) NH_2NH_2, KOH ; HCl ; Mg in ether ; H_2O

d) CH_3Li, H_2O

825. What is the best method for the following conversion?

$$CH_3CH_2COOH \longrightarrow CH_3CH_2COCH_3$$

a) $SOCl_2$; $(CH_3)_2Cd$

b) $SOCl_2$; $LiAlH[OC(CH_3)_3]_3$

c) $SOCl_2$; $LiAlH[OC(CH_3)_3]_3$; CH_3CH_2MgBr ; $H_3\overset{\oplus}{O}$, $K_2Cr_2O_7$

d) HCl; $(CH_3CH_2)_2Cd$

826. What are the best conditions for the following conversion?

$$C_6H_5\text{-}COOH \longrightarrow C_6H_5\text{-}CH_2\text{-}C_6H_5$$

a) $SOCl_2$; C_6H_6, $AlCl_3$; Zn(Hg), HCl

b) $SOCl_2$; C_6H_5MgBr, ether

c) $SOCl_2$; C_6H_6, $AlCl_3$; $(C_6H_5)Cd$

d) $LiAlH_4$; $SOCl_2$; Mg in ether ; H_2O ; C_6H_6, $AlCl_3$

TIP Answers a), b) and c) contain as a first step the conversion to an acid chloride, so we should begin our thinking there. Answer b) leads to a tertiary alcohol so is ruled out based on the structure of the product. In a) and c), the second product is benzophenone. The final step in c), however, involves incorrect chemistry. Answer d) has different flaws. The reduction before reaction with thionyl chloride leads to an alcohol and then benzyl chloride. Grignard formation is then followed by protonation to toluene and then the final step just doesn't go. The final step in answer a) incorporates the Clemmensen reaction for successful completion of the intended synthesis.

827. What are the best conditions for the following conversion?

$$CH_3CH_2\overset{\overset{O}{\|}}{C}OH \longrightarrow CH_3CH_2\overset{\overset{O}{\|}}{C}CH_2CH_3$$

a) $SOCl_2$; $(CH_3)_2Cd$ b) HCl ; $(CH_3CH_2)_2Cd$ c) $SOCl_2$; $LiAlH[OC(CH_3)_3]_3$

d) $SOCl_2$; $LiAlH[OC(CH_3)_3]_3$; CH_3CH_2MgBr ; $H_3\overset{\oplus}{O}$, $K_2Cr_2O_7$

828. What are the best conditions for the following conversion?

C_6H_5–COOH → C_6H_5–CH_2O–(cyclopentyl)

a) $SOCl_2$; $NaBH_4$; NaH ; (cyclopentyl)–Br

b) PCl_5 ; $LiAlH[OC(CH_3)_3]_3$; (cyclopentyl)–MgBr , $H_3O^{\oplus}$

c) $SOCl_2$; (cyclopentyl)–Cd–(cyclopentyl) ; $LiAlH_4$; $H_3O^{\oplus}$

d) $LiAlH_4$; $H_3O^{\oplus}$; $SOCl_2$; (cyclopentyl)–ONa

829. Which of the following conditions best describes the hydrolysis of ethyl acetate in the pH range of 1-14?

a) The rate increases with increasing pH.
b) The rate decreases with increasing pH.
c) The rate reaches a minimum with increasing pH.
d) The rate reaches a maximum with increasing pH.

TIP The ester hydrolysis is both acid catalyzed and base promoted. Therefore in a neutral solution (pH of 7) reaction is slowest, and the rate increases at both lower and higher pH. This translates to a high rate at pH 1, going to a minimum at pH 7, back to a high rate at pH 14. This activity gives a hint as to what is happening at physiological pH.

830. The following ester is labeled with ^{18}O at the indicated positions. Where does the label appear in the hydrolyzed products?

CH_3CH_2OOC–C_6H_4–SO_2OCH_3 (arrows indicate the labeled O in CH_3CH_2**O**OC and in SO_2**O**CH_3)

a) methanol
b) ethanol
c) both methanol and ethanol
d) neither methanol nor ethanol

831. What is the major product from the following reactions?

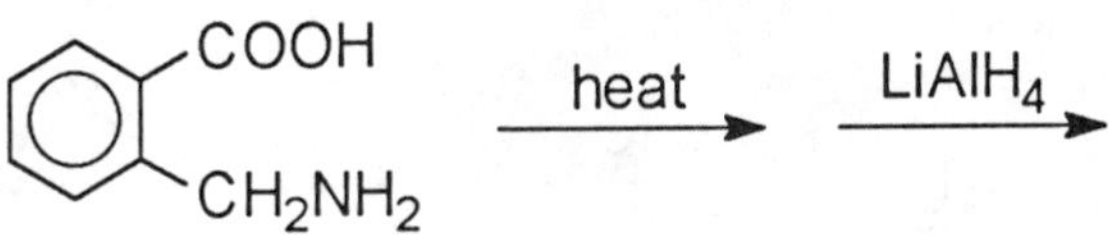
COOH
CH2NH2
heat
LiAlH4

a) 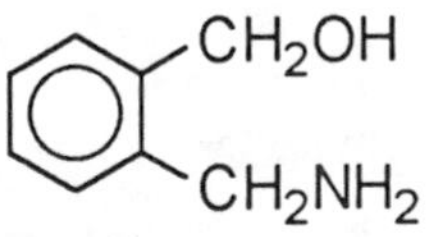
CH2OH
CH2NH2
b) 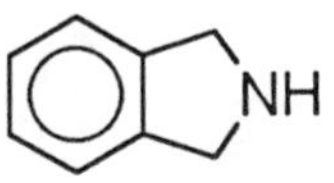
NH
c) 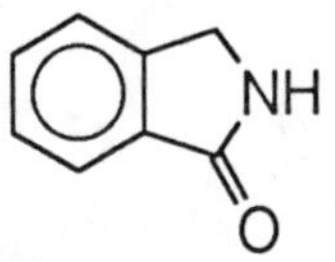
NH
O
d) 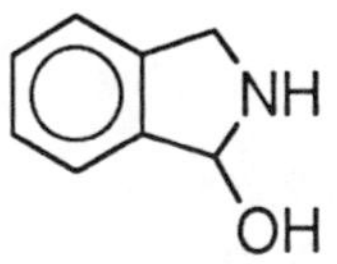
NH
OH

832. What is the major product from the following reactions?

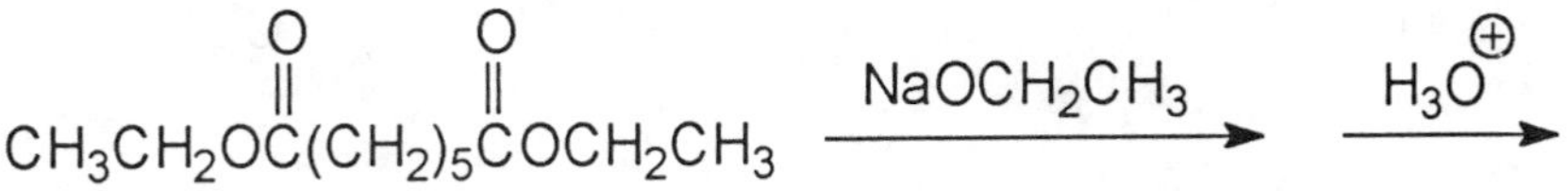
CH3CH2OC(CH2)5COCH2CH3
NaOCH2CH3
H3O⊕

a) 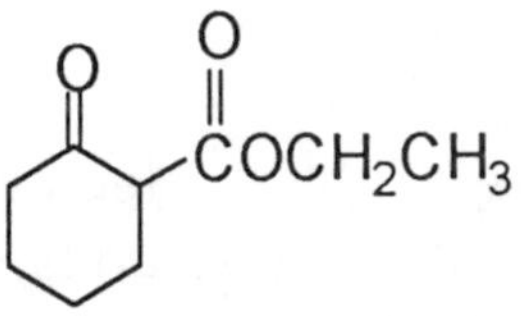
COCH2CH3
b) 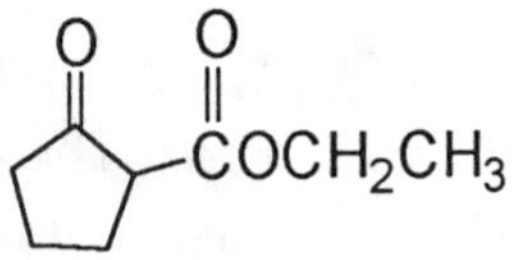
COCH2CH3

c) 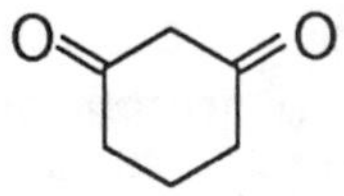
d) 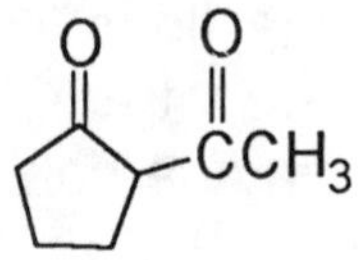
CCH3

833. What is the major product from the following reactions?

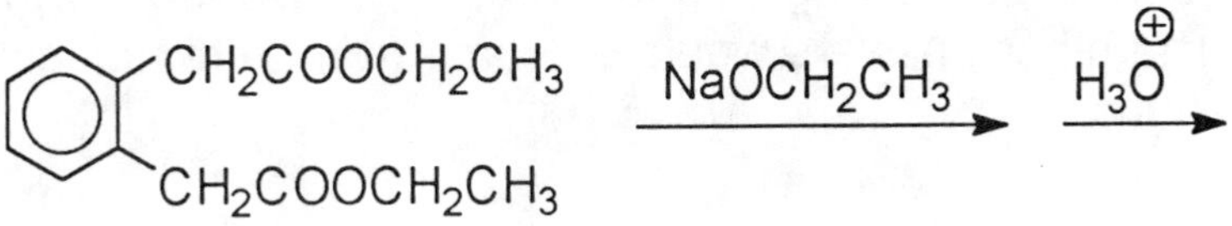
CH2COOCH2CH3
CH2COOCH2CH3
NaOCH2CH3
H3O⊕

a) 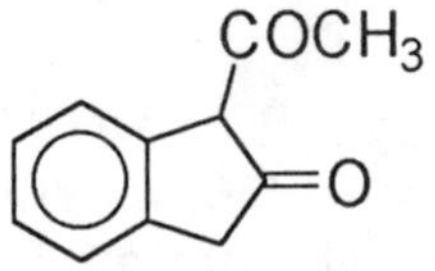
COCH3
b) 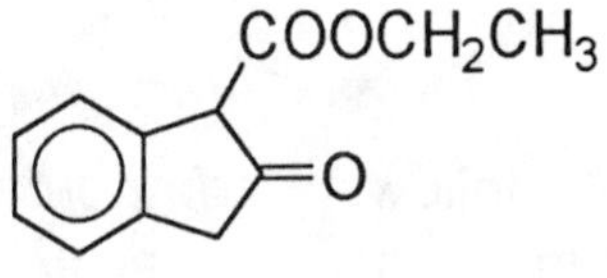
COOCH2CH3

c) 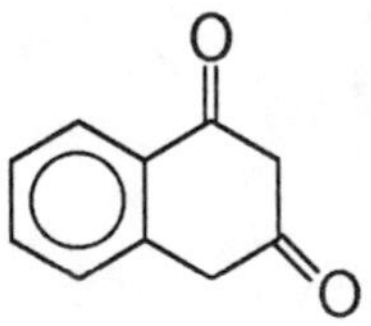
d) 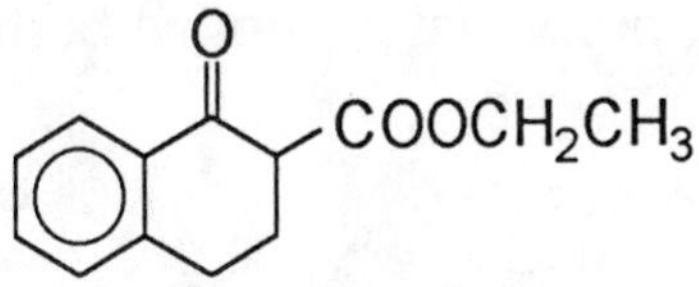
COOCH2CH3

834. What is the major product from the following reactions?

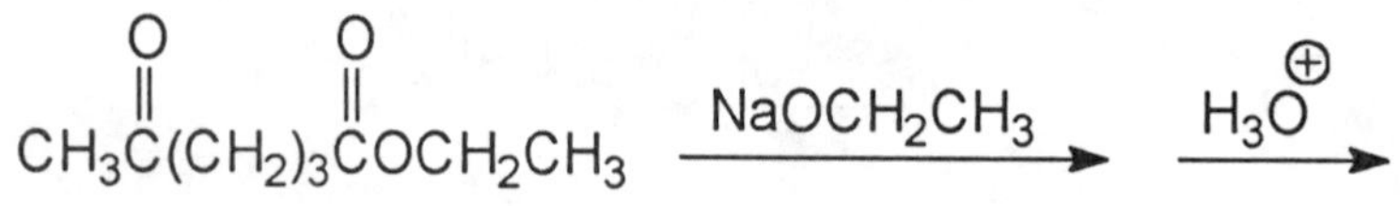

$$CH_3\overset{O}{\overset{\|}{C}}(CH_2)_3\overset{O}{\overset{\|}{C}}OCH_2CH_3 \xrightarrow{NaOCH_2CH_3} \xrightarrow{H_3O^{\oplus}}$$

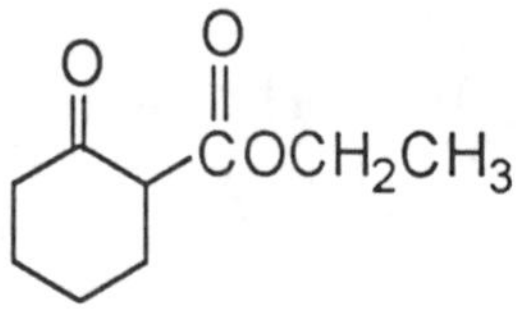

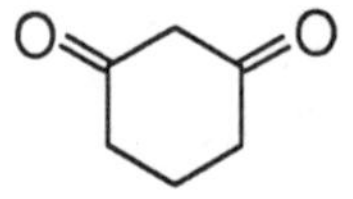

a) 2-($\overset{O}{\overset{\|}{C}}OCH_2CH_3$)cyclohexanone

b) cyclohexane-1,3-dione

c) 2-($\overset{O}{\overset{\|}{C}}OCH_3$)cyclohexanone

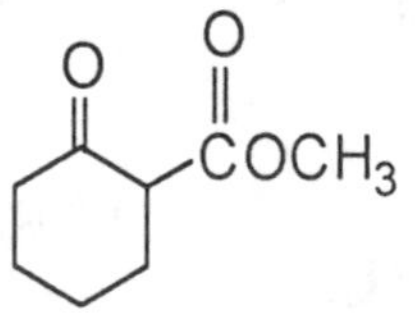

d) 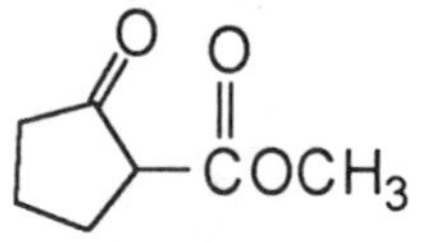 2-($\overset{O}{\overset{\|}{C}}OCH_3$)cyclopentanone

TIP For questions 832- 834. These three reactions are all intramolecular ester condensations. The first two are Claisen reactions (two ester units) and the third is a Claisen-Schmidt (a ketone and an ester).

For the first two (questions 832 and 833), the product will be a cyclic beta-keto ester since these are intramolecular reactions. Ring size is determined by the structure, so counting carbon atoms between the two ester functions leads to a six membered ring in 832 and a five membered ring in 833.

In question 834, the reaction is an intramolecular Claisen-Schmidt in which the ketone serves as the enolate source and the ester furnishes the carbonyl. The product then will be 1,3 diketone and, in a cyclic system when there is a choice, the most stable ring size is formed. Only answer b) fulfills all the criteria. The other answers would all come from intramolecular Claisen condensations.

835. Which of the following compounds is least likely to be a useful synthetic compound in a crossed Claisen condensation with ethylacetate?

a) ethyl propionate b) ethyl benzoate c) ethyl formate d) ethyl oxalate

836. What is the major product from the crossed Claisen condensation reaction of acetone and diethyl carbonate?

a) $CH_3\overset{\overset{O}{\|}}{C}CH_2\overset{\overset{O}{\|}}{C}COOCH_2CH_3$ b) $H\overset{\overset{O}{\|}}{C}CH_2\overset{\overset{O}{\|}}{C}OCH_2CH_3$

c) $CH_3\overset{\overset{O}{\|}}{C}CH_2\overset{\overset{O}{\|}}{C}OCH_2CH_3$ d) $CH_3CH_2O\overset{\overset{O}{\|}}{C}CH_2\overset{\overset{O}{\|}}{C}OCH_2CH_3$

837. What is the major product from the crossed Claisen condensation reaction of acetone and diethyl oxalate?

a) $CH_3\overset{\overset{O}{\|}}{C}CH_2\overset{\overset{O}{\|}}{C}COOCH_2CH_3$ b) $CH_3\overset{\overset{O}{\|}}{C}CH_2\overset{\overset{O}{\|}}{C}OCH_2CH_3$

c) $CH_3\overset{\overset{O}{\|}}{C}CH_2\overset{\overset{O}{\|}}{C}H$ d) $CH_3CH_2O\overset{\overset{O}{\|}}{C}CH_2\overset{\overset{O}{\|}}{C}OCH_2CH_3$

E. "ROADMAP REVIEW"

838. The aldehyde, propanal, is reacted with a Grignard reagent, phenyl magnesium bromide, to give an addition product upon workup. This product is reacted further with acetic anhydride. What is the most reasonable structure for the final product?

a) $C_6H_5-CH_2CH_2CH_2O\overset{\overset{O}{\|}}{C}CH_3$ b) $C_6H_5-\underset{\underset{CH_2CH_3}{|}}{C}HO\overset{\overset{O}{\|}}{C}CH_3$

c) $C_6H_5-\underset{\underset{CH_2CH_3}{|}}{C}H\overset{\overset{O}{\|}}{C}OCH_3$ d) $CH_2CH_2\overset{\overset{C_6H_5}{|}}{\underset{\underset{C_6H_5}{|}}{C}}O\overset{\overset{O}{\|}}{C}CH_3$

839. Which of the following series of reactions lead to acrylamide?

a) acetone $\xrightarrow{\text{HCN}}$ $\xrightarrow[\text{room temp.}]{\text{conc. } H_2SO_4}$

b) acetaldehyde $\xrightarrow{\text{HCN}}$ $\xrightarrow[\text{room temp.}]{\text{conc. } H_2SO_4}$

c) acetone $\xrightarrow[\text{NaOH}]{Br_2}$ $\xrightarrow{NH_3}$

d) acetaldehyde $\xrightarrow[\text{NaOH}]{Br_2}$ $\xrightarrow{NH_3}$

840. Ethyl benzoate is reacted with DIBALH in an intermediate step in a synthesis. The product is analyzed by infrared spectroscopy and a standard chemical test. Which chemical test would you use and which band would you look for in the IR spectrum?

Chemical Test	IR Band
A. Lucas	I. 1610 cm^{-1}
B. Br_2 in CCl_4	II. 1710 cm^{-1}
C. Tollens	III. 2200 cm^{-1}
D. Baeyer	IV. 3300 cm^{-1}

a) A and III b) C and II c) D and I d) B and IV

TIP A DIBALH reduction of an ester leads to an aldehyde. Aldehydes are classified chemically by the Tollens reagent and spectroscopically by the sharp band at 1710 cm^{-1} in the infrared.

841. An unknown compound, X, has the molecular formula of C_4H_7NO and a band in the infrared spectrum in the region of 3300 cm^{-1}. The compound reacts with bromine in sodium hydroxide to give Compound Y which has the molecular formula, C_3H_7N. Compound Y also has an IR band in the 3300 cm^{-1} region. What is the most reasonable structure for Compound X?

a) [structure: NH_2-substituted dihydrofuran ring with O] b) [structure: six-membered ring with O and N] c) cyclopropyl–$CONH_2$ d) $CH_3CH_2CH_2CN$

842. Toluene is reacted with sulfuric acid and nitric acid, and the product is reduced with zinc and hydrochloric acid to give an intermediate compound, A. Compound A is treated with sodium nitrite and hydrochloric acid followed by reaction with cuprous cyanide to give Compound B. Compound B is treated with lithium aluminum hydride to give the final product, Compound C. Which of the following structures is the product, Compound C, that the synthesis was designed to give?

a) 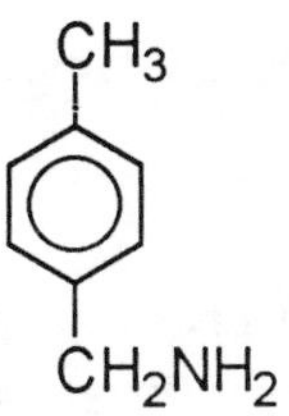 CH_3 / CH_2NH_2 b) 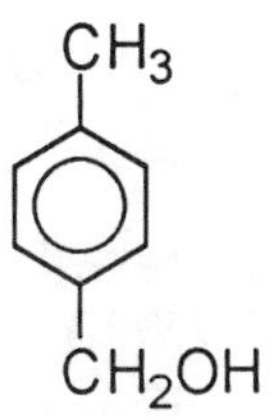 CH_3 / CH_2OH c) 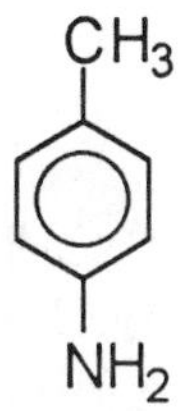 CH_3 / NH_2 d) 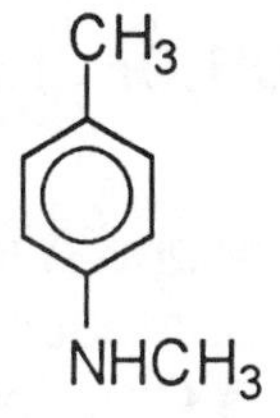 CH_3 / $NHCH_3$

TIP The reactions are: nitration, reduction, diazotization, Sandemeyer with CuCN and finally reduction of the nitrile. One way to see through the maze is to note that the product must add one carbon atom and must be a primary amine. Answers b), c) and d) violate those requirements.

843. Toluene is reacted with NBS and then treated with aqueous base. The product is reacted with chromium trioxide in pyridine and then with zinc and α-bromoethyl acetate. This product is treated first with base and then neutralized with acid. What is the most likely structure for the product?

a) 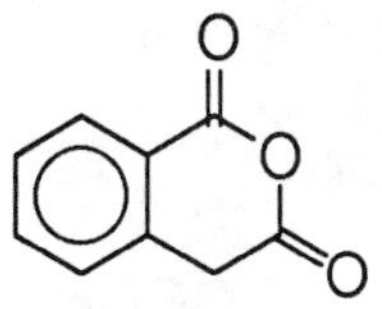 O O O b) 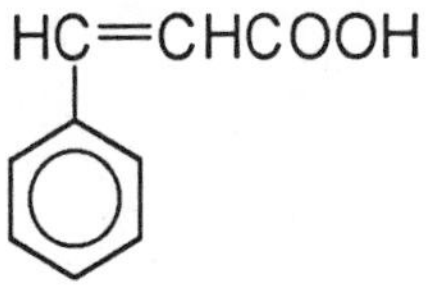 HC=CHCOOH

c) 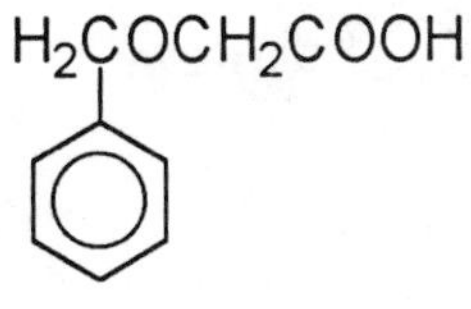 H_2COCH_2COOH d) 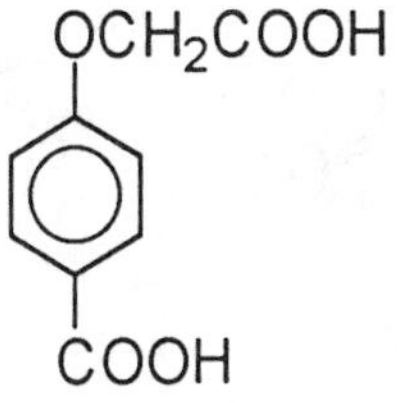 OCH_2COOH / COOH

844. What is the final product of the reaction of benzoic acid with thionyl chloride and ammonia, followed by reaction with sodium hydroxide and bromine, and finally reaction with bromine?

a) 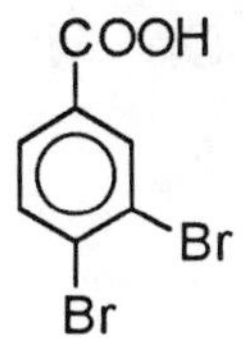 COOH Br Br b) 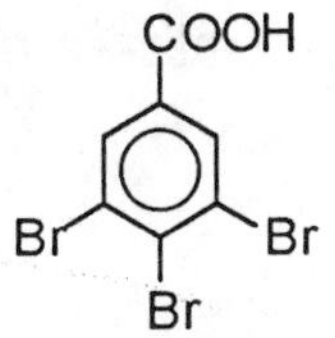 COOH Br Br Br c) 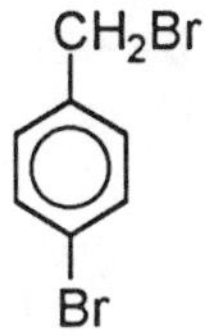 CH_2Br Br d) 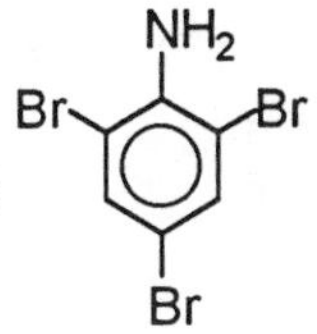 NH_2 Br Br Br

845. Alpha bromopropionic acid is reacted with ethanol and hydrochloric acid. The product is reacted with zinc and acetaldehyde in toluene. Which of the following is most likely to be the final product?

a) $HC(=O)CH_2C(=O)CH(OCH_3)OCH_2CH_2CH_3$

b) $CH_3CH(Br)CH(OH)CH_2OH$

c) $CH_3CH(OH)CH(CH_3)CH_2OH$

d) $CH_3CH(OH)CH(CH_3)COOCH_2CH_3$

846. Compound A is treated with sodium dichromate in sulfuric acid and water to give Compound B. Compound B is reacted with ethanol and hydrochloric acid to give Compound C. Reaction of Compound C with sodium ethoxide followed by treatment with dilute acid gives Compound D, the final product. Given the structure of Compound A as follows, what is the most likely structure for Compound D?

Compound A: benzene ring bearing two ortho CH_2CH_2OH groups

Compound A

a) indane with CHO substituent

b) indene with $COOCH_2CH_3$ substituent

c) tetralin-1,3-dione (O, =O)

d) indanone with $COOCH_2CH_3$ substituent and =O

TIP The oxidation to Compound B yields a dicarboxylic acid, and then esterification leads to Compound C, a diester. Reaction of the diester with ethoxide is then an intramolecular Claisen condensation. Such a reaction always leads to a beta-keto ester and in this case a cyclic beta-keto ester. Only d) is such a structure.

847. Chlorobenzene is added to a mixture of magnesium in ether and then treated with cadmium chloride to give Compound A. Compound A is reacted with butanoyl chloride, followed by reaction with hydrazine to give Compound B. Compound B is treated with potassium hydroxide and then heated to give the final product, C. What is the most likely structure for Compound C?

a) $C_6H_5CH_2CH_2CH_2CH_3$

b) $CH_3CH_2CH_2CH(OH)C_6H_5$

c) $C_6H_5C(=NNH_2)CH_2CH_2CH_2CH_3$

d) $C_6H_5CH_2CH_2C(=O)NH_2$

848. Compound A is treated with thionyl chloride and ammonia to give Compound B. Compound B is then reacted with bromine and sodium hydroxide to give Compound C. Compound C is heated to give the final product, Compound D. Given the structure for Compound A as follows, what is the most likely structure for Compound D?

Cyclohexane ring bearing CCH_3 (C=O) and CH_2COOH on adjacent carbons

Compound A

a) Cyclohexane ring bearing COOH and CH_2CONH_2 on adjacent carbons

b) Cyclohexane ring bearing CCH_3 (C=O) and CH_2NH_2 on adjacent carbons

c) Cyclohexane fused to a five-membered ring: NH, C=O, NH

d) Cyclohexane fused to a five-membered lactam ring: C=O, NH, CH_2

TIP The formation of the amide B is unexceptional. The really tricky part of this problem is to detect that two different functional groups will react with bromine and sodium hydroxide. The amide will undergo a Hoffmann reaction to give the amine. The methyl ketone will undergo a bromoform reaction to give the carboxylic acid. In this molecule the amine and carboxylic acid react intramolecularly with heat to lead to the lactam in answer d).

CHAPTER 21 ENOLATE ANIONS and ENAMINES

A. DIRECTED and MIXED ALDOL REACTIONS

849. What are the major products when the following ketones are converted to the enolate anions under thermodynamic control, and then reacted with trimethylsilyl chloride?

$$CH_3\overset{\overset{O}{\|}}{C}CH_2(CH_2)_2CH_3 \qquad\qquad \text{(cyclohexanone ring, C=O)}\!-\!CH(CH_3)_2$$

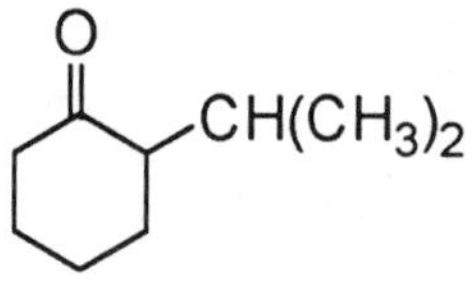

I. $CH_2{=}\overset{\overset{OSi(CH_3)_3}{|}}{C}CH_2(CH_2)_2CH_3$

II. $CH_3\overset{\overset{OSi(CH_3)_3}{|}}{C}{=}CH_2(CH_2)_2CH_3$

III. 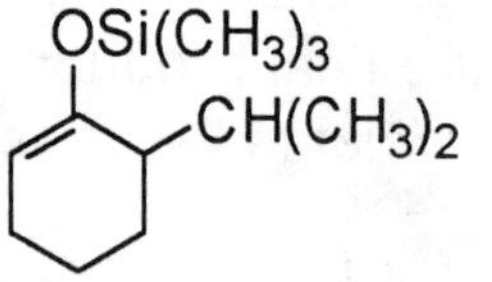 $OSi(CH_3)_3$, $CH(CH_3)_2$ (cyclohexene ring)

IV. 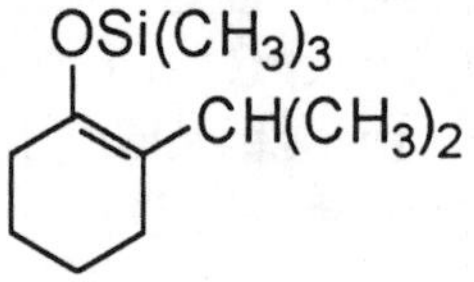 $OSi(CH_3)_3$, $CH(CH_3)_2$ (cyclohexene ring)

a) I, III b) II, III c) I, IV d) II, IV

850. Which of the following reactions will proceed as written?

I. $CH_3CH_2\overset{\ominus}{O} + CH_3\overset{\overset{O}{\|}}{C}CH_2\overset{\overset{O}{\|}}{C}OEt \longrightarrow CH_3\overset{\overset{O}{\|}}{C}\overset{\ominus}{C}H\overset{\overset{O}{\|}}{C}OEt + EtOH$

II. $CH_3CH_2\overset{\ominus}{O} + CH_3\overset{\overset{O}{\|}}{C}OEt \longrightarrow \overset{\ominus}{C}H_2\overset{\overset{O}{\|}}{C}OEt + EtOH$

III. $CH_3CH_2OH + CH_3\overset{\overset{O}{\|}}{C}\overset{\ominus}{C}H_2 \longrightarrow CH_3CH_2\overset{\ominus}{O} + CH_3\overset{\overset{O}{\|}}{C}CH_3$

IV. $CH_3CH_2OH + CH_3\overset{\overset{O}{\|}}{C}\overset{\ominus}{C}H\overset{\overset{O}{\|}}{C}OEt \longrightarrow CH_3CH_2\overset{\ominus}{O} + CH_3\overset{\overset{O}{\|}}{C}CH_2\overset{\overset{O}{\|}}{C}OEt$

a) I, II b) III, IV c) II, IV d) I, III

851. What are the major products when the following ketones are converted to the enolate anions under kinetic control, and then reacted with trimethylsilyl chloride?

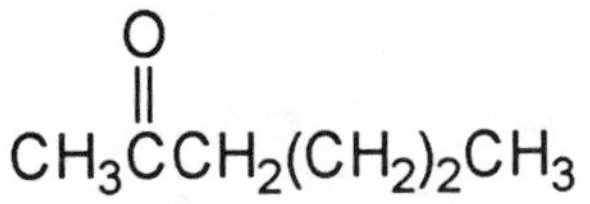

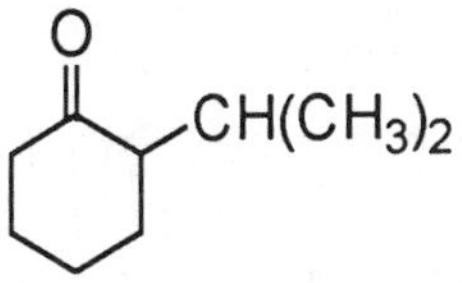

$CH_3C(=O)CH_2(CH_2)_2CH_3$ (2-isopropylcyclohexanone structure, $CH(CH_3)_2$)

I. 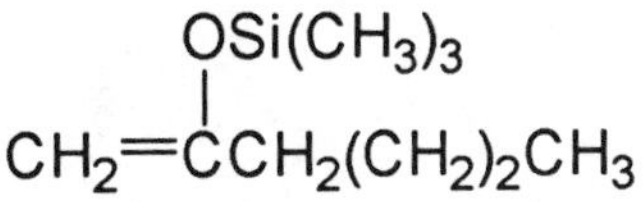 $CH_2{=}C(OSi(CH_3)_3)CH_2(CH_2)_2CH_3$

II. 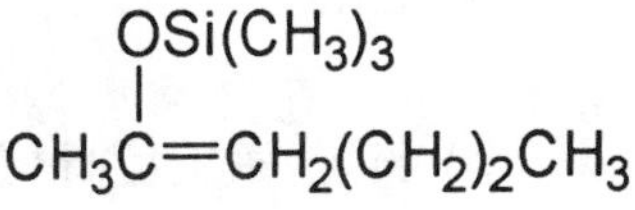 $CH_3C(OSi(CH_3)_3){=}CH_2(CH_2)_2CH_3$

III. 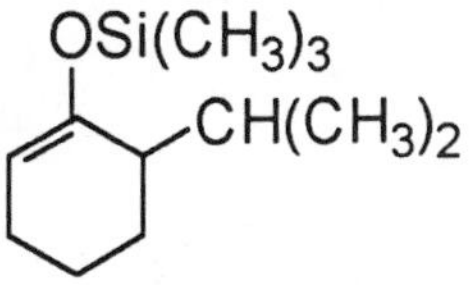 (cyclohexene with $OSi(CH_3)_3$ and $CH(CH_3)_2$)

IV. 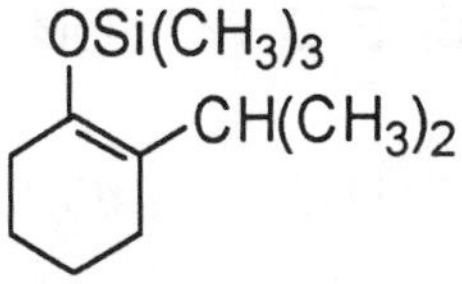 (cyclohexene with $OSi(CH_3)_3$ and $CH(CH_3)_2$)

a) I, IV b) II, III c) I, III d) II, IV

TIP When two different enolate anions can be formed by proton removal from two different alpha hydrogens, often the direction can be predicted by whether the conditions call for kinetic or thermodynamic control. For kinetic control, the more available alpha hydrogens dominate. Thus in question 851, the less sterically hindered methyl for the first compound and methylene for the second compound will lead to the silyl products I and III. Under thermodynamic control, the more stable enolate ion dominates, and since the enolate ion with the more highly substituted double bond is more stable, the order is opposite to kinetic control. Thus in question 849, the silyl products, II and IV, arise from the more stable enolate ions.

852. What is the major product from the following reaction?

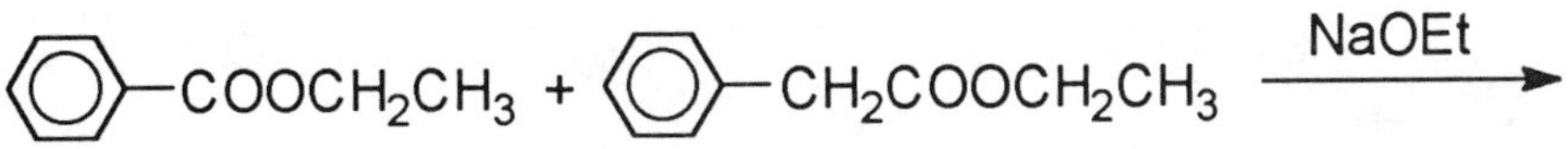

$C_6H_5{-}COOCH_2CH_3 + C_6H_5{-}CH_2COOCH_2CH_3 \xrightarrow{NaOEt}$

a) 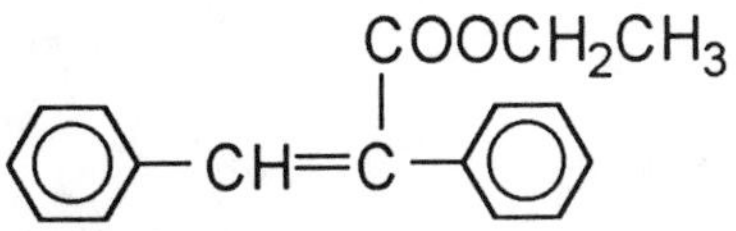 $C_6H_5{-}CH{=}C(COOCH_2CH_3){-}C_6H_5$

b) $C_6H_5{-}C(=O)CH_2{-}C_6H_5$

c) $C_6H_5{-}C(=O)CH(COOCH_2CH_3){-}C_6H_5$

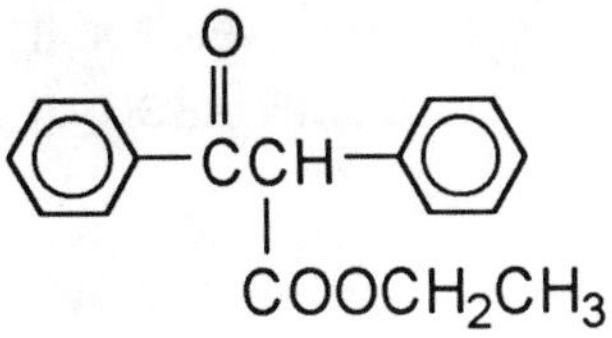

d) 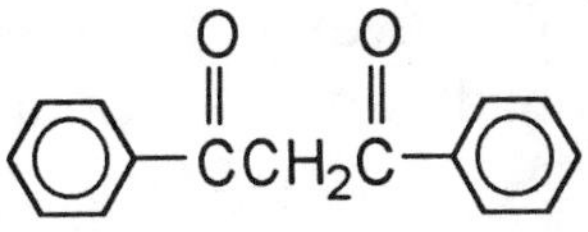 $C_6H_5{-}C(=O)CH_2C(=O){-}C_6H_5$

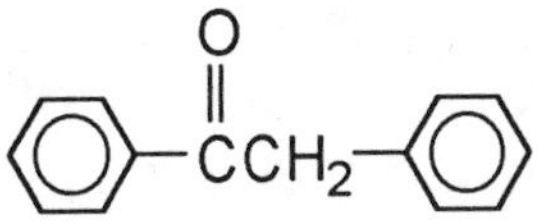

853. What are the best conditions for the following transformation?

$$HO_2C{-}CO_2H \longrightarrow CH_3CH_2O\overset{O}{\overset{\|}{C}}CH_2\overset{O}{\overset{\|}{C}}COOEt$$

a) $EtO^{\ominus}$ + EtOH ; $(CH_3)_2C{=}O$ + $EtO^{\ominus}$

b) $H_3O^{\oplus}$ + EtOH ; CH_3CO_2Et + $EtO^{\ominus}$

c) $EtO^{\ominus}$ + EtOH ; CH_3CO_2Et

d) $H_3O^{\oplus}$ + EtOH ; $(CH_3)_2C{=}O$ + $EtO^{\ominus}$

854. What is the best reaction route for the following transformation?

$$CH_3\overset{O}{\overset{\|}{C}}CH_3 \longrightarrow CH_3\overset{O}{\overset{\|}{C}}CH_2\overset{O}{\overset{\|}{C}}CH_3$$

a) $\xrightarrow{\text{NaOEt}}$ $\xrightarrow{CH_3\overset{O}{\overset{\|}{C}}CH_2COOEt}$ $\xrightarrow{H_3O^{\oplus}}$

b) $\xrightarrow{\text{NaOEt}}$ $\xrightarrow{CH_3\overset{O}{\overset{\|}{C}}CH_3}$ $\xrightarrow{H_3O^{\oplus}}$

c) $\xrightarrow{\text{NaOEt}}$ $\xrightarrow{CH_3\overset{O}{\overset{\|}{C}}CH_2CH_3}$ $\xrightarrow{H_3O^{\oplus}}$

d) $\xrightarrow{\text{NaOEt}}$ $\xrightarrow{CH_3\overset{O}{\overset{\|}{C}}OCH_2CH_3}$ $\xrightarrow{H_3O^{\oplus}}$

TIP A helpful insight comes from the structure of the product which is a beta diketone. This rules out aldol reactions b) and c) which lead to beta hydroxy ketones. This leaves the two esters in a) and d) as candidates for the required Claisen-Schmidt reaction. Counting carbons in the product rules out answer a).

B. ENAMINES

855. What are the reactants in the enamine synthesis of the following compound?

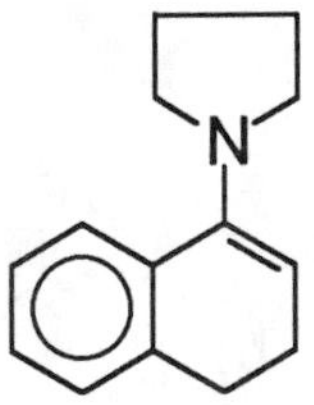

N

a) 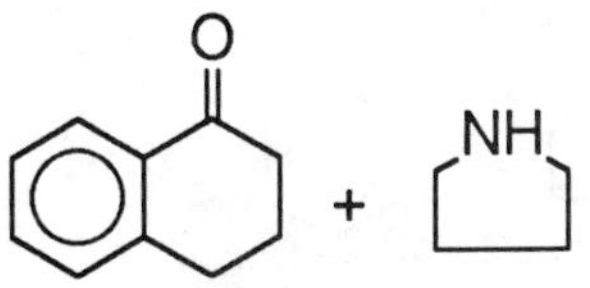 O + NH

b) 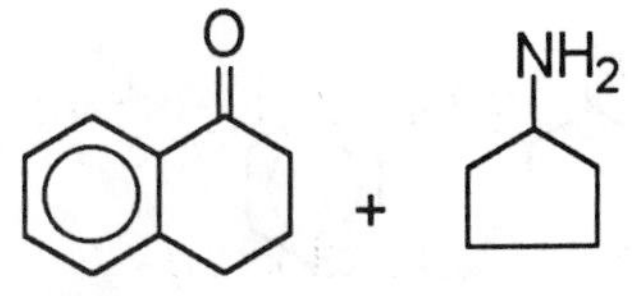 O + NH_2

c) 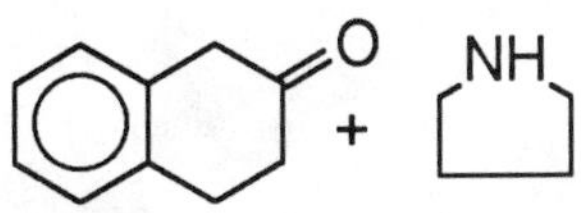 O + NH

d) 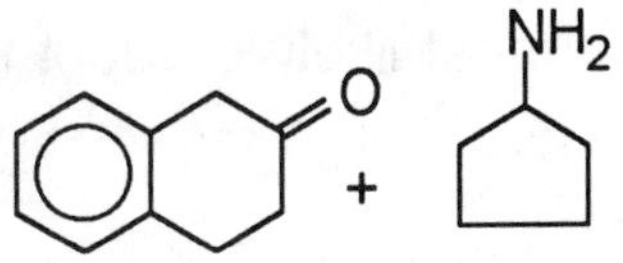 O + NH_2

856. Which carbolamine is formed in the following reaction?

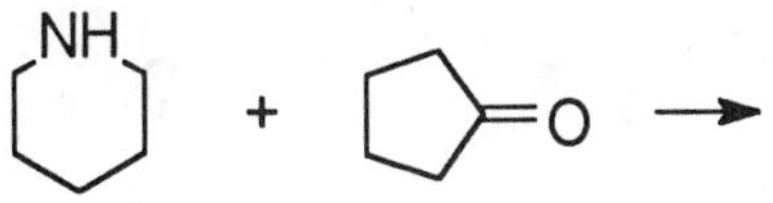

NH + O →

a) 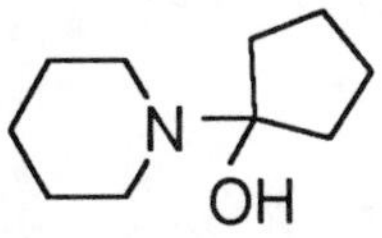 N OH

b) 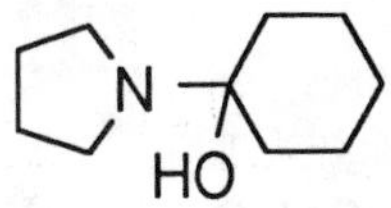 N HO

c) 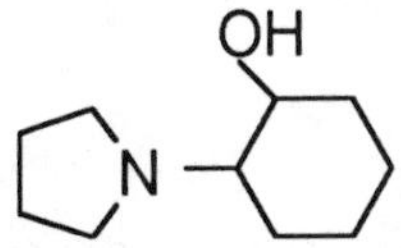 OH N

d) 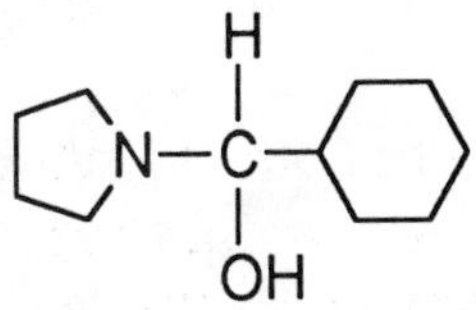 H N—C— OH

857. What is the major product from the following reaction?

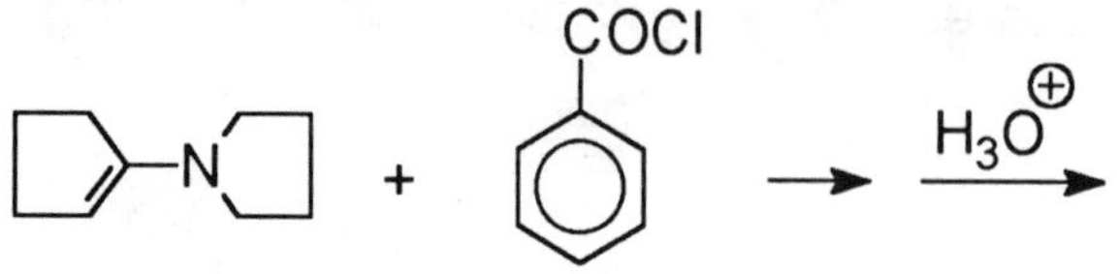

COCl

-N + → $H_3O^{\oplus}$ →

a) 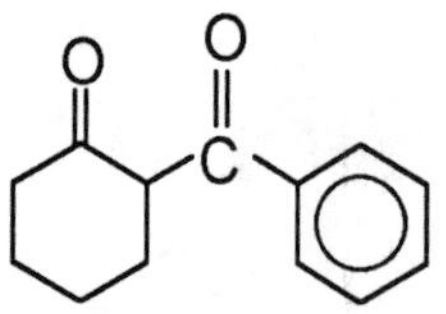 C

b)

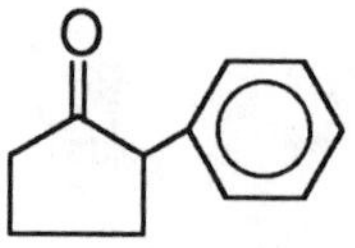

c) 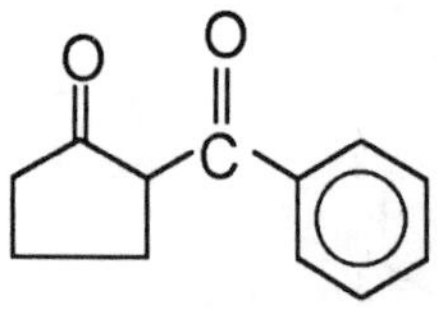 C

d) 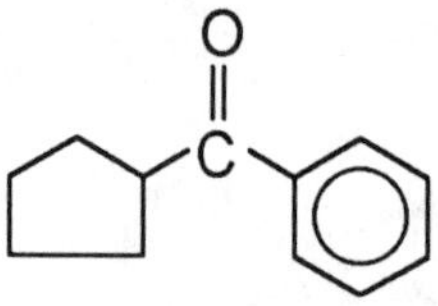 C

858. What is the most likely product from the following series of reactions?

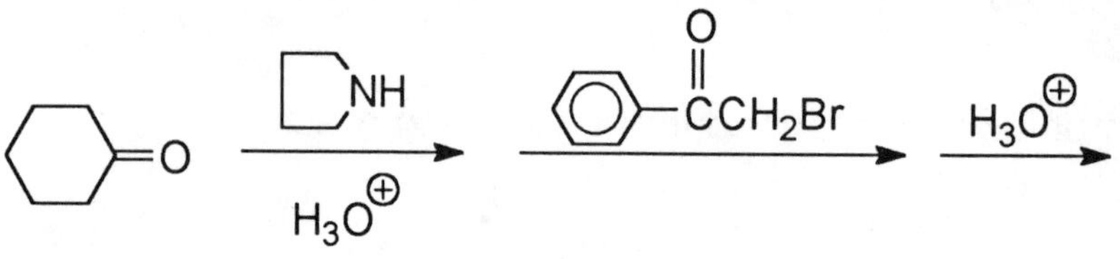

NH

=O → ($H_3O^{\oplus}$) → $\overset{O}{\overset{\|}{C}}CH_2Br$ → $H_3O^{\oplus}$ →

a) 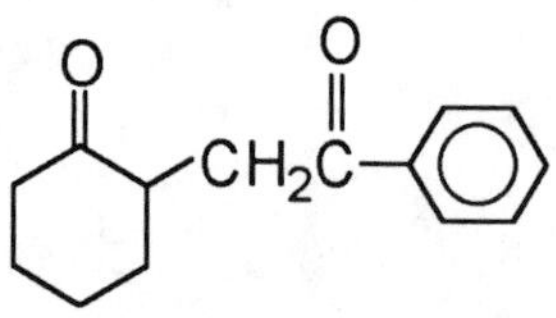 $CH_2\overset{O}{\overset{\|}{C}}$–

b) 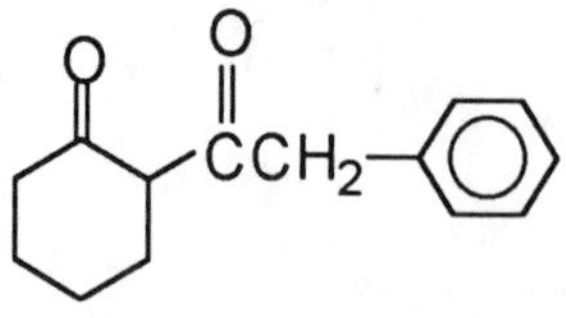 $\overset{O}{\overset{\|}{C}}CH_2$–

c) $CH_2\overset{O}{\overset{\|}{C}}$–

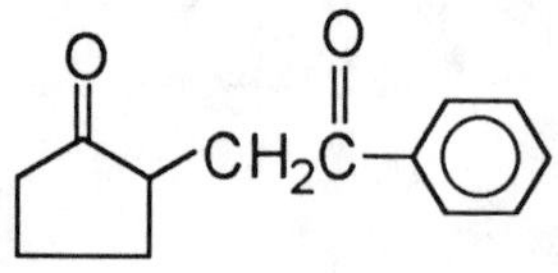

d) 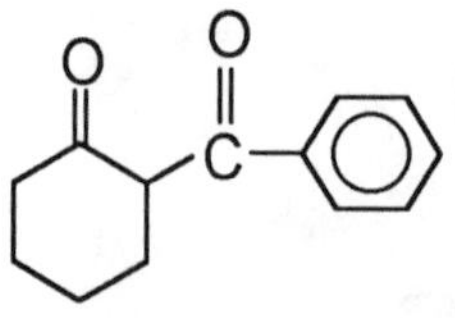 C–

859. What is the major product from the following series of reactions?

Cyclohexanone $\xrightarrow[H_3O^{\oplus}]{\text{pyrrolidine (NH)}}$ $\xrightarrow{CH_2{=}CHCO_2Et}$ $\xrightarrow{H_3O^{\oplus}}$

a) 2-(CH=CHCO$_2$Et)cyclohexanone b) 2-($CH_2CH_2CO_2Et$)cyclohexanone

c) 2-($CH_2CH_2CO_2Et$)cyclopentanone d) bicyclic ketone (C=O bridge)

TIP Enamines react effectively as nucleophiles in the Michael reactions, and since the reactant is a classic Michael acceptor, this route is suggested. Answer a) is incorrect because it does not include the addition to the double bond that is characteristic of the Michael reaction. Answer c) is incorrect since the reacting enamine will be a cyclohexanone derivative. Answer d) is the wrong chemistry.

C. ACETOACETIC ESTER SYNTHESIS

860. Which of the following compounds undergoes the fastest exchange of hydrogen for deuterium when treated with D_2O and a trace of OD^- ?

a) 2,2-dimethylcyclohexane-1,3-dione (H_3C, CH_3) b) cyclohexane-1,2-dione c) cyclohexane-1,3-dione d) cyclohexane-1,4-dione

861. Which of the following halides is NOT effective in the alkylation reaction of sodioacetic ester?

I: $CH_2{=}CHBr$ II: $(CH_3)_3CBr$ III: $C_6H_5{-}CH_2Cl$ IV: $CH_3C(=O)CH_2Br$

a) I, III b) II, IV c) III, IV d) I, II

862. What is the most likely product from the following series of reactions?

$$CH_3C(=O)\overset{\ominus}{C}HCOOEt\ Na^{\oplus} \xrightarrow{C_6H_5{-}CH_2Br} \xrightarrow{NaOH} \xrightarrow{H_3O^{\oplus}} \xrightarrow{heat}$$

a) $C_6H_5{-}CH_2C(=O)CH_3$ b) $C_6H_5{-}C(=O)CH_2CH_3$

c) $CH_3C(=O)CH_3$ d) $C_6H_5{-}CH_2CH_2C(=O)CH_3$

863. What is the final product from the following series of reactions?

$$CH_3C(=O)CH_2COOCH_2CH_3 \xrightarrow[2)\ C_6H_5{-}C(=O)CH_2Br]{1)\ NaOEt} \xrightarrow[2)\ CH_3I]{1)\ NaOEt} \xrightarrow[2)\ H_3O^{\oplus}\ 3)\ heat]{1)\ KOH}$$

a) $C_6H_5{-}C(=O)CH(CH_3)CH_2COCH_3$ b) $C_6H_5{-}C(=O)CH(CH_3)COCH_2COCH_3$

c) $C_6H_5{-}C(=O)CH_2CH(CH_3)COCH_3$ d) $C_6H_5{-}C(=O)CH_2CH_2C(=O)CH_3$

864. What is the final product from the following series of reactions?

$$CH_3\overset{O}{\overset{\|}{C}}CH_2COOCH_2CH_3 \xrightarrow[2)\ CH_2{=}CHCH_2Br]{1)\ NaOEt} \xrightarrow[2)\ CH_3I]{1)\ NaOEt} \xrightarrow[2)\ H_3O^{\oplus}\ \ 3)\ heat]{1)\ KOH}$$

a) $CH_3CH_2\overset{O}{\overset{\|}{C}}CH_2CH_2CH{=}CH_2$

b) $CH_3\overset{O}{\overset{\|}{C}}\underset{CH_3}{\underset{|}{C}}HCH_2CH{=}CH_2$

c) $CH_2{=}\overset{CH_3}{\overset{|}{C}}HCH_2CHCOOH$

d) $CH_2{=}CHCH_2CH_2COOH$

TIP For many problems, knowing what kind of product can be expected is often a great shortcut. A helpful tip for the acetoacetic ester synthesis is that the product will always be a monosubstituted acetone or a disubstituted acetone where both substitutions are on the same carbon. This simple device leads immediately to the correct answer b) in this problem.

865. What is the major product from the following series of reactions?

$$CH_3\overset{O}{\overset{\|}{C}}CH_2COOCH_2CH_3 \xrightarrow[2)\ CH_3COCl]{1)\ NaOEt} \xrightarrow[2)\ CH_3I]{1)\ NaOEt} \xrightarrow[2)\ H_3O^{\oplus}\ \ 3)\ heat]{1)\ KOH}$$

a) $CH_3COCH_2CH(CH_3)COCH_3$

b) $CH_3COCH_2COCH_3$

c) $CH_3COCH(CH_3)COCH_3$

d) $CH_3COCH(CH_3)CH_2COCH_3$

866. In order to prepare 4-methyl-5-phenyl-2-pentanone, ethyl acetoacetate is alkylated with which halide?

a) 2-bromo-1-phenylpropane
b) 1-bromo-2-methyl-3-phenylpropane
c) benzyl bromide + methyl bromide
d) 1-bromo-2-phenylethane + methyl bromide

D. MALONIC ESTER SYNTHESIS

867. Diethyl malonate and which halide can be used to prepare butyric acid?

a) methyl bromide
b) ethyl bromide
c) 1-bromopropane
d) isopropyl chloride

868. The alkylation reaction of sodiomalonate ester is a useful reaction for which of the following halides?

I. allyl chloride
II. tert. butyl bromide
III. ortho- bromotoluene
IV. bromoacetone

a) I, III
b) I, IV
c) III, IV
d) II, IV

869. Diethyl malonate and which halide can be used to prepare isobutyric acid?

a) methyl bromide
b) ethyl bromide
c) 1-bromopropane
d) isopropyl chloride

870. What is the major product from the following series of reactions?

$$CH_3CH_2O\overset{O}{\overset{\|}{C}}CH_2\overset{O}{\overset{\|}{C}}OCH_2CH_3 \xrightarrow[2)\ CH_3CH_2I]{1)\ NaOCH_2CH_3} \xrightarrow[2)\ CH_3I]{1)\ NaOCH_2CH_3} \xrightarrow[\substack{2)\ H_3O^{\oplus} \\ 3)\ heat}]{1)\ KOH}$$

a) $CH_3CH_2\underset{}{\overset{CH_3}{\overset{|}{C}}}HCOOH$

b) $CH_3\overset{CH_3}{\overset{|}{C}}HCH_2COOH$

c) $CH_3CH_2\overset{CH_2CH_3}{\overset{|}{C}}HCOOH$

d) $CH_3CH_2CH_2CH_2COOH$

871. What is the major product from the following reaction?

$$BrCH_2CH_2Br + 2\ CH_3CH_2OOC\overset{\ominus}{C}HCOOCH_2CH_3\ Na^{\oplus} \longrightarrow \xrightarrow[\text{2) } H_3O^{\oplus}\text{; 3) heat}]{\text{1) KOH}}$$

a) cyclohexane with COOH at positions 1 and 4

b) cyclohexane with $COOCH_2CH_3$ at positions 1 and 4

c) $HOOC(CH_2)_4COOH$

d) $CH_3\overset{O}{\overset{\|}{C}}CH_2CH_2\overset{O}{\overset{\|}{C}}CH_3$

872. Which compound is the reactant in the following reaction scheme?

$$\overset{\ominus}{C}H(COOEt)_2\ Na^{\oplus} + ? \longrightarrow \xrightarrow[\text{2) } H_3O^{\oplus}\text{; 3) heat}]{\text{1) KOH}} \xrightarrow{LiAlH_4} HO(CH)_5OH$$

a) $CH_2{=}\overset{CH_3}{\overset{|}{C}}COOEt$

b) $CH_2{=}CH\overset{O}{\overset{\|}{C}}CH_3$

c) $BrCH_2CH_2CH_2OEt$

d) $CH_2{=}CHCN$

TIP This is another problem where knowing what kind of product can be expected is often a great shortcut. A helpful tip for the malonic ester synthesis is that the initial product will always be a monosubstituted or a disubstituted acetic acid. In this problem, that means that the product before the reduction with $LiAlH_4$ is the five carbon glutaric acid. The reactant then must contribute three carbon atoms and lead to a 1,5 dicarboxylic acid. This rules out answers a), b) and c).

E. THE MICHAEL REACTION

873. Which of the following molecules can become effective nucleophiles for the Michael reaction?

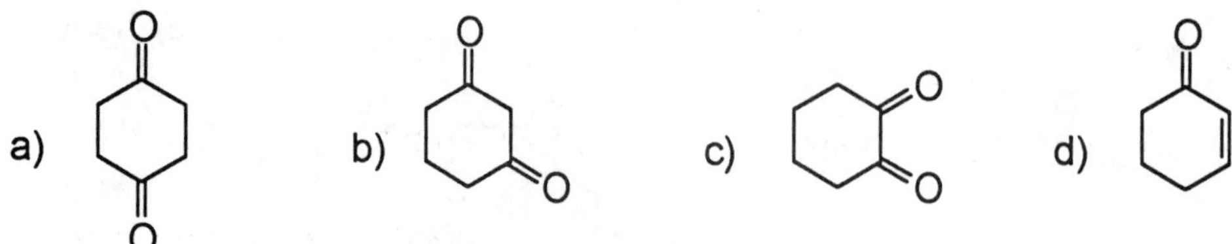

874. Which of the following can react as nucleophilic acceptors for the Michael reaction?

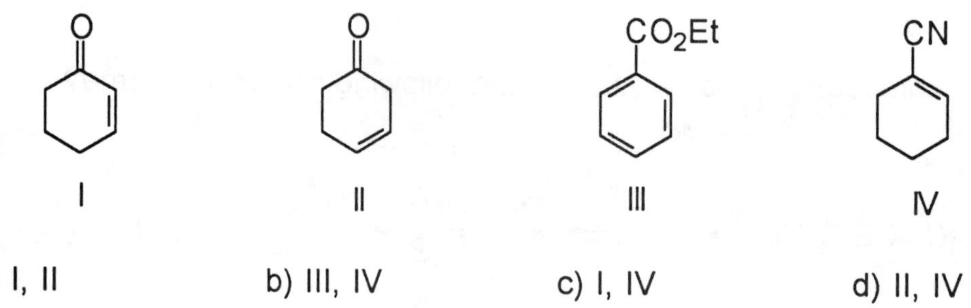

a) I, II   b) III, IV   c) I, IV   d) II, IV

TIP The Michael reaction involves the unusual nucleophilic addition to a carbon-carbon double bond. The double bond must have electron withdrawing groups in conjugation for the reaction to proceed. Thus in question 874, compounds II and III are easily discarded. The effective Michael nucleophiles are anions with extensive resonance delocalization. Of the set in question 873, only b) will lead to such an anion.

875 What is the major product from the following series of reactions?

$$CH_2(CO_2Et)_2 + CH_2{=}CHC{\equiv}N \xrightarrow{\text{NaOEt}} \xrightarrow[\text{2) } H_3O^{\oplus} \text{ 3) heat}]{\text{1) KOH}}$$

a) $HOOC(CH_2)_2COOH$ b) $HOOC(CH_2)_3COOH$

c) $HOOC(CH_2)_4COOH$ d) $HOOC(CH_2)_5COOH$

876. What is the major product from the Michael addition of sodiomalonic ester to 3-butene-2-one, followed by saponification, acidification and heat?

a) $CH_3C(=O)(CH_2)_3C(=O)CH_3$ b) $HOOC(CH_2)_3C(=O)CH_3$

c) $HOOC(CH_2)_3COOH$ d) $HOOCCH_2CH_2C(=O)CH_2CH_3$

877. What is the major product from the Michael addition of sodiomalonic ester to acrylonitrile, followed by saponification, acidification and heat?

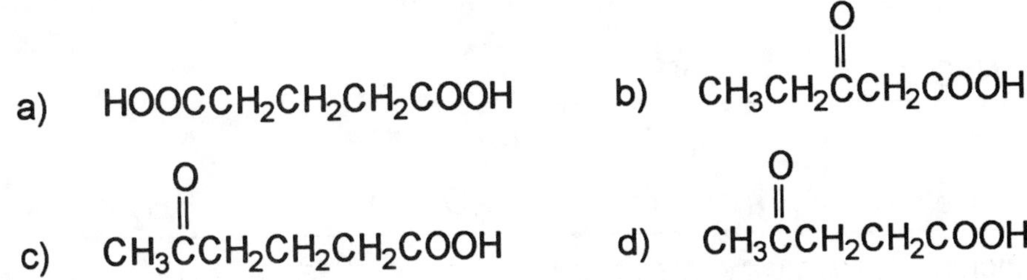

TIP The combination of a Michael addition with a malonic ester synthesis can be predicted from the kinds of diagnostics described for questions 873 and 874. A substituted acetic acid will come from the malonic ester molecule and a nucleophile addition to the double bond will come from the Michael acceptor part. In question 876, this leads to the expectation that the product will be a six carbon atom substituted acetic acid with a terminal methyl ketone. In question 876, this leads to the expectation that the product will be a substitued acetic acid with five carbon atoms and a terminal carboxyl group.

F. COMBINATIONS OF REACTION TYPES

878. Addition and/or condensation reactions are not useful for preparing which of the following kinds of dicarbonyl compounds?

a) β-dicarbonyl compounds
b) γ-dicarbonyl compounds
c) δ-dicarbonyl compounds
d) α-dicarbonyl compounds

879. Ethyl benzoate is reacted with ethyl acetate in a solution of sodium ethoxide. The product is treated with sodium ethoxide and ethyl bromide, followed by reaction with aqueous acid and then heated. What is the most likely final product of these reactions?

a) $C_6H_5-CH_2CH_2COCH_3$
b) $C_6H_5-COCH_2COOCH_2CH_3$
c) $C_6H_5-COCH_2CH_2CH_3$
d) (benzo-fused lactone structure with O and =O)

880. What is the best reaction scheme for the following transformation?

CH_3COCH_2, $CH_2COOCH_2CH_3$ (on cyclopentane) → (spiro bicyclic compound)

a) NaOEt → ; $H_3O^{\oplus}$ → ; Zn(Hg)/HCl →

b) $NaBH_4$ → ; HBr → ; Mg/ether → ; H_2O → ; H_2SO_4/heat → ; H_2/Pt →

c) Br_2/P → ; Zn → ; $LiAlH_4$ → ; heat → ; H_2/Pt →

d) Br_2/NaOH → ; $SOCl_2$ → ; CH_3CH_2OH → ; NaOEt → ; NH_2NH_2/heat →

TIP This is good problem to work backwards. Both a) and d) involve conversion of carbonyls to methylenes, and b) and c) involve hydrogenation of alkenes to make alkanes. But the earlier chemistry of b) and c) has serious mistakes. Reaction sequence d) will involve the loss of a carbon atom in the initial haloform reaction and is therefore not a good candidate. This leaves answer a) which is an intramolecular Claisen-Schmidt reaction which leads to beta diketones and a cyclic one in this case. The final step is a reduction.

CHAPTER 22 AMINES

A. NOMENCLATURE

881. Which of the following structures are named correctly?

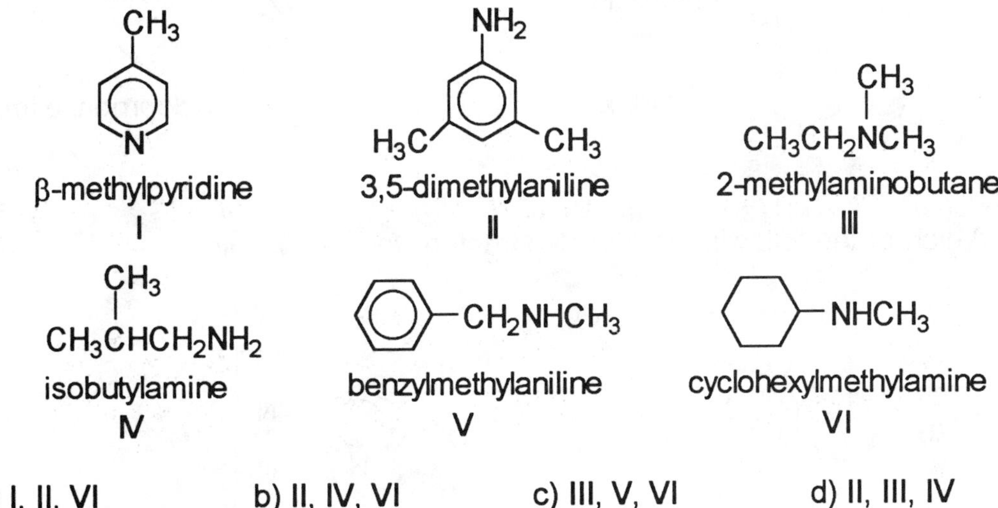

a) I, II, VI   b) II, IV, VI   c) III, V, VI   d) II, III, IV

882. What is the correct structure for 3-(N,N-dimethylamino)hexane?

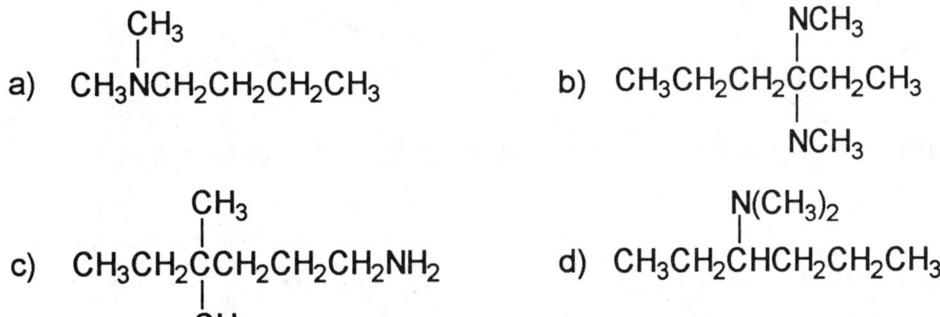

883. Which of the following structures are named correctly?

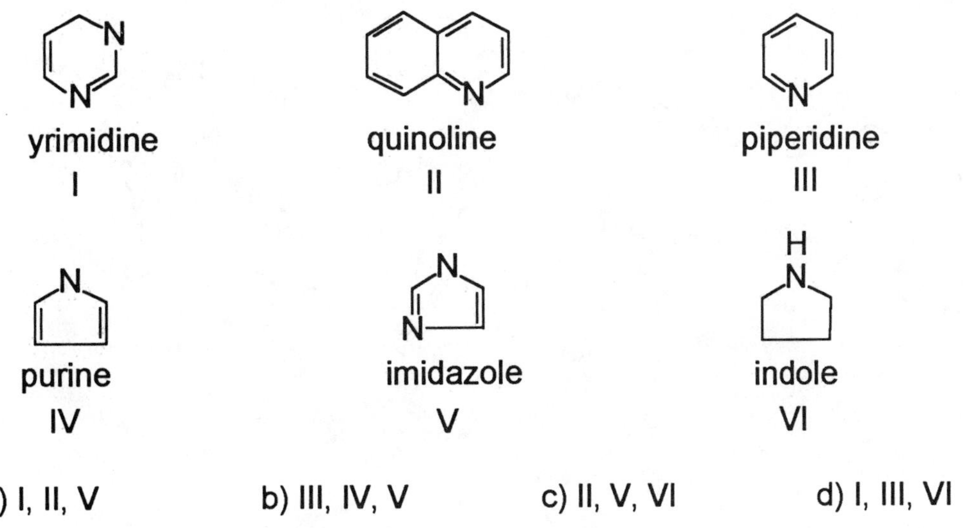

a) I, II, V   b) III, IV, V   c) II, V, VI   d) I, III, VI

884. Which of the following structures has the correct name?

a) 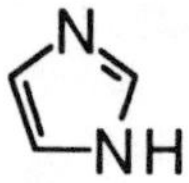 imidazole
b)  piperidine
c)  pyrrolidine
d) 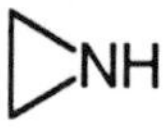 cyclopropane imine

885. Which of the following is a likely structure for an alkaloid?

a) 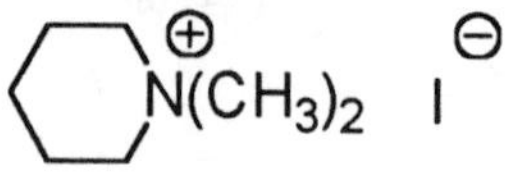 $N(CH_3)_2$ I
b) 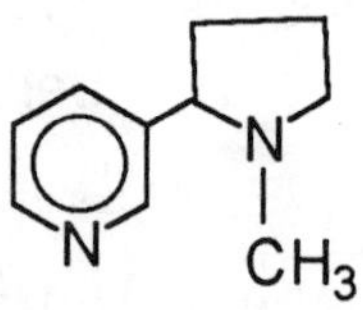 N CH_3
c) 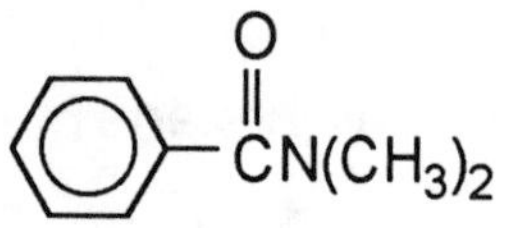 $CN(CH_3)_2$
d) 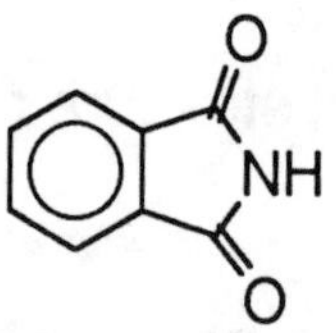 NH

886. Which of the following compounds is most likely found in coffee?

a) 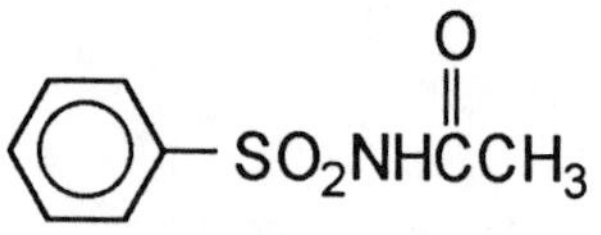 SO_2NHCCH_3
b) 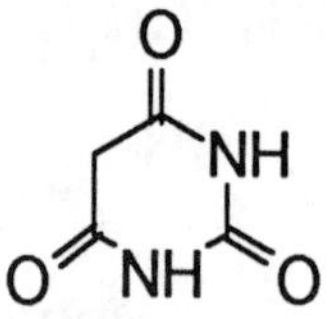 NH NH
c) 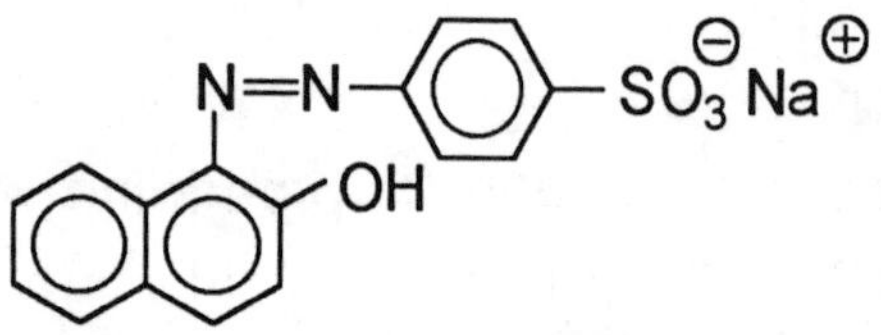 N=N SO_3 Na OH
d) 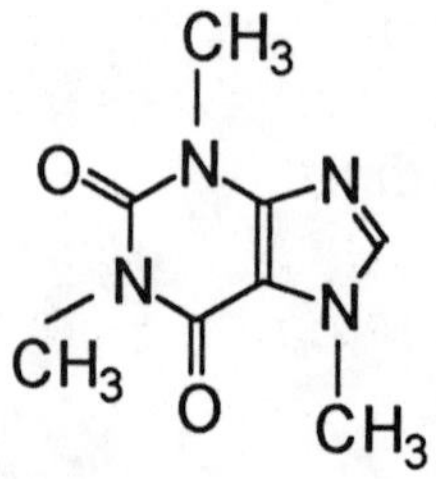 CH_3 N N N N CH_3 CH_3

B. PHYSICAL PROPERTIES

887. How can benzene and pyridine be distinguished chemically?

a) concentrated H_2SO_4 b) dilute NaOH c) dilute HCl d) $NaHCO_3$

888. Which reagents can be used to separate a mixture of aniline and toluene?

a) ether and dilute H_2SO_4 b) $NaHCO_3$
c) concentrated H_2SO_4 d) $Ag(NH)_4OH$

889. Which chemical test can be used to distinguish between benzylamine and benzamide?

a) decolorizing of bromine in carbon tetrachloride at room temperature
b) solubility in dilute hydrochloric acid
c) formation of a stable diazonium compound with nitrous acid at 0-5°C
d) solubility in dilute sodium bicarbonate

890. Which of the following is insoluble in water but is soluble in dilute HCl?

a) $CH_3CH_2CH_2NH_2$ b) 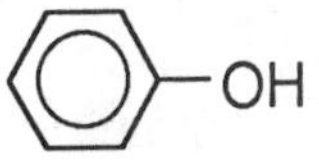 OH

c) 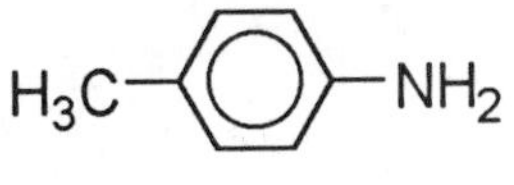 H_3C NH_2 d) 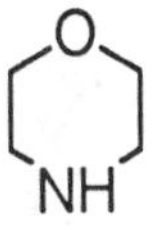 O NH

TIP For questions 887- 890. Solubility properties (determined by fundamental chemistry) can be useful for distinguishing between compounds. Starting with questions 887 and 888, all aromatic compounds are soluble in concentrated sulfuric acid, so benzene and pyridine, and aniline and toluene cannot be separated by this reagent. However, basic compounds (amines) are soluble in dilute acid. Since pyridine (887), aniline (888), benzyl amine (889) and para-toluidine (890) are all basic compounds, separations are possible with dil HCl or dil H_2SO_4. Water solubility is restricted to polar molecules with four or less carbon atoms. Thus in question 890, a) and d) are water soluble and can be ruled out as answers.

891. How can cyclohexylamine and piperidine be distinguished?

a) benzene sulfonyl chloride in KOH b) dilute HCl
c) concentrated H_2SO_4 d) acetic anhydride

892. Which of the following explanations can account for the fact that pyrrole has a low basicity, an enhanced acidity, and a significant dipole moment?

a) the 4n + 2 aromatic rule
b) the heterocyclic ring system
c) a reduced steric effect
d) the nitrogen atom electronegativity

893. Which of the following compounds has the lowest heat of hydrogenation?

a) 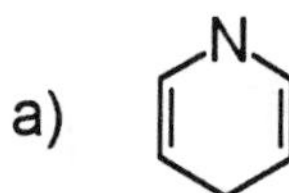
b)
c)
d)

894. What is the order of increasing value for the dipole moment for the following compounds (lowest first)?

I II III IV

a) I, II, III, IV
b) III, II, I, IV
c) II, IV, I, III
d) III, IV, I, II

895. What is the product from the reaction of one equivalent of acid with one equivalent of the following compound?

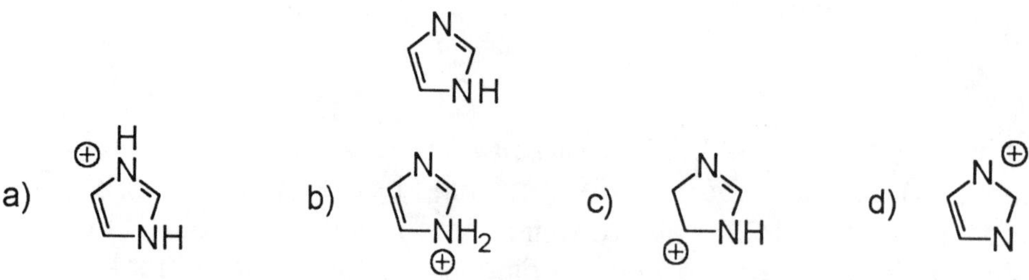

a) b) c) d)

TIP When the number of pi electrons in the conjugated cyclic system is 4n+2, there is considerable aromatic resonance stabilization. In pyrrole and imidazole, a 4n+2 system is obtained when the lone pair of electrons on the nitrogen is used as part of the aromatic six-electron system. This has dramatic ramifications on stability, charge, difficulty of protonation and results in much lower heats of hydrogenation. In question 893 only b) and d) fulfill the rule, but resonance structures with a positive charge on oxygen are less favorable than resonance structures for b). The resonance structures which include charges will lead to dipole moments, and in question 894, the charge is stabilized most in IV. Finally the protonation of the nitrogen atom that contributes 2 electrons to the aromatic sextet will be greatly resisted as in question 895 b), whereas protonation of the other nitrogen atom (a) does not disrupt the aromatic sextet.

896. Which of the following amines have a chiral center?

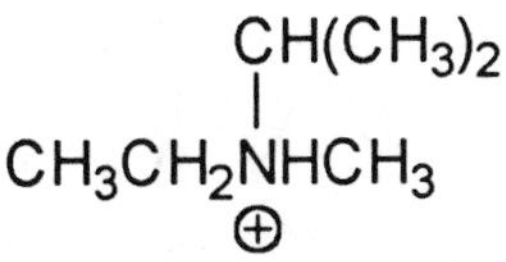

I: $CH_3CH_2\overset{\oplus}{N}H(CH(CH_3)_2)CH_3$

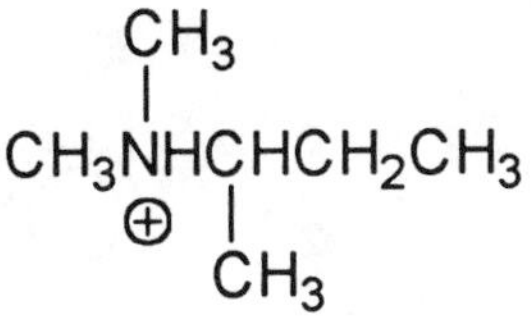

II: $CH_3\overset{\oplus}{N}H(CH_3)CH(CH_3)CH_2CH_3$

III: $CH_3CH(NH_2)CH_2CH_3$

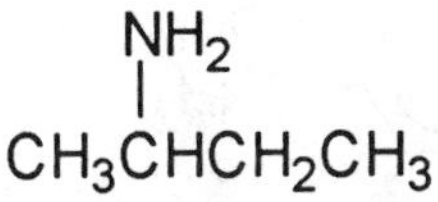

IV: 4,4-dimethylpiperidine (H_3C, H_3C on ring carbon; NH)

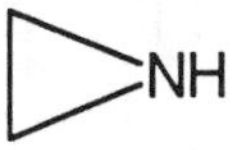

V: aziridine (NH)

a) I, II, III b) III, IV, V c) I, III, IV d) II, III, IV

897. Which of the following amines can be resolved into enantiomers?

I: $CH_3CH_2NHCH_3$

II: $CH_3CH_2CH(CH_3)NHCH_3$

III: 4,4-dimethyl-1-methylpiperidine (H_3C, H_3C; NCH_3)

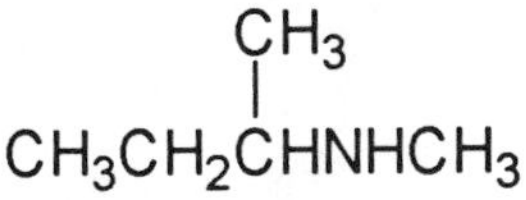

IV: 2,2-dichloroaziridine (Cl, Cl; NH)

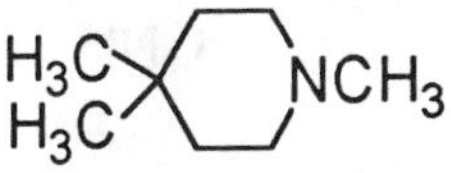

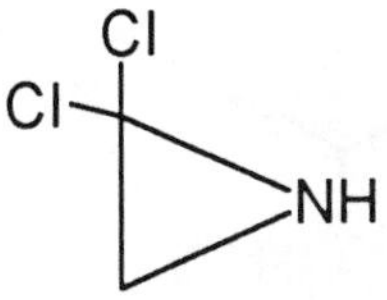

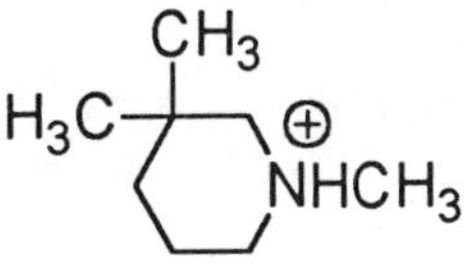

V: 3,3-dimethyl-1-methylpiperidinium (CH_3, H_3C; $\overset{\oplus}{N}HCH_3$)

a) I, II, III b) III, IV, V c) II, III, IV d) II, IV, V

C. ACIDITY and BASICITY

898. Which of the following compounds is the strongest base?

a) 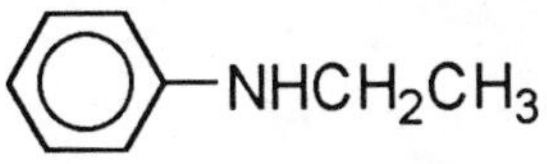 $C_6H_5-NHCH_2CH_3$ b) $C_6H_5-NHCCH_3$ (C=O)

c) $C_6H_5-CNHCH_3$ (C=O) d) 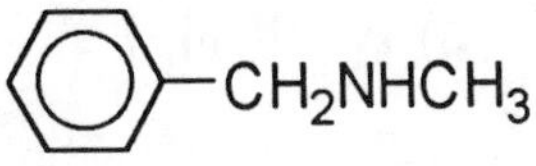 $C_6H_5-CH_2NHCH_3$

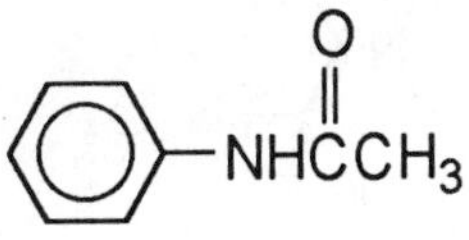

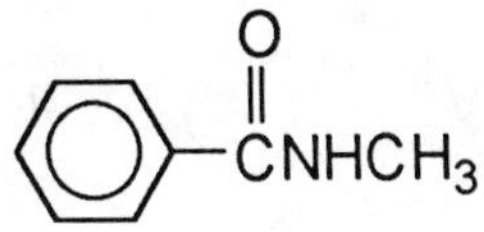

899. What is the order of increasing base strength for the following compounds (weakest first)?

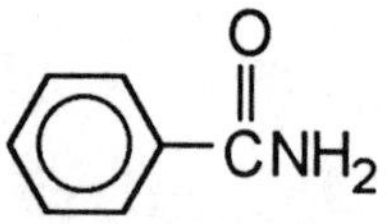

I: C_6H_5–C(=O)NH_2 II: C_6H_5–NH_2 III: C_6H_5–S(=O)(=O)NH_2 IV: C_6H_5–NH–C_6H_5

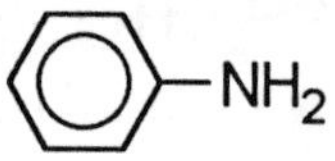

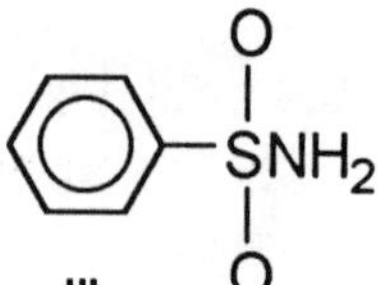

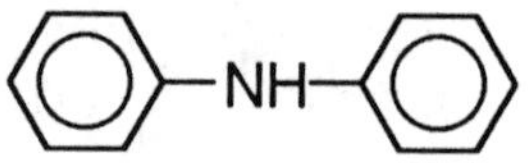

a) II, III, IV, I b) II, I, III, IV c) I, II, IV, III d) III, IV, I, II

900. Which of the following compounds is the weakest base?

a) 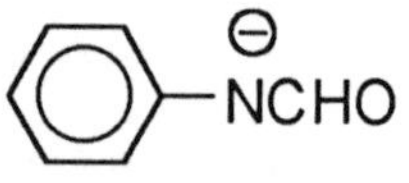 C_6H_5–$\overset{\ominus}{N}$CHO b) C_6H_5–$\overset{\ominus}{N}CH_3$

c) C_6H_5–C(=O)$\overset{\ominus}{N}$H d) 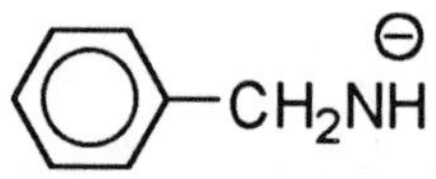 C_6H_5–$CH_2\overset{\ominus}{N}H$

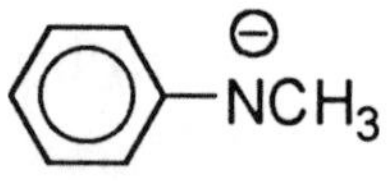

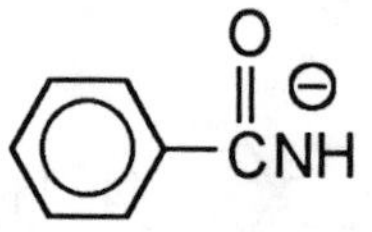

901. Which of the following compounds is the strongest base?

a) 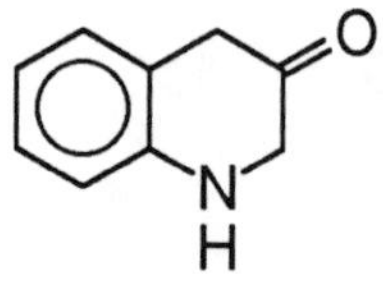 3-oxo ring, N–H b) 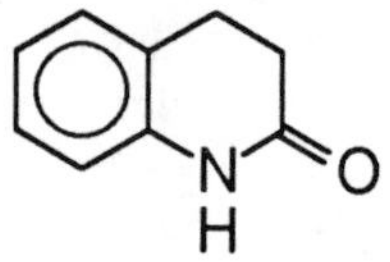 ring with N–H and C=O c) ring with C=O and NH d) 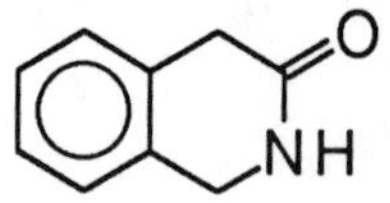 ring with C=O and NH

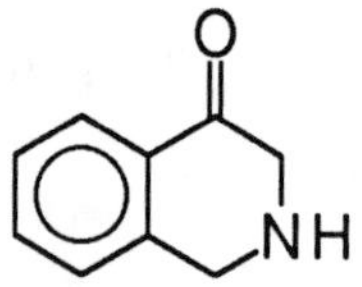

902. What is the order of increasing base strength for the following compounds (weakest first)?

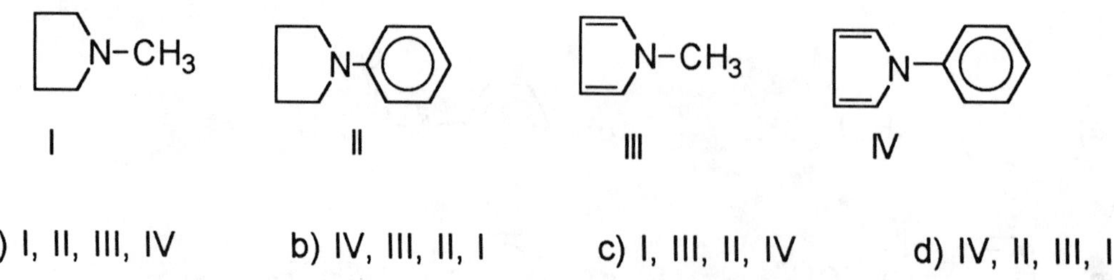

a) I, II, III, IV b) IV, III, II, I c) I, III, II, IV d) IV, II, III, I

TIP The major factors affecting amine basicity are, in decreasing order of importance, (1) when the amine is part of a 4n +2 aromatic system, (2) when the amine is adjacent to a carbonyl, i.e an amide, and (3) when the amine is attached to an aromatic ring. The solution to the question is then to recognize the environment of the amine nitrogen. In question 900 a), the basicity of the amine is lowered by both aromatic and carbonyl groups. In this question, is it clear that d) is the strongest base? Similar reasoning applies to question 901 where b) is the weakest base with the same two base lowering effects. In question 902, there are two factors as well but here the factors are 4n +2 aromaticity and an adjacent aromatic ring. Careful reasoning should lead to I being many orders of magnitude more basic than IV.

903. In which of the following reactions does the nitrogen atom act as a nucleophile rather than a Bronsted base?

I. $C_6H_5-NH_2 + CH_3CH_2Br \longrightarrow C_6H_5-\overset{\oplus}{N}H_2CH_2CH_3\ \overset{\ominus}{Br}$

II. $CH_3NH_2 + H_2O \longrightarrow CH_3\overset{\oplus}{N}H_3 + \overset{\ominus}{O}H$

III. $NH_3 + HCl \longrightarrow NH_4Cl$

IV. $C_5H_5N + HOC(=O)CH_3 \longrightarrow C_5H_5\overset{\oplus}{N}H\ \overset{\ominus}{O}C(=O)CH_3$

V. $C_6H_5-NH_2 + Br_2 \longrightarrow$ 2,4,6-tribromoaniline ($Br_3C_6H_2-NH_2$) $+ HBr$

VI $C_6H_5-COCl + CH_3NH_2 \longrightarrow C_6H_5-CONHCH_3 + HCl$

a) I, IV b) I, VI c) II, III d) IV, V

904. In which of the following reactions does the nitrogen atom act as a Bronsted base?

I.  $C_6H_5-NH_2 + CH_3CH_2Br \rightarrow C_6H_5-\overset{\oplus}{N}H_2CH_2CH_3\ \overset{\ominus}{Br}$

II. $CH_3NH_2 + H_2O \rightarrow CH_3\overset{\oplus}{N}H_3 + \overset{\ominus}{O}H$

III. $NH_3 + HCl \rightarrow NH_4Cl$

IV.  $C_5H_5N + HOCCH_3$ (with C=O) $\rightarrow C_5H_5\overset{\oplus}{N}H\ \ \overset{\ominus}{O}CCH_3$ (with C=O)

V. 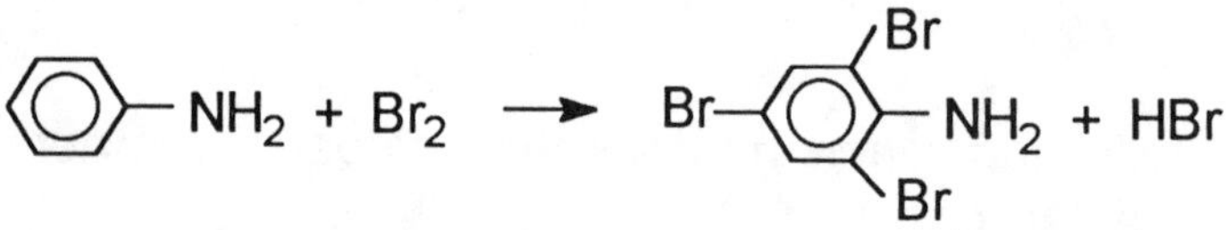 $C_6H_5-NH_2 + Br_2 \rightarrow$ (2,4,6-tribromo) $Br_3C_6H_2-NH_2 + HBr$

VI $C_6H_5-COCl + CH_3NH_2 \rightarrow C_6H_5-CONHCH_3 + HCl$

a) I, II, III b) II, III, IV c) II, IV, V d) III, IV, VI

905. Which methyl hydrogen atoms are most acidic in the following compounds?

a) 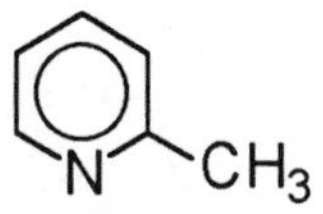 2-methylpyridine (CH_3 on the carbon next to N) b) 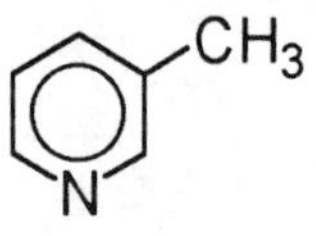 3-methylpyridine (CH_3) c) 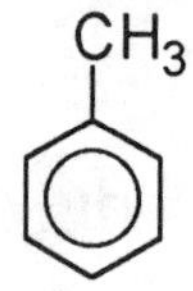 toluene (CH_3 on benzene) d) 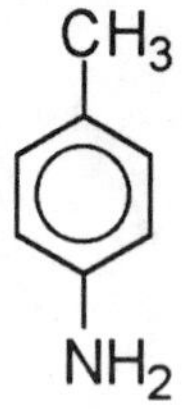 p-toluidine (CH_3 and NH_2 para on benzene)

D. PREPARATIONS

906. Which conditions are best for the following transformation?

$$CH_3CH_2CH_2NH_2 \longrightarrow CH_3CH_2CH_2NHCH_2CH_2CH_3$$

a) H_2CO + Ni / H_2

b) $CH_3CH_2CH_2CH_2NH_2$ + heat

c) CH_3COCl ; $LiAlH_4$; $H_3O^{\oplus}$

d) $CH_3CH_2CH_2Br$ + heat ; $\overset{\ominus}{O}H$

907. Which conditions are best for the following preparation?

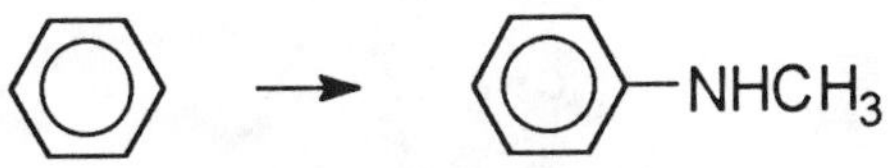

$$C_6H_6 \longrightarrow C_6H_5\text{-}NHCH_3$$

a) Cl_2 + $FeCl_3$; NH_2CH_3

b) CH_3COCl + $AlCl_3$; NH_3 + Ni / H_2

c) HNO_3 + H_2SO_4 ; Fe + HCl ; CH_3I (at low concentration)

d) CH_3CONH_2 + PCl_5

908. Which compound is prepared from the following reaction scheme?

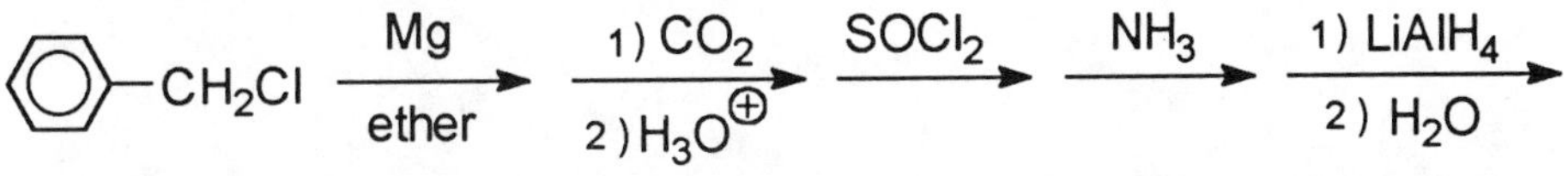

$$C_6H_5\text{-}CH_2Cl \xrightarrow[\text{ether}]{Mg} \xrightarrow[2)\,H_3O^{\oplus}]{1)\,CO_2} \xrightarrow{SOCl_2} \xrightarrow{NH_3} \xrightarrow[2)\,H_2O]{1)\,LiAlH_4}$$

a) 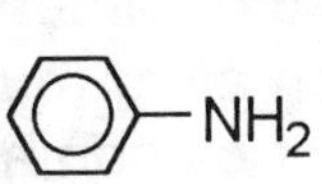 $C_6H_5\text{-}NH_2$

b) $C_6H_5\text{-}CH_2NH_2$

c) $C_6H_5\text{-}CH_2CH_2NH_2$

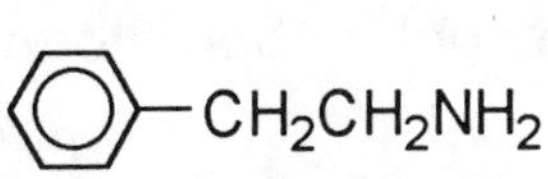

d) 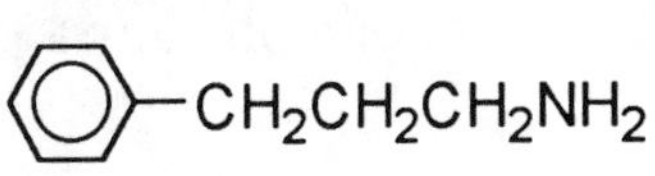 $C_6H_5\text{-}CH_2CH_2CH_2NH_2$

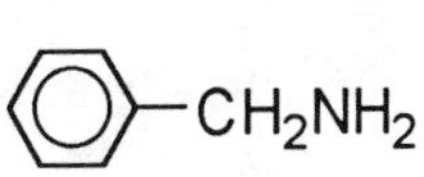

909. The same amine is prepared by which of the following reaction sequences?

I. $CH_3CH_2CH_2C(=O)NH_2 \xrightarrow[Br_2]{NaOH}$

II. $CH_3CH_2CH_2CH_2Br \xrightarrow{LiN_3} \xrightarrow{LiAlH_4}$

III. $CH_3CH_2CH_2CH_2Br \xrightarrow{LiCN} \xrightarrow{Pt / H_2}$

IV. Potassium phthalimide ($N^{\ominus}$ $K^{\oplus}$) $\xrightarrow{CH_3CH_2CH_2CH_2I} \xrightarrow{NH_2NH_2}$

a) I, II b) III, IV c) I, IV d) II, IV

910. Which of the following reaction schemes is not suitable for the preparation of 1-aminopentane?

a) $CH_3CH_2CH_2CH_2CH_2I \xrightarrow[EtOH]{NaN_3} \xrightarrow{LiAlH_4}$

b) $CH_3CH_2CH_2CH_2Br \xrightarrow[EtOH]{NaCN} \xrightarrow[NH_3]{H_2/Pt}$

c) $CH_3CH_2CH_2CH_2COOH \xrightarrow{NH_3} \xrightarrow{heat} \xrightarrow{LiAlH_4}$

d) $CH_3CH_2CH_2CH_2COOH \xrightarrow{SOCl_2} \xrightarrow{NH_3} \xrightarrow[H_2O]{NaOH, Br_2}$

TIP For questions 908-910. Amine synthetic methods can add a carbon atom (questions 908, 909 III and 910 b); they can lose a carbon atom (questions 909 I and 910 d); and finally they can leave the number of carbon atoms unchanged (questions 909 II and IV, and 910 a and c). Knowing the chemistry, the solution to the problem then is to count carbon atoms.

911. What is the principal product from the following series of reactions?

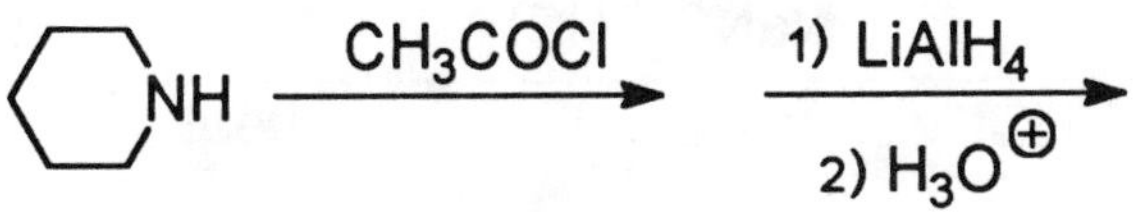

Piperidine (NH) $\xrightarrow{CH_3COCl}$ $\xrightarrow[2)\ H_3O^{\oplus}]{1)\ LiAlH_4}$

a) 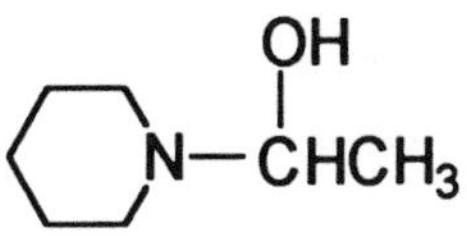 N–CH(OH)CH$_3$ (piperidinyl)
b) 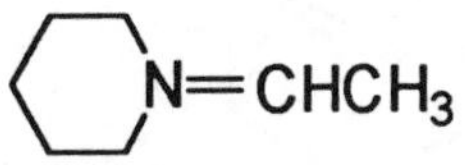 N=CHCH$_3$ (piperidinyl)
c) 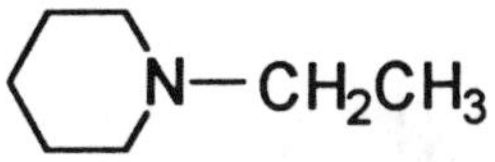 N–CH$_2$CH$_3$ (piperidinyl)
d) 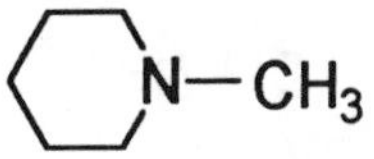 N–CH$_3$ (piperidinyl)

912. What is the principal product from the following series of reactions?

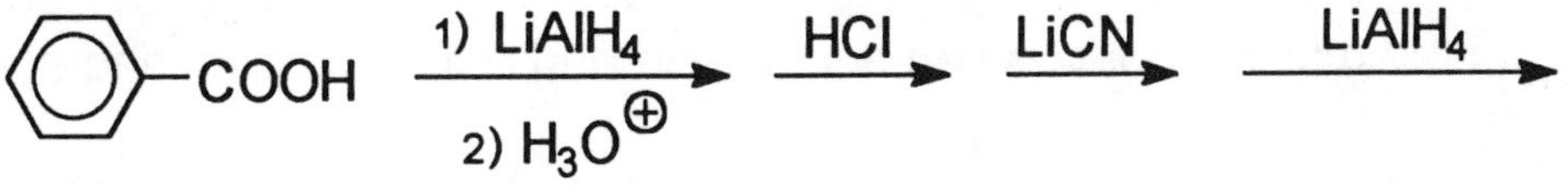

C_6H_5–COOH $\xrightarrow[2)\ H_3O^{\oplus}]{1)\ LiAlH_4}$ $\xrightarrow{HCl}$ $\xrightarrow{LiCN}$ $\xrightarrow{LiAlH_4}$

a) 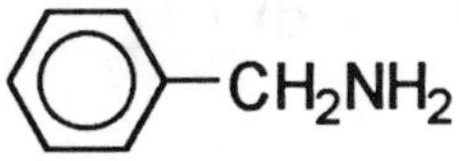 C_6H_5–CH_2NH_2
b) 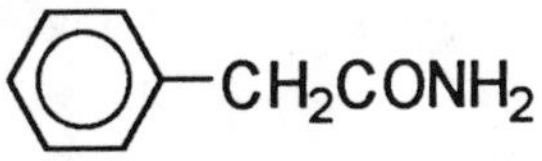 C_6H_5–CH_2CONH_2
c) 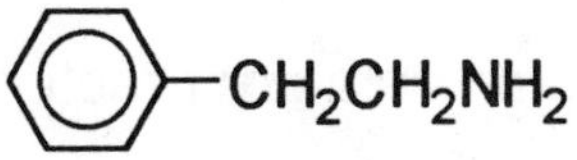 C_6H_5–$CH_2CH_2NH_2$
d) 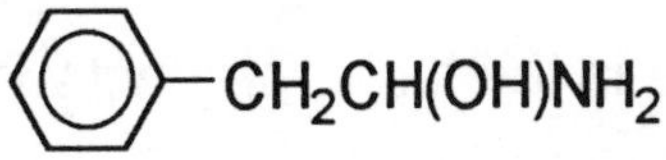 C_6H_5–$CH_2CH(OH)NH_2$

913. What is the major product from the following reaction?

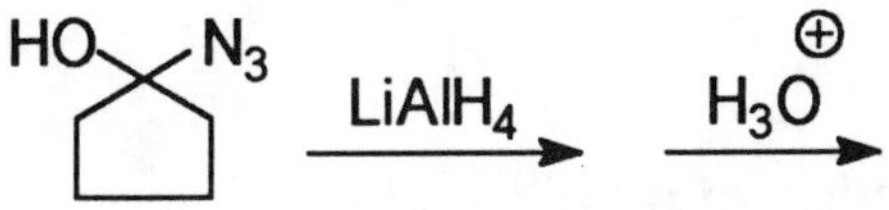

1-hydroxy-1-azidocyclopentane (HO, N_3) $\xrightarrow{LiAlH_4}$ $\xrightarrow{H_3O^{\oplus}}$

a) 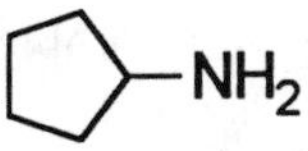 cyclopentyl–NH_2
b) 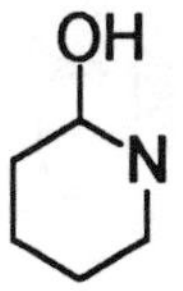 2-hydroxypiperidine (OH, N)
c) 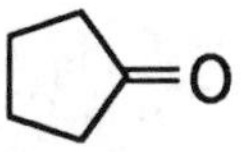 cyclopentanone (=O)
d) 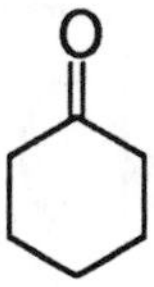 cyclohexanone (O)

914. What is the most likely product from the following series of reactions?

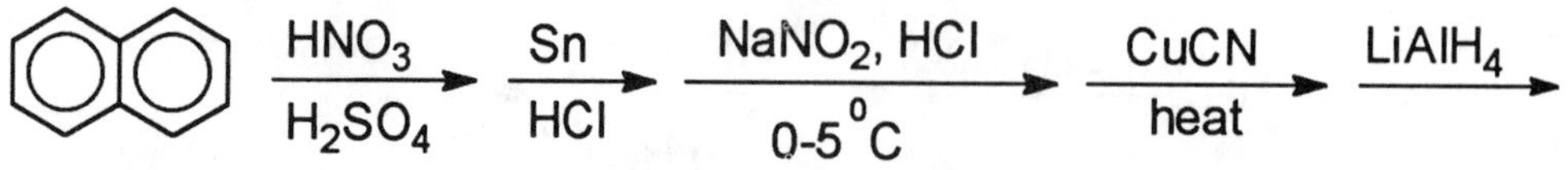

Naphthalene $\xrightarrow[H_2SO_4]{HNO_3}$ $\xrightarrow[HCl]{Sn}$ $\xrightarrow[0\text{-}5^{\circ}C]{NaNO_2,\ HCl}$ $\xrightarrow[heat]{CuCN}$ $\xrightarrow{LiAlH_4}$

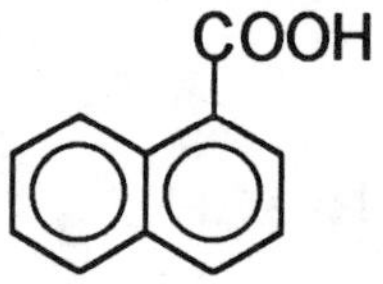

a) COOH (1-naphthoic acid)

b) 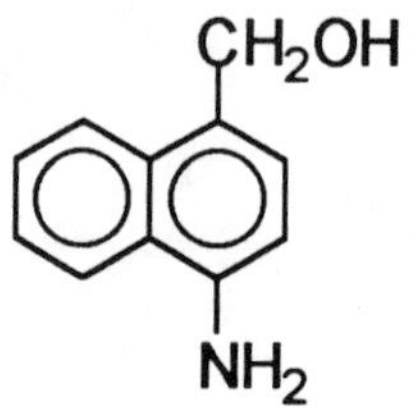 CH_2OH, NH_2 (4-amino-1-naphthalenemethanol)

c) CH_2NH_2 (1-naphthalenemethylamine)

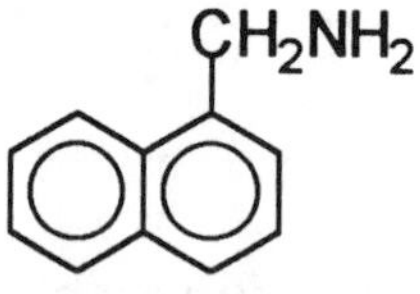

d) 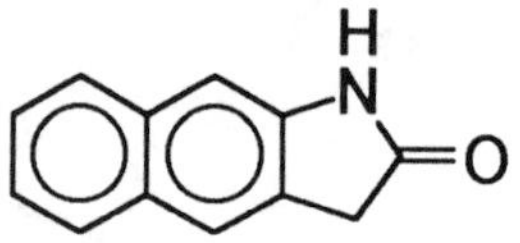 H, N, =O (naphtho-fused lactam)

E. REACTIONS with ACID DERIVATIVES

915. What is the major product when acetic acid and ammonia are reacted at 25°C?

a) $H_2NOC(=O)CH_3$ b) $H_2NC(=O)CH_3$ c) $CH_3C(NH_2)(OH)OH$ d) $CH_3C(=O)O^{\ominus}\ NH_4^{\oplus}$

916. What is the major product when acetic acid and ammonia are reacted at 200°C?

a) $H_2NOC(=O)CH_3$ b) $H_2NC(=O)CH_3$ c) $CH_3C(NH_2)(OH)OH$ d) $CH_3C(=O)O^{\ominus}\ NH_4^{\oplus}$

917. What is the product from the reaction of pyrole and succinic anhydride?

a) 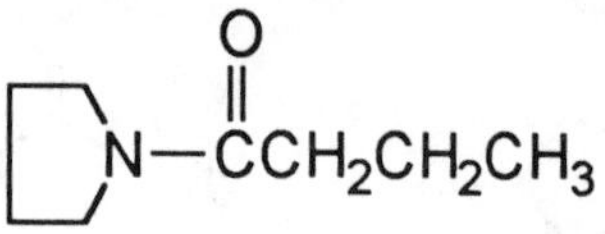 $N-C(=O)CH_2CH_2CH_3$

b) $N-C(=O)-CH_2CH_2COOH$

c) $N-C(=O)-C_6H_4$, COOH

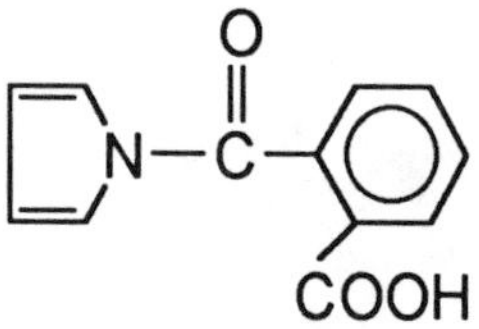

d) 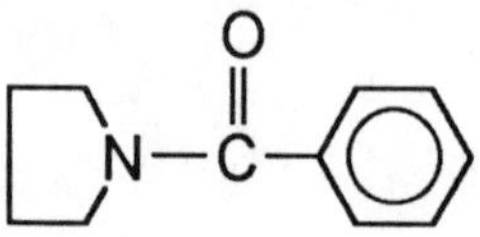 $N-C(=O)-C_6H_5$

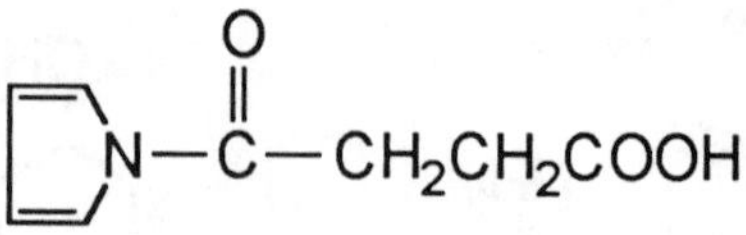

918. What is the product from the reaction of pyrrolidine and benzoic anhydride?

a) 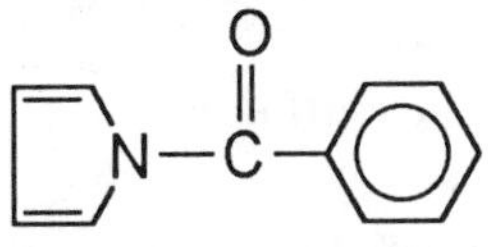 b) 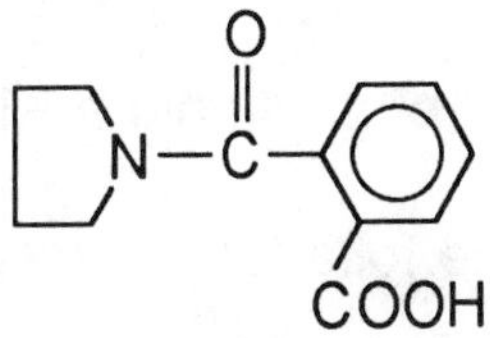 c) 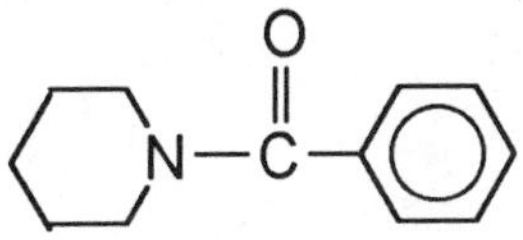 d)

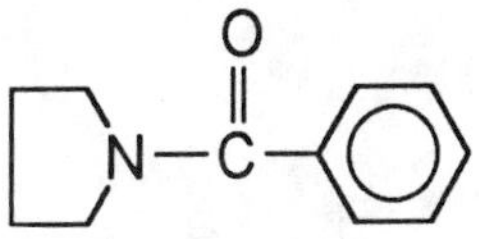

919. What is the most likely product from heating the following compound?

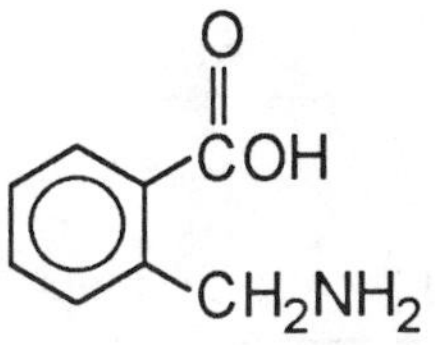

a) 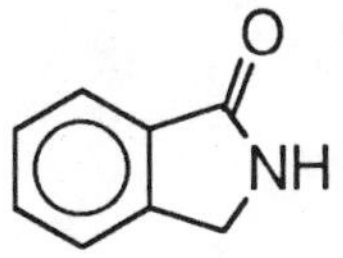 b) 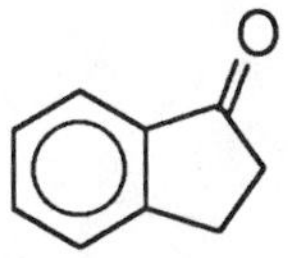 c) 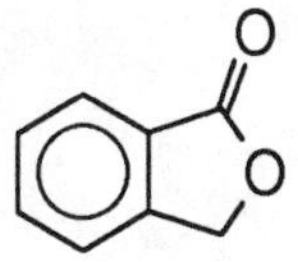 d)

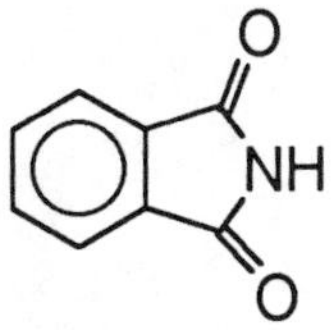

F. REACTIONS with NITROUS ACID

920. Which reagents are best for the following reaction?

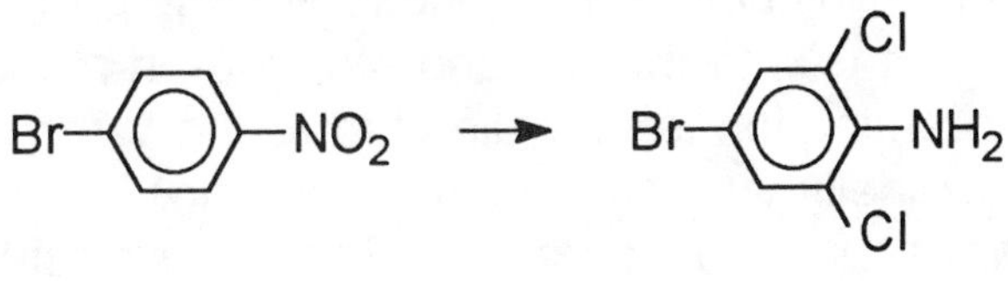

a) Ni / H_2 ; Cl_2 + $FeCl_3$

b) Sn / HCl ; $(CH_3CO)_2O$; Cl_2 + $FeCl_3$; $LiAlH_4$; $H_3O^{\oplus}$

c) $(CH_3CO)_2O$; Cl_2 + $FeCl_3$; $H_3O^{\oplus}$

d) Sn / HCl ; HCl + $NaNO_2$; $COCl$; $FeCl_3$ + Cl_2

921. Which of the following is the best starting material for the synthesis of para-iodonitrobenzene?

a) chlorobenzene
b) nitrobenzene
c) nitrobenzenesulfonic acid
d) para-nitroaniline

922. Which of the following reactions is best for the synthesis of 3,5-dibromo-toluene?

a) toluene $\xrightarrow[\text{heat}]{Br_2,\ Fe}$

b) para-methylaniline $\xrightarrow[H_2O]{Br_2}$ $\xrightarrow{HONO}$ $\xrightarrow{H_3PO_2}$

c) toluene $\xrightarrow[H_2SO_4]{\text{conc } HNO_3}$ $\xrightarrow[H_2S]{NH_3}$ $\xrightarrow{HONO}$ $\xrightarrow{CuBr}$

d) 3-bromotoluene $\xrightarrow{\text{conc } HNO_3}$ $\xrightarrow{NH_3}$ $\xrightarrow{HONO}$ $\xrightarrow{CuBr}$

TIP The tips for aromatic diazonium syntheses are that reaction always occurs exactly where the diazonium nitrogen is and that there are several well defined and useful reactions of these diazonium compounds. It is as if we had a pair of tweezers to selectively change a substituent. Bromination of toluene gives substituents in the wrong positions (ortho-para not meta) and so a simple reaction as in answer a) will not work. Instead we rely on the diazonium synthesis. The best start is with the needed structure in place, namely para-methylaniline. The bromination is now directed to the ortho positions relative to the para NH_2 group. Finally we can get rid of the para NH_2 group with the convienent diazonium and H_3PO_2 reactions. The other strategies, c) and d), are good chemistry, but the positions for the bromo groups will not be what was asked for at 3 and 5.

923. What is the expected product from the following reaction sequence starting with nitrobenzene?

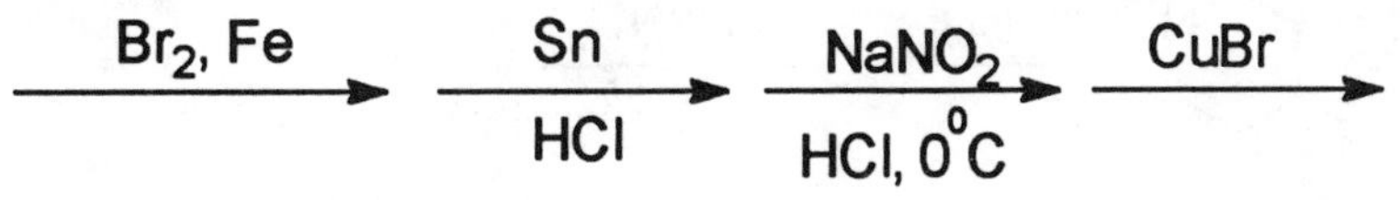

$\xrightarrow{Br_2, Fe}$ $\xrightarrow[HCl]{Sn}$ $\xrightarrow[HCl,\ 0^0C]{NaNO_2}$ $\xrightarrow{CuBr}$

a) 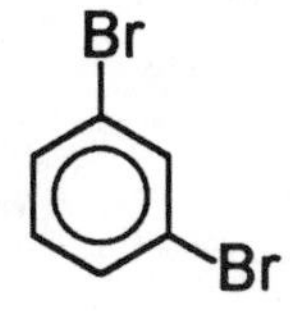 (1,3-dibromobenzene: Br, Br) b) 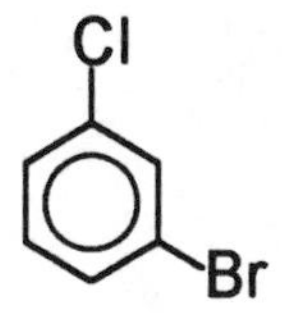 (Cl, Br at meta) c) 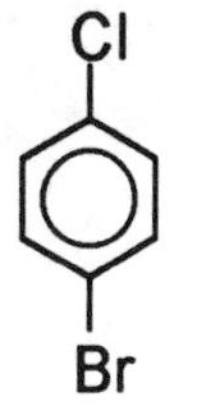 (Cl, Br at para) d) 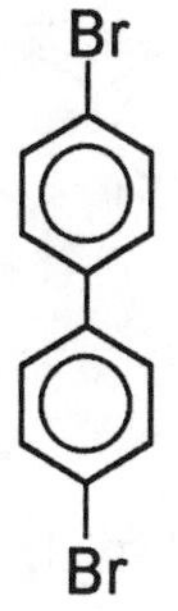 (4,4'-dibromobiphenyl: Br, Br)

924. What is the principal product from the following series of reactions?

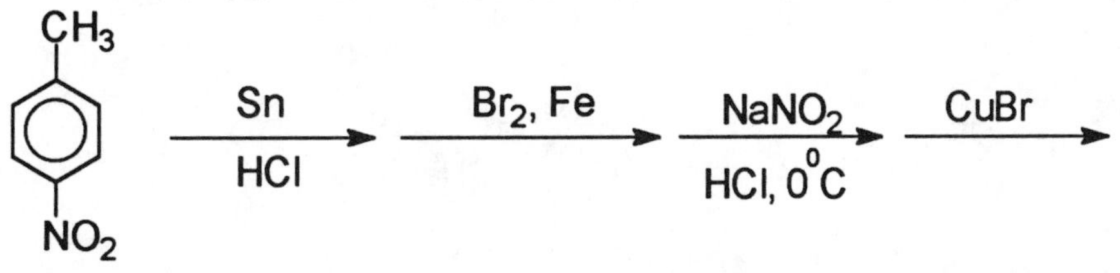

(CH_3, NO_2 para-substituted benzene) $\xrightarrow[HCl]{Sn}$ $\xrightarrow{Br_2, Fe}$ $\xrightarrow[HCl,\ 0^0C]{NaNO_2}$ $\xrightarrow{CuBr}$

a) 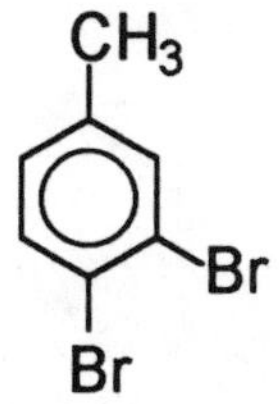 (CH_3, Br, Br) b) 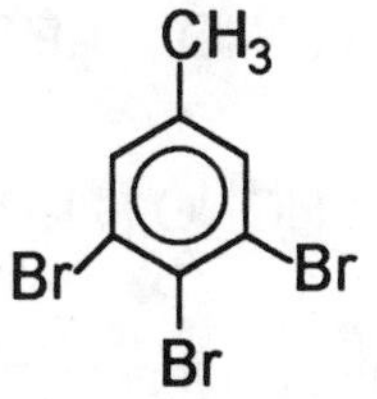 (CH_3, Br, Br, Br) c) 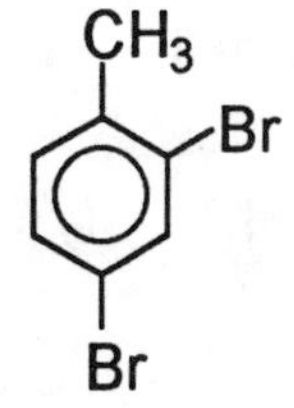 (CH_3, Br, Br) d) 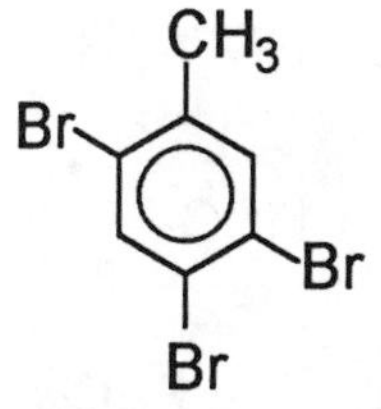 (CH_3, Br, Br, Br)

925. In the Schlieman reaction, a diazonium compound is converted to an aromatic fluorine compound by using which reagent?

a) KF b) CuF c) HBF_4 d) NaF

926. Which compounds can be synthesized by using the Sandmeyer reaction?

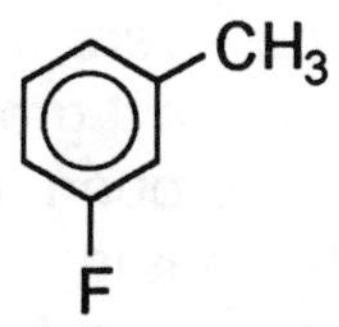

(CH_3, F) I (CH_3, Cl) II (CH_3, Br) III (CH_3, I) IV

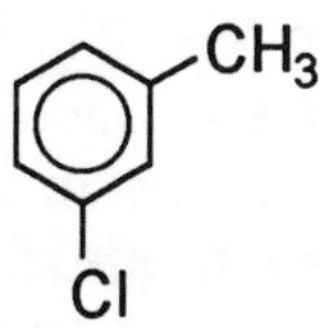

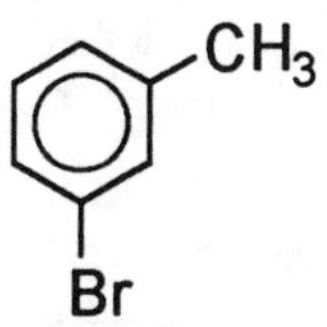

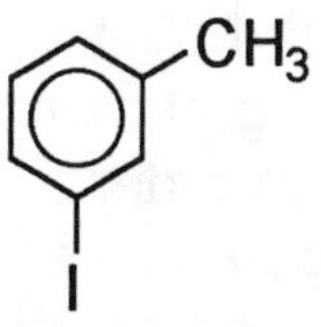

a) I, II b) II, III c) III, IV d) I, IV

927. What is the major product from the following reactions?

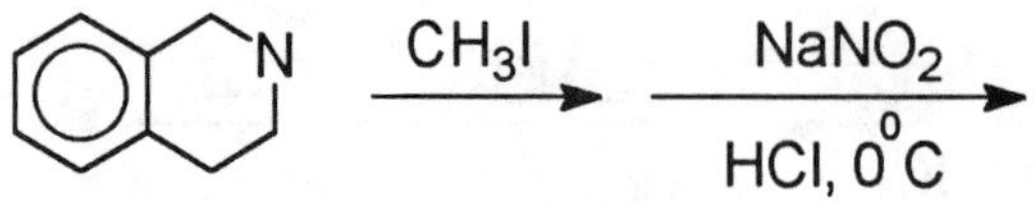

$\xrightarrow{CH_3I}$ $\xrightarrow[HCl,\ 0^{0}C]{NaNO_2}$

a) 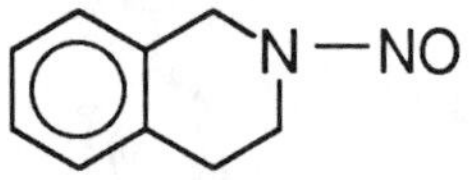 N—NO

b) 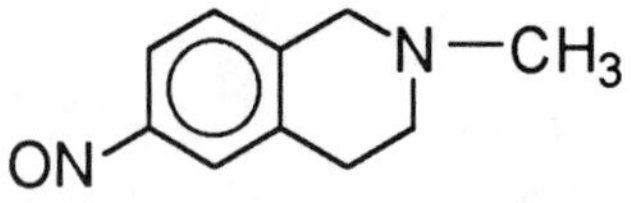 N—CH_3, ON

c) $\overset{\ominus}{Cl}\ \overset{\oplus}{N_2}$, N—$CH_3$

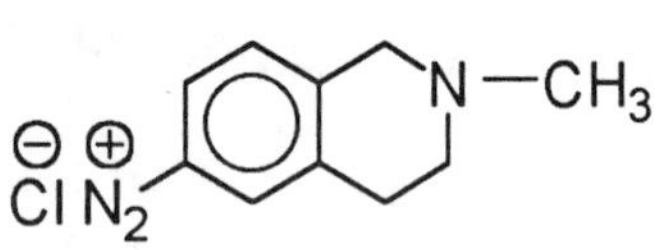

d) 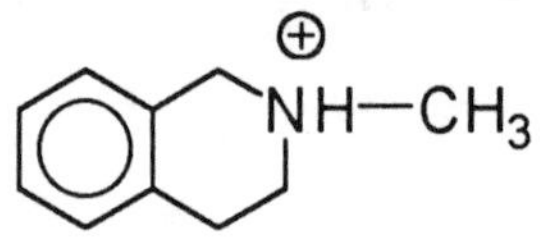 $\overset{\oplus}{N}H$—CH_3

928. What are the best conditions for the following synthesis?

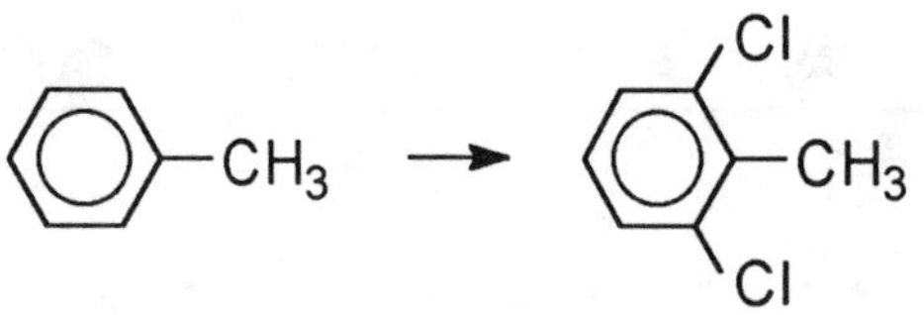

—CH_3 → Cl, —CH_3, Cl

a) HNO_3, H_2SO ; Cl_2, $FeCl_3$; Sn, HCl ; HONO ; H_3PO_2

b) HNO_3, H_2SO_4 ; Sn, HCl ; Cl_2. $FeCl_3$; HONO ; H_3PO_2

c) Cl_2, $FeCl_3$; HNO_3, H_2SO_4 ; Sn, HCl ; HONO ; H_3PO_2

d) HNO_3, H_2SO_4 ; Sn, HCl ; HONO ; H_3PO_2 ; Cl_2, $FeCl_3$

TIP This is a variation of the strategy we used in question 922. We first nitrate toluene and then take advantage of the meta-directing nitro group in the position para to the methyl group. So, now is the time for putting on the chloro groups. If the nitro group is first reduced to the NH_2 group (answers b and d), the chloro groups now go in the 3,5 positions relative to the methyl group--not wanted. In order to leave the position para to the methyl group open, it must be blocked (answer c just won't do this). As in question 922, that is accomplished by the strategy to reduce the nitro group, diazotize and replace the diazonium with a hydrogen by using H_3PO_2.

G. HOFMANN ELIMINATION REACTIONS

929. What is the product from the following reaction?

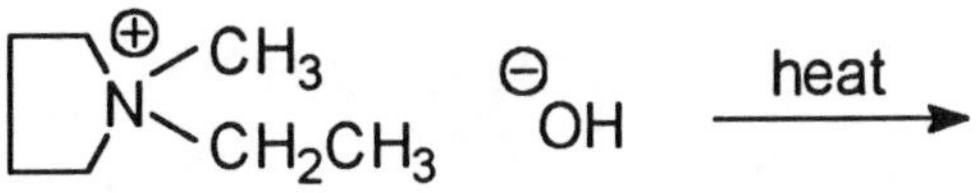

(pyrrolidinium ring) $N^{\oplus}(CH_3)(CH_2CH_3)$ $^{\ominus}OH$ $\xrightarrow{\text{heat}}$

a) $CH_2{=}CHCH_2CH_2N(CH_3)CH_2CH_3$ b) $CH_2{=}CHCH{=}CH_2$

c) 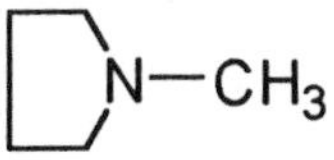 (pyrrolidine ring) $N{-}CH_3$ d) 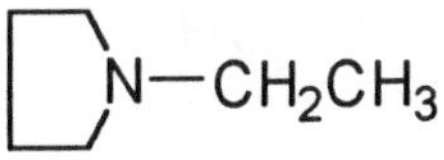 (pyrrolidine ring) $N{-}CH_2CH_3$

930. What is the product from the following reaction sequence?

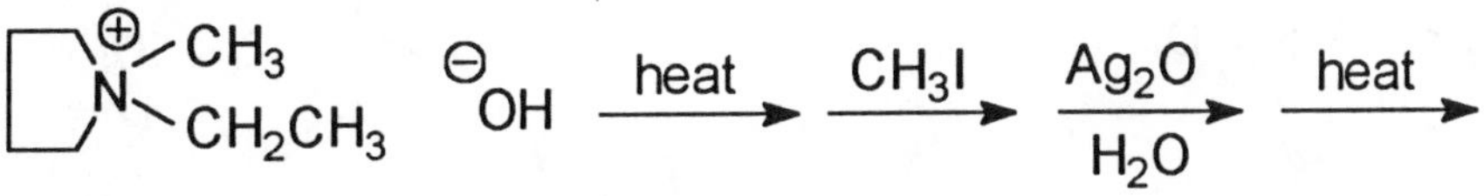

(pyrrolidinium ring) $N^{\oplus}(CH_3)(CH_2CH_3)$ $^{\ominus}OH$ $\xrightarrow{\text{heat}}$ $\xrightarrow{CH_3I}$ $\xrightarrow[H_2O]{Ag_2O}$ $\xrightarrow{\text{heat}}$

a) $CH_2{=}CHCH{=}CH_2$ b) $CH_2{=}CHCH_2CH{=}CH_2$

c) 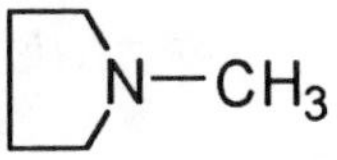 (pyrrolidine ring) $N{-}CH_3$ d) $(CH_3)_2NCH_2CH_2CH{=}CH_2$

931. What is the product from the following pyrrolysis?

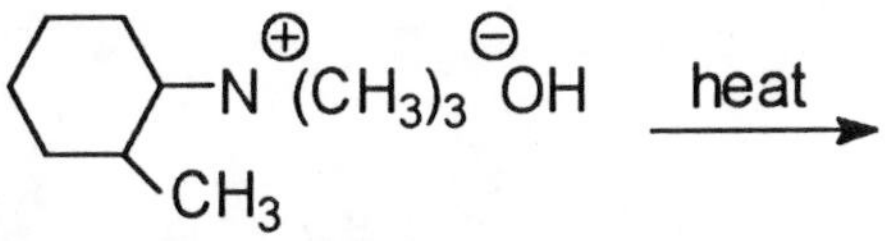

(2-methylcyclohexyl) $-\overset{\oplus}{N}(CH_3)_3$ $^{\ominus}OH$ $\xrightarrow{\text{heat}}$

a) 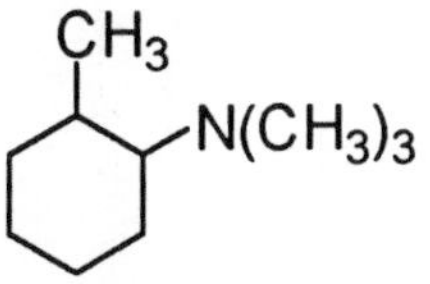 (2-methylcyclohexyl) $-N(CH_3)_3$ b) 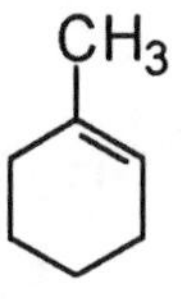 1-methylcyclohexene (CH_3) c) 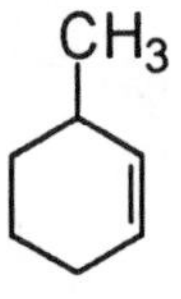 3-methylcyclohexene (CH_3) d) 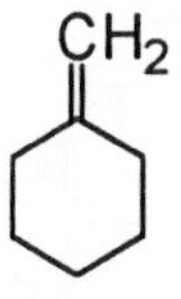 methylenecyclohexane (CH_2)

TIP The Hofmann elimination reaction is an E2 reaction that differs markedly from others in that the product is always the **least** highly substituted alkene. Thus in question 931, the beta hydrogen at carbon atom six will be removed to give c) as a prduct. Is it clear that b) would be the answer for a Saytzeff-like elimination?

H. COPE ELIMINATION REACTIONS

932. Which of the following amines will react with hydrogen peroxide and then undergo a Cope elimination?

a) $(CH_3)_3CCH_2N(CH_3)_2$ b) $CH_3CH_2NHCH_3$

c) $(CH_3)_3N$ d) $(CH_3)_3CN(CH_3)_2$

933. What is the major product from the following reaction?

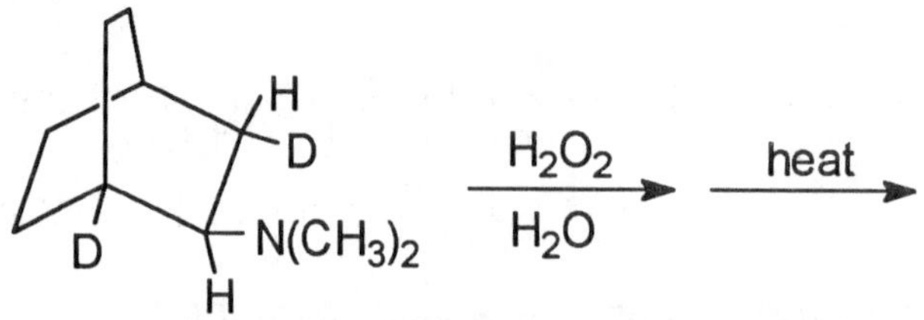

H, D, D, $N(CH_3)_2$, H $\xrightarrow[H_2O]{H_2O_2}$ $\xrightarrow{heat}$

a) 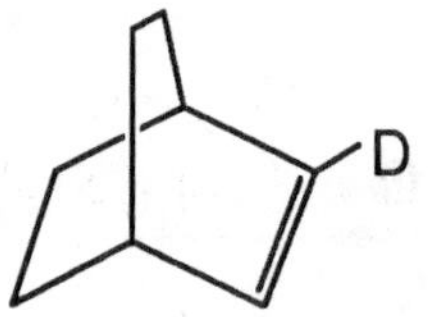 D b) 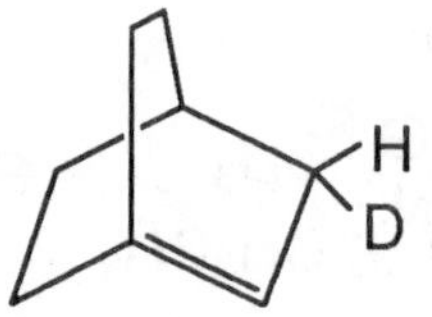 H, D

c) 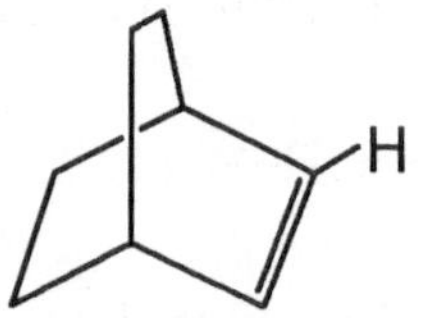 H d) D, $N(CH_3)_2$

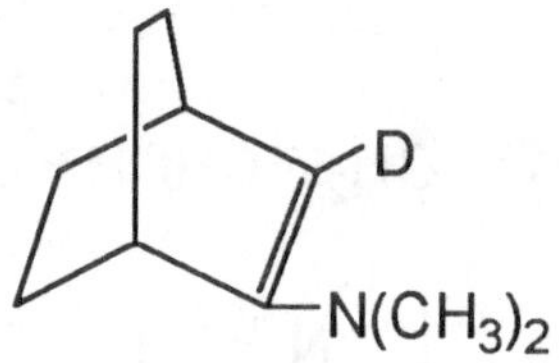

934. What is the major product from the following reaction?

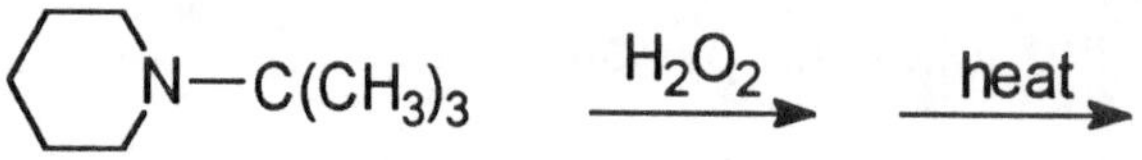

N—$C(CH_3)_3$ $\xrightarrow{H_2O_2}$ $\xrightarrow{heat}$

a) 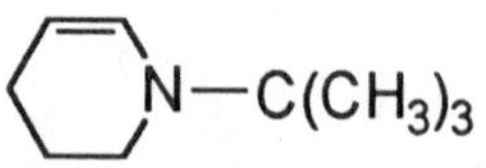 N—$C(CH_3)_3$ b) $CH_2=C(CH_3)_2$

c) 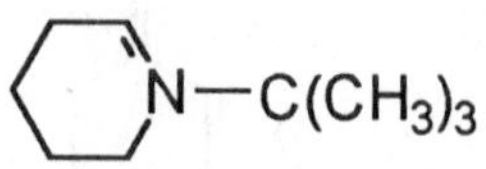 N—$C(CH_3)_3$ d) $CH_3CH=CHCH_3$

TIP The Cope elimination for tertiary amines involves a cyclic transition state and the formation of the **least** highly substituted alkene. Thus in question 933, it is the cis hydrogen atom that is removed to give a), and in question 934, the methyl hydrogen atom is removed to give answer b).

CHAPTER 23 LIPIDS

935. Which of the following saturated fatty acids are named correctly?

$CH_3(CH_2)_{10}COOH$	$CH_3(CH_2)_{12}COOH$	$CH_3(CH_2)_{14}COOH$
lauric acid	palmitic acid	myristic acid
I	II	III

$CH_3(CH_2)_{16}COOH$	$CH_3(CH_2)_{18}COOH$
stearic acid	arachidic acid
IV	V

a) I, II, III b) III, IV, V c) II, III, IV d) I, IV, V

936. Which of the following fatty acids has the highest degree of unsaturation?

a) stearic acid b) linoleic acid c) linolenic acid d) palmitic acid

937. What properties are characteristic for the most abundant fatty acids found in plants and animals?

I. They contain an even number of carbon atoms, in the range of 10-20.
II. The E isomer predominates.
III. The unsaturated fatty acids have a higher melting point than the corresponding saturated acids.
IV. The most abundant fatty acids are palmitic, stearic and oleic.

a) I, II b) III, IV c) I, III d) I, IV

938. Which physical property of oils is responsible for the fact that they are usually liquids at room temperature?

a) They have long, unbranched side chains.
b) The carbon-carbon double bonds have the Z configuration.
c) There is a large degree of hydrogen bonding.
d) There are many van der Waals attractions.

TIP These two problems, 937 and 938, contain the defining features of fats and oils, namely that fatty acids are unbranched with an even number of carbon atoms mostly between 10 and 20 carbon atoms. Moreover unsaturated fatty acids (mostly with the Z configuration) do not pack as well and hence have higher melting points.

939. Which of the following is NOT characteristic of phospholipids?

a) phosphate esters b) fatty acid esters
c) glyceride esters d) polyamides

940. Which property of phospholipids accounts for their ability to form micelles and bilayers?

a) nonpolarity
b) unsaturation
c) lipophilicity
d) hydrophilicity and lipophilicity

941. Which property of phospholipids accounts for their ability to form membranes that are fluid?

a) nonpolarity
b) unsaturation
c) lipophilicity
d) hydrophilicity and lipophilicity

942. What is the structural feature of terpenes that distinguishes them from other natural products?

a) phosphate esters
b) multi-ring systems
c) isoprene units
d) polyunsaturated chains

943. Terpenes are intermediates in the biosynthesis of which natural products?

a) prostaglandins b) steroids c) phospholipids d) glycerides

944. Which of the following structures is a prostaglandin?

a) CH_3 CH_3 CH_3 CH_2OH CH_3 CH_3

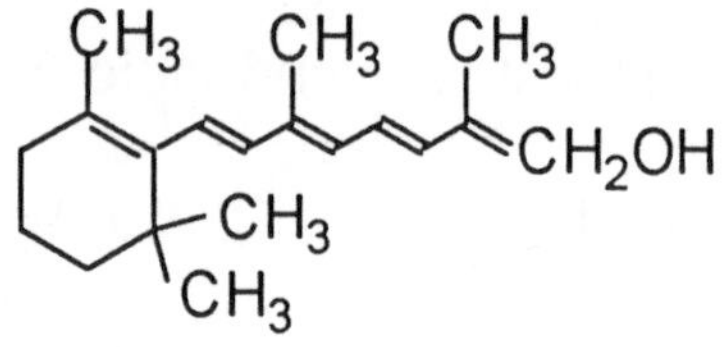

b) HO COOH HO HO H

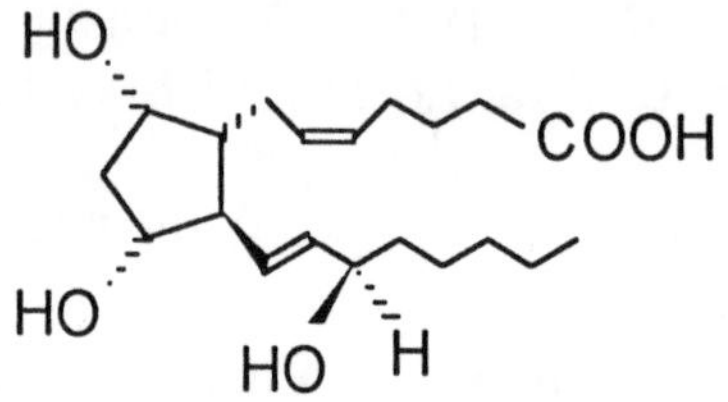

c) HO H_3C $COCH_2OH$ OH H_3C H H H O

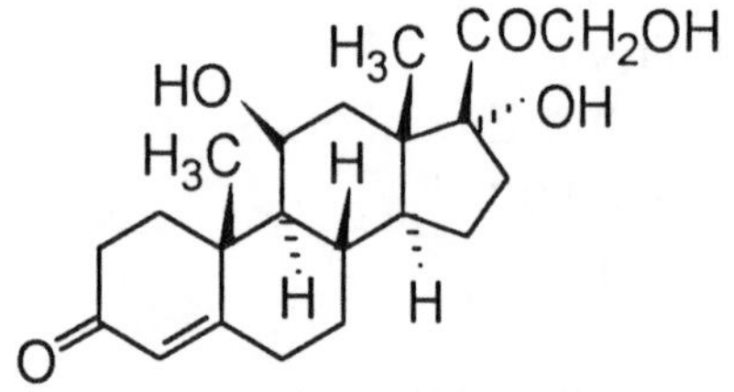

d) HO

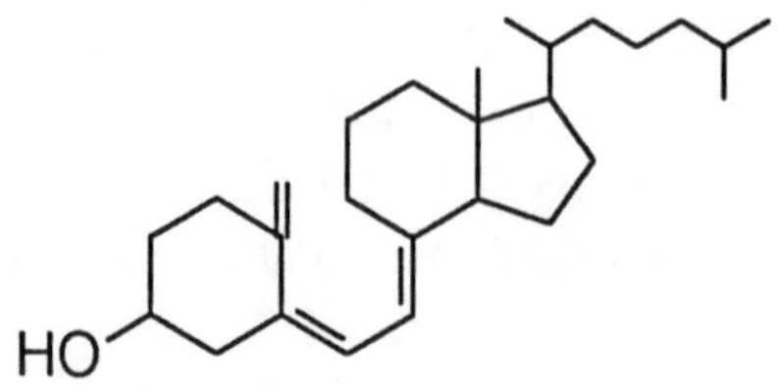

TIP For questions 942-944. The key to the biosynthesis of many cyclic and branched chain hydrophobic groups is the isoprene unit. A combination of two of these isoprene units leads to the 10 carbon terpenes (question 942). Combinations of terpenes lead to larger units, some of which are cyclic, and often these fundamental building blocks of isoprene units can be discerned. For example, in question 944, answers a), c) and d) clearly highlight that they are derivatives of isoprene. Answer c) is a steroid and b) is a prostaglandin but not a steroid nor an isoprene derivative.

945. Which of the following are derivatives of the parent steroid, pregnane?

a) estrone and progesterone
b) progesterone and cholesterol
c) cortisone and progesterone
d) cholic acid and cortisone

946. Which of the following steroids is most acidic?

a) testerone
b) estradiol
c) progesterone
d) cortisone

947. Which reagent is best for separating estrone and testosterone by extraction?

a) aqueous HCl
b) aqueous $NaHCO_3$
c) aqueous KOH
d) Br_2 in CCl_4

948. Which functional group accounts for the ability to separate estradiol from progesterone?

a) carboxylic acid b) phenolic OH c) amide d) alcoholic OH

TIP For questions 946-948. A key difference between the male and female sex hormones is that ring A is aromatic for the female sex hormone and therefore the attached hydroxyl is a phenol. Since phenols are acidic, we can utilize this acidity as a means of separation or identification. Clearly an important part of getting to the answers here is knowledge of the structures and names of the steroids. The female sex hormones are estrone and estradiol.

949. Which of the following structures is a steroid?

a) 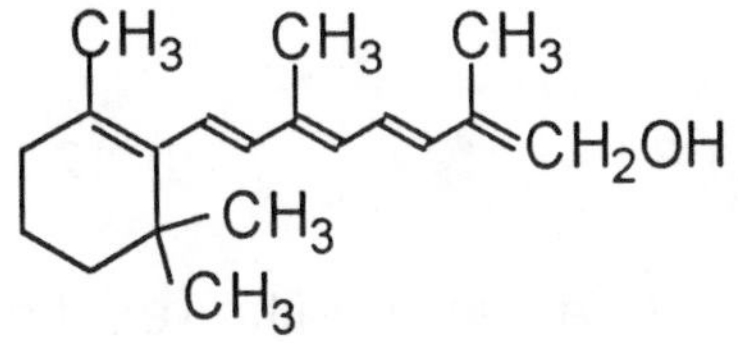 CH_3 CH_3 CH_3 CH_2OH CH_3 CH_3

b) 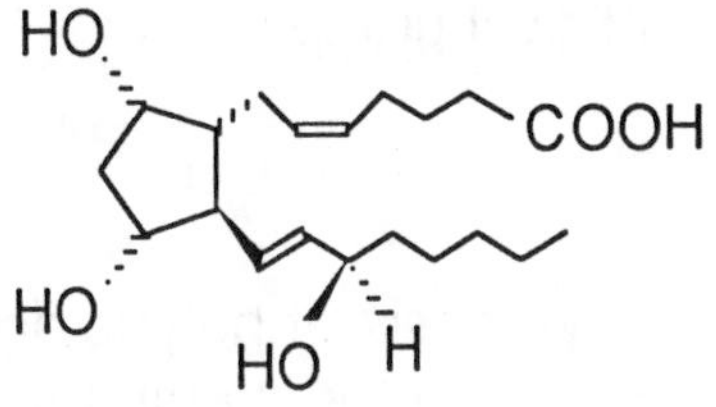 HO COOH HO HO H

c) 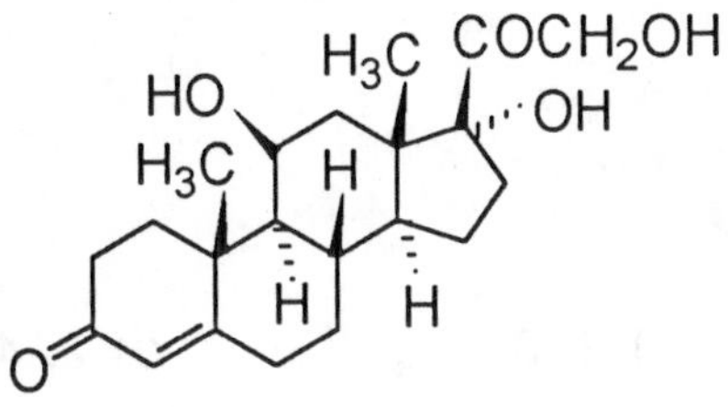 H_3C $COCH_2OH$ HO OH H_3C H H H O

d) 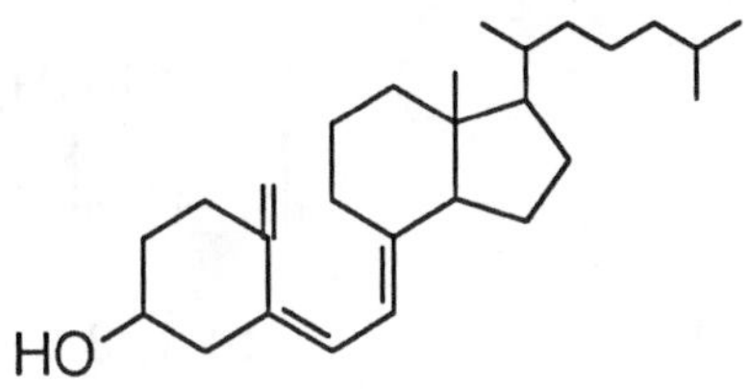 HO

950. The following are key steps in fatty acid metabolism. What is the best description for the third step (C)?

$$RCH{=}CHC(=O)SCoA \xrightarrow{A} RCH(OH)CH_2C(=O)SCoA \xrightarrow{B} RC(=O)CH_2C(=O)SCoA \xrightarrow{C} RC(=O)SCoA + CH_3C(=O)SCoA$$

a) oxidation
b) Michael addition
c) retro-Claisen reaction
d) selective reduction

TIP For all retro reactions (Aldol, Claisen, Diels Alder, etc.) the trick is to think backwards. Sometimes it helps to think of the general structure of the product. Since a beta keto ester is the product of the Claisen reaction of two esters, then a retro-Claisen must generate two esters from a beta keto ester. Thus answer c) is the thio ester analog of the retro-Claisen. The forward reaction (not given here) is critical for the formation of the alkyl chain, as for example, the R group in the starting material.

951. The average molecular weight for a fat is 730. What is the saponification number for this?

a) 77 b) 195 c) 230 d) 690

952. How many triglycerides, including stereoisomers, are possible that have three different acyl groups?

a) 4 b) 6 c) 8 d) 12

953. The biosynthesis of butyric acid proceeds via a Claisen-like reaction of acetylSCoA. If the starting material is labeled at the methyl carbon atom, which positions will have the label in the product?

$CH_3CH_2CH_2COOH$ (positions 1, 2, 3, 4 indicated by arrows under CH_3, CH_2, CH_2, $COOH$ respectively)

a) 1, 2 b) 1, 3 c) 2, 3 d) 2, 4

CHAPTER 24 AMINO ACIDS and PROTEINS

A. STRUCTURE and NOMENCLATURE

954. Which of the following amino acids are named correctly?

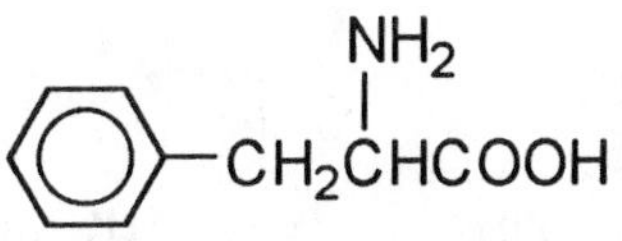

C_6H_5–$CH_2CH(NH_2)COOH$
phenylalanine
I

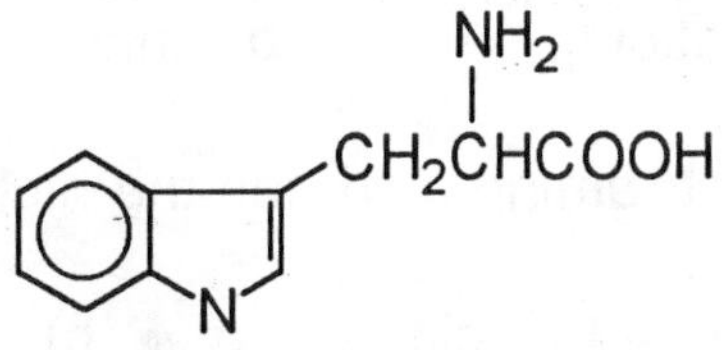

(indol-3-yl)–$CH_2CH(NH_2)COOH$
tyrosine
II

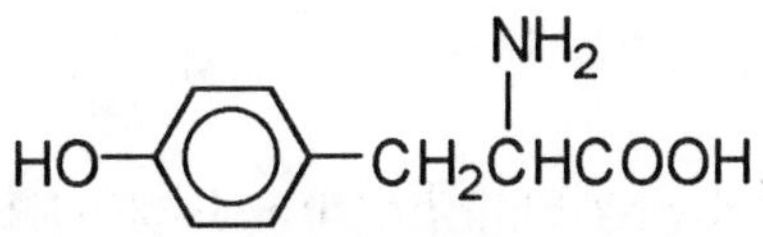

HO–C_6H_4–$CH_2CH(NH_2)COOH$
tryptophan
III

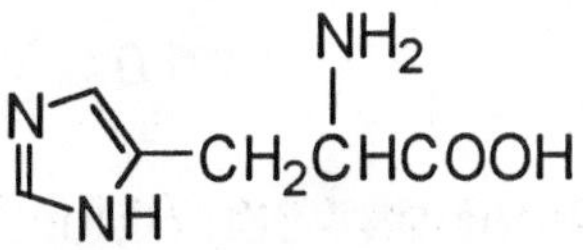

(imidazolyl)–$CH_2CH(NH_2)COOH$
histidine
IV

a) I, II b) III, IV c) I, IV d) II, III

955. Which of the following amino acids are named correctly?

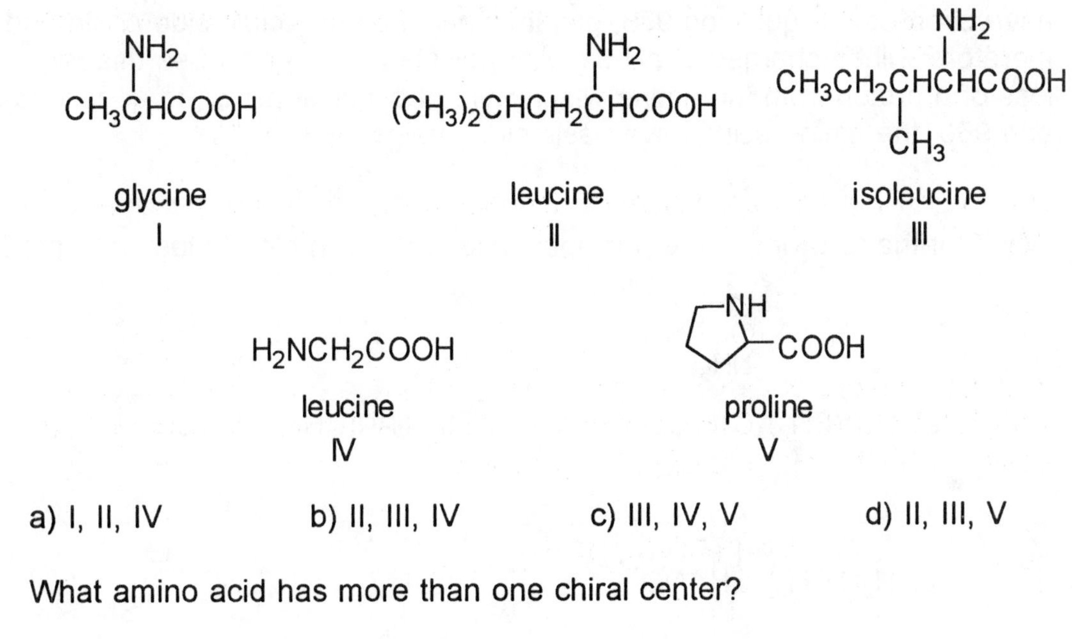

a) I, II, IV b) II, III, IV c) III, IV, V d) II, III, V

956. What amino acid has more than one chiral center?

a) leucine b) isoleucine c) valine d) phenylalanine

957. What form of leucine is present in an aqueous solution at pH 6?

a) cationic b) anionic c) dianionic d) zwitterionic

958. What form of glutamic acid is present in an aqueous solution at pH 6?

a) cationic b) anionic c) dianionic d) zwitterionic

959. Which amino acid has the highest isoelectric point?

a) aspartic acid b) arginine c) proline d) glutamine

960. What is the isoelectric point of lysine?

a) 2.8 b) 5.0 c) 6.3 d) 9.7

TIP For questions 957-960. Most amino acids are cations at low pH (the amine is protonated and the carboxyl is not ionized) and anions at high pH (the carboxyl is ionized and the amine is deprotonated). At the intermediate pH, defined as the isoelectric point, they carry no charge. For most amino acids this occurs at around pH 6. Some amino acids have either acidic or basic side chains which modify the above. With an acidic side chain, the isolectric point will occur at pH values lower than 6 and with a basic side chain, the isoelectric point will occur at pH values higher than 6. In question 957, leucine is a typical amino acid and thus has no charge at pH 6. This rules out answers a), b) and c) which all have charges. In question 958, glutamic acid has an acidic side chain and therefore will be charged at pH 6. Answer c) is wrong because this requires loss of a proton from the amine which will not occur at pH 6. In questions 959 and 960, the amino acids have basic side chains.

961. Which of the following is the principal ionic species in blood plasma at pH 7.4?

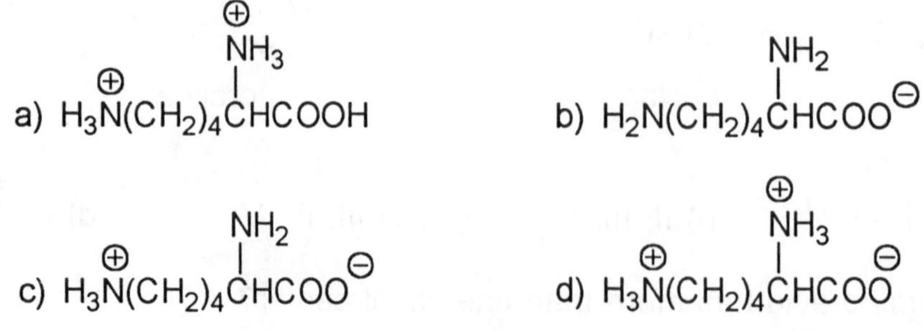

962. At what pH is the cationic form of glutamic acid present as the dominant species?

a) 2 b) 3.2 c) 5.6 d) 9.7

963. What charge does tyrosine have at pH 7 and which electrode does it migrate toward during electrophoresis?

a) positive, anode
b) positive, cathode
c) negative, anode
d) negative, cathode

964. Which mixture of amino acids can be separated by electrophoresis?

a) H_2NCH_2COOH $C_6H_5-CH_2CH(NH_2)COOH$ $H_2NC(=O)CH_2CH_2CH(NH_2)COOH$

b) $HOOCCH_2CH(NH_2)COOH$ $H_2N(CH_2)_4CH(NH_2)COOH$ H_2NCH_2COOH

c) $CH_3S(CH_2)_2CH(NH_2)COOH$ $C_6H_5-CH_2CH(NH_2)COOH$ $HOCH_2CH(NH_2)COOH$

d) H_2NCH_2COOH $CH_3CH(NH_2)COOH$ $(CH_3)_2CHCH(NH_2)COOH$

TIP For questions 961-964. In question 961, the amino acid, lysine, has a basic side chain. Thus, since the isoelectric point will be higher than 7.4, we require a charged species. Moreover the carboxyl group must be ionized at this pH, and answer a) is ruled out. In question 962, we need similar reasoning but applied to an amino acid with an acidic side chain. The isoelectric point in this case will be at low pH so the cation form will predominate at very low pH. In question 963, we have another amino acid with an acidic side chain and thus at pH 7 we have an anionic species. If cations migrate to the cathode, does that lead to answer c) as correct? In question 964, the charge on amino acids and therefore the migration during electrophoresis can separate a cation, an anion, and a neutral species. This suggests that the answer should have amino acids with an acidic group, a basic group and a neutral side chain.

B. REACTIONS of AMINO ACIDS

965. What is the role of benzoyl chloride in the resolution of amino acid racemates?

a) a protective group for the acid group
b) an enzyme resolving agent
c) a protective group for the amino group
d) a diastereomeric salt

966. Which amino acid does not give nitrogen upon treatment with $NaNO_2$ and HCl?

a). glycine b) tryptophan c) valine d) proline

967. The ninhydrin color test will NOT work for which amino acid?

a) leucine b) glutamic acid c) serine d) proline

968. What is the major product from the reaction of phenylalanine with ninhydrin?

a) 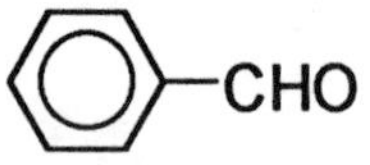 C_6H_5-CHO

b) C_6H_5-$CH_2CH(NH_2)OOCH_3$

c) (phthalimide: benzene ring fused with C(=O)–NH–C(=O)) NH

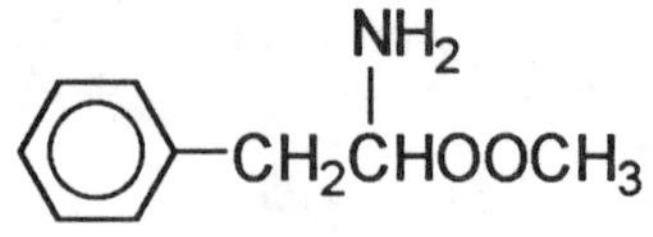

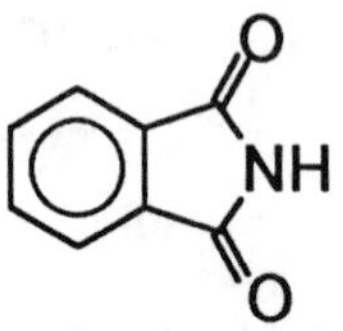

d) 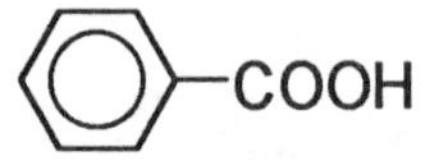 C_6H_5-COOH

969. What is the major product from the following reaction?

$$HOOCCH_2CH_2CH_2CH(NH_2)COOH + CH_3OH + HCl \longrightarrow$$

a) $HOOC(CH_2)_3CH(NHOCH_3)COOH$

b) $HOOC(CH_2)_3CH(NH_2)COOCH_3$

c) $CH_3OOC(CH_2)_3CH(NH_2)COOCH_3$

d) $CH_3OOC(CH_2)_3CH(NHOCH_3)COOCH_3$

970. Which amino acid yields lactic acid upon treatment with $NaNO_2$ and HCl?

a) glycine b) alanine c) valine d) leucine

971. Which amino acid reacts with bromine water and gives off N_2 upon reaction with $NaNO_2$ and HCl?

a) proline b) serine c) tyrosine d) threonine

C. POLYPEPTIDES

972. Multiple protective groups are likely to be required for polypeptide synthesis with which amino acids?

I. arginine II. proline III. glutamic acid IV. tryptophan

a) I, II b) I, III c) II, IV d) III, IV

973. Which of the following reaction schemes is best for the synthesis of ala-gly?

a) $H_2NCH(CH_3)COOH + H_2NCH_2COOH \xrightarrow{\text{heat}}$

b) $H_2NCH(CH_3)COOH \xrightarrow{SOCl_2} \xrightarrow{H_2NCH_2COOH}$

c) $H_2NCH(CH_3)COOH \xrightarrow{CH_3COCl} \xrightarrow{SOCl_2} \xrightarrow{H_2NCH_2COOH} \xrightarrow{H_3O^{\oplus}}$

d) $H_2NCH(CH_3)COOH \xrightarrow{C_6H_5CH_2OCOCl} \xrightarrow{SOCl_2} \xrightarrow{H_2NCH_2COOH} \xrightarrow[Pd]{H_2}$

974. Which of the following statements about a Merrifield peptide synthesis is NOT true?

a) The BOC protecting groups are used.
b) The DCC agents are used.
c) It begins with an N-terminal amino acid.
d) It begins with a C-terminal amino acid.

975. What is the best reagent to remove the peptide from the solid support in a Merrifield synthesis?

a) H_2 / Pd b) HF

c) $OH^{\ominus} + H_2O ; H_3O^{\oplus}$ d) conc HCl + heat

976. What dipeptides are expected from the reaction of alanine with dicyclohexyl carbodiimide and glycine?

I. ala-gly   II. gly-ala   III. ala-ala   IV. gly-gly

a) I, III   b) II, IV   c) I, III, IV   d) I, II, III, IV

977. Which methods can be used to sequence a tripeptide, assuming the total amino acid composition is known?

I. the Sanger method
II. successive Edman degradations
III. partial hydrolysis and the Sanger method
IV. a single Edman degradation followed by the Sanger method

a) I, II   b) III, IV   c) II, IV   d) I, III

978. A peptide gave the following proportions of amino acids upon hydrolysis:
try (1)   pro (1)   leu (2)   val (1)   ala (3)
Treatment with 2,4-dinitrofluorobenzene, followed by hydrolysis, gave 2,4-dinitrophenylleucine. Partial hydrolysis of the peptide gave the following fragments:
pro-val-ala   ala-leu-try   try-ala-pro
What is the sequence of amino acids in the peptide?

a) leu-ala-leu-try-ala-pro-val-ala   b) try-ala-pro-val-ala-leu-try-leu
c) ala-leu-try-ala-pro-val-ala-leu   d) leu-pro-val-ala-leu-try-ala-pro

979. Partial hydrolysis of a peptide gave the following tripeptides among others:
ala-leu-gly   leu-gly-val   gly-val-leu
Treatment of the the peptide with 2,4-dinitrofluorobenzene, followed by hydrolysis, gave 2,4-dinitrophenylalanine. What is the most likely sequence for the peptide?

a) gly-ala-leu-gly-val-leu   b) ala-leu-ala-gly-leu-ala
c) ala-leu-gly-val-leu-ala   d) ala-leu-gly-gly-val-leu

TIP The trick with this kind of problem is to use all the information that is given. The information from partial hydrolysis alone, or from the Sanger reaction alone, does not lead to unambiguous answers, but taken together they do. In the partial hydrolysis data, look for the common factor, which is gly in this problem. Hydrolysis of answers of b) and d) cannot give these gly fragments, but a) and c) can. The Sanger method identifies ala as the N terminal amino acid and thus rules out answer a).

980. The secondary structure of proteins depends primarily on which property of the amino acids?

a) disulfide bonds
b) hydrogen bonds
c) amide bonds
d) polar side chains

981. The α-helical structure of fibrous proteins depends primarily on which property?

a) disulfide bonds
b) hydrogen bonds
c) amide bonds
d) polar side chains

982. The solubility of globular proteins in aqueous solutions depends primarily on which property?

a) disulfide linkages
b) hydrogen bonds
c) pleated sheets
d) polar side chains

983. Which amino acid side chains are probably on the surface of a globular protein in an aqueous enviornment?

HO–C_6H_4–CH_2 (I)

C_6H_5–CH_2 (II)

$HSCH_2$ (III)

indol-3-yl–CH_2 (IV)

imidazol-4-yl–CH_2 (V)

$(CH_3)_2CHCH_2$ (VI)

a) I, IV, V
b) II, IV, VI
c) I, III, IV
d) I, III, V

TIP In an aqueous environment, we would expect the nonpolar side chains to be in the hydrophobic interior of the globular protein (as in a micelle), and the polar or ionized side chains to be at the hydrophilic surface. Side chains II, IV and VI are clearly nonpolar, so a correct answer does not include these side chains.

CHAPTER 25 NUCLEIC ACIDS

984. Which of the following structures are named correctly?

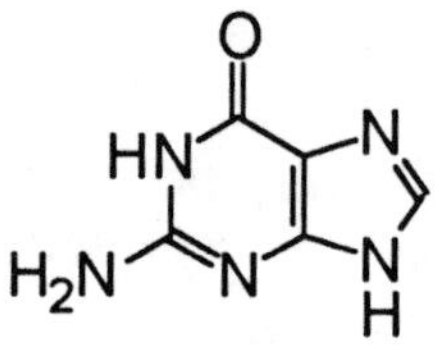

guanine I

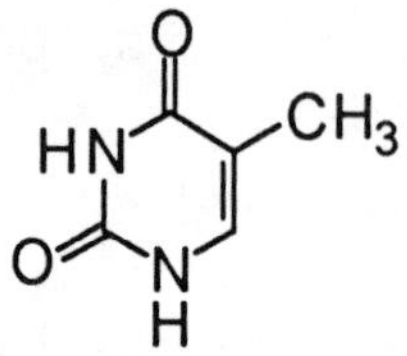

adenine II

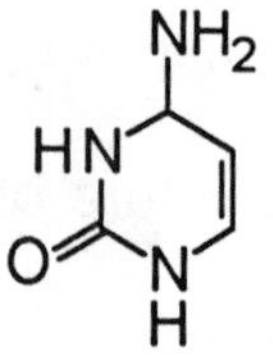

thymine III

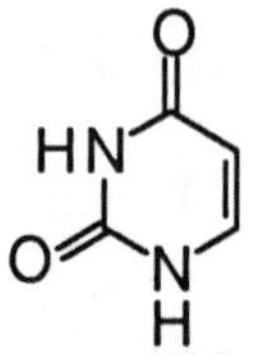

uracil IV

a) I, II b) III, IV c) I, IV d) II, III

985. Which of the following structures is a nucleotide?

a)

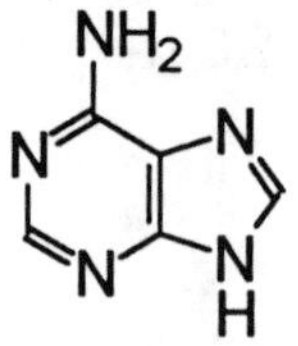

b)

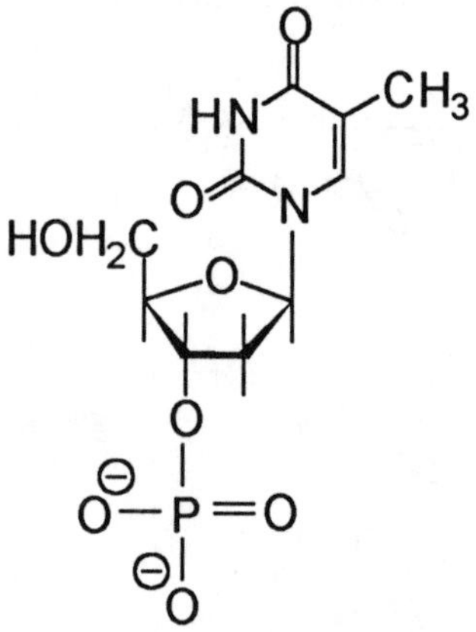

c)

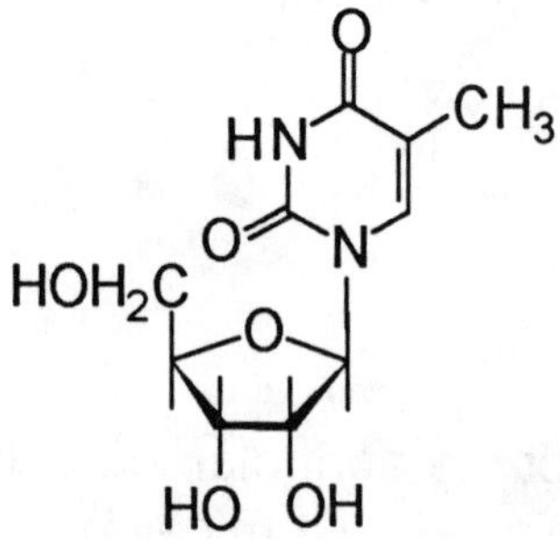

d)

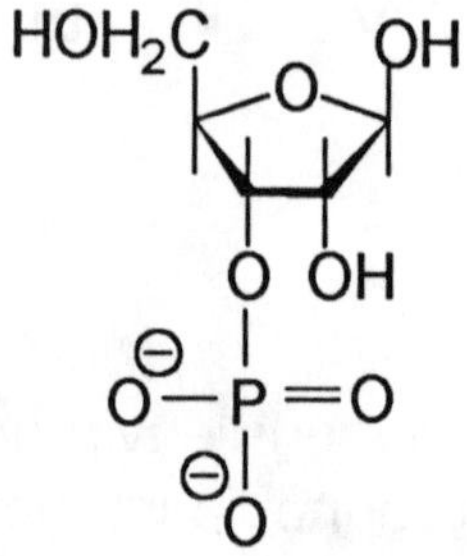

986. DNA and RNA differ accordingly to which of the following?

I. the position of attachment of phosphate groups
II. the position of attachment of base groups
III. the sugar structure at C 2'
IV. the structure of the bases

a) I, II b) II, III c) III, IV d) II, IV

987. What is the correct pairing of bases bound to RNA?

A—C	G—U	A—G	G—C	A—U	C—U
I	II	III	IV	V	VI

a) I, II b) III, V c) III, IV d) IV, V

TIP In the strands of DNA, the base pairs are A-T and G-C. It is important to remember however, that in RNA, U replaces T, which leads to A-U and G-C.

988. What is the charge on adenosine phosphate at pH 3?

a) 0 b) -1 c) -2 d) -3

989. Which functional groups of DNA do the following structures protect during synthesis?

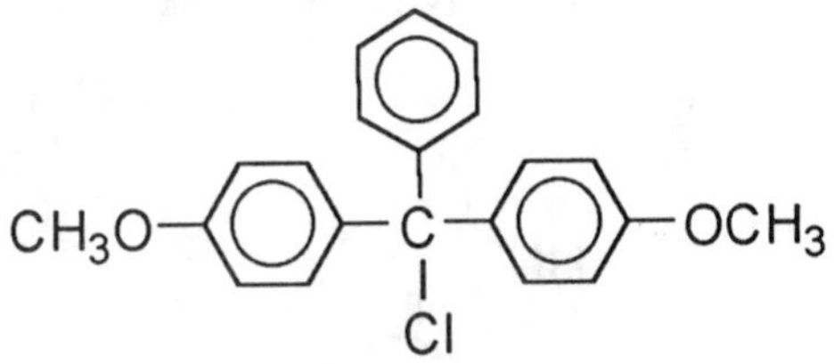

CH_3O–C_6H_4–C(C_6H_5)(Cl)–C_6H_4–OCH_3 (I)

C_6H_5–COCl (II)

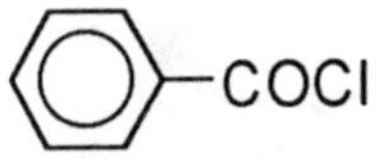

a) I for amine, II for 5' hydroxyl
b) I for 5' hydroxyl, II for amine
c) I for 3' nucleoside, II for 5' nucleoside
d) I for 5' nucleoside, II for 3' nucleoside

990. Which of the following is true for DNA?

a) (A+G) = (C+T)
b) A=G, C=T
c) (A+T) = (G+C)
d) (A+C) = (G+T)

991. What is the complementary tetranucleotide for 5'-AGCT-3'

a) 3'-TCGA-5'
b) 3'-AGCT-5'
c) 3'-GATC-5'
d) 3'-CTAG-5'

992. How many amino acids can be coded for using only two bases?

a) 2 b) 4 c) 16 d) 64

TIP This problem illustrates the essentials of using a code. If we have a two 'word' code using the four bases, then we have a total of 16 'words'. If each 'word' is assigned to one amino acid, then the maximum number is 16. In actual fact there are 21 amino acids and the actual code utilizes 3 bases for the 'word'.

993. Addition of one carbon atom will cause which of the following base conversions?

a) T to U b) U to T c) T to C d) C to T

994. How can cytosine be converted to uracil?

a) reaction with dil HCl
b) reaction with $NaNO_2$ / HCl
c) reaction with NaOH
d) addition of one carbon atom

995. Hydrogen bonding is strongest between which two structures?

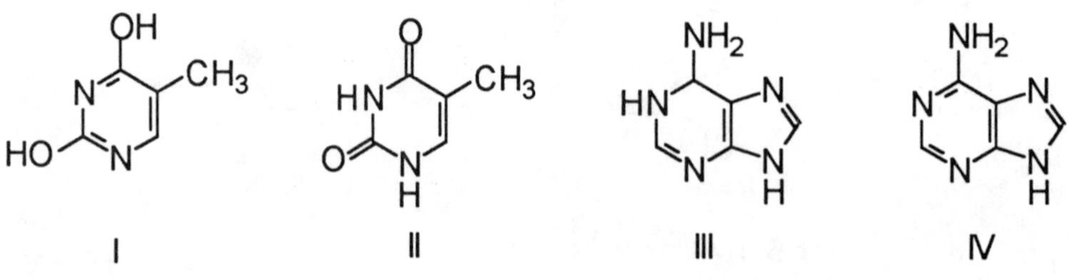

a) I and II b) I and IV c) II and III d) II and IV

996. What is the most likely site of attack of alkylating agents on the DNA structure?

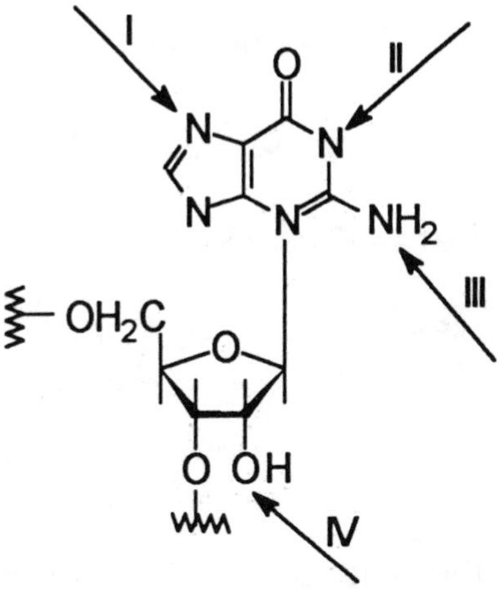

a) I b) II c) III d) IV

TIP Most alkylating agents are carcinogenic and it is thought that they attack the DNA bases at basic sites. For guanidine, we can easily rule out the alcohol site (IV) and the amide nitrogen (II) as basic sites. In contrast to alkylation at site (III), alkylation at site (I) leads to a resonance stabilized form.

997. What is the diameter of the double helix form of DNA?

a) 34 Å b) 10 Å c) 20 Å d) 4 m

998. What is the complementary sequence in RNA for the DNA sequence 5'-AATCAGTT-3'?

a) 5'-AAUCAGUU-3'
b) 5'-AACUGAUU-3'
c) 5'-UUAGUCAA-3'
d) 5'-CCAUCGAA-3'

999. When RNA is hydrolyzed, which of the following is true for the concentration of the products?

a) A = G = C = U
b) (A+G) = (C+U)
c) (A+U) = (G+C)
d) there is no relationship

1000. What is the order of increasing reactivity in a nucleophilic reaction for the following base positions (least reactive first)?

I. N-7 of guanine II. N-3 of adenine III. N-1 of cytosine

a) I, II, III b) II, I, III c) III, I, II d) III, II, I

1001. The backbone of DNA is best described as:

a) pairs of complementary bases
b) polyphosphate esters of 1,3-glycols
c) pyrophosphates
d) sugar glycosides

TIP The trick here is to focus on the difference between the continuous chain (the backbone) and other features of DNA. The complementary pairs are what hold the already formed chains together in the double helix. The sugar glycosides are the structures that incorporate the bases. The phosphate (not pyro) linkages at C5 and C3 are key to the continuous chain. How many carbon atoms separate the repeating unit?